T0189660

Measuring Scholarly Impact

Ying Ding • Ronald Rousseau • Dietmar Wolfram
Editors

Measuring Scholarly Impact

Methods and Practice

 Springer

Editors
Ying Ding
School of Informatics and Computing
Indiana University
Bloomington, IN, USA

Ronald Rousseau
University of Antwerp
Antwerp, Belgium

Dietmar Wolfram
University of Wisconsin-Milwaukee
Milwaukee, WI, USA

ISBN 978-3-319-34863-6   ISBN 978-3-319-10377-8 (eBook)
DOI 10.1007/978-3-319-10377-8
Springer Cham Heidelberg New York Dordrecht London

Preface

The measurement and assessment of scholarly impact have been experiencing rapid changes over the past two decades thanks to developments in how scholarship is communicated and advances in the tools and techniques that may be used to study scholarly communication. Measures of research impact play an increasingly important role in how individuals, research groups, journals, academic departments, institutions, and countries are ranked in their respective areas of contribution to scholarship. From the beginnings of metrics-related research in the early twentieth century, when small-scale quantitative studies of scholarly communication first revealed distinct patterns in the way publications are produced, research approaches have evolved to today's methods that employ a range of tools and techniques on large-scale datasets. Research contributions from statistical sciences, scientific visualization, network analysis, text mining, and information retrieval have provided tools and techniques to investigate metric phenomena and to assess scholarly impact in new ways. The core and complementary interests of metric studies are reflected in the names that are used to describe the field, which are also used in this edited book. Authors may have preferred terms to describe what they research. Regardless of the preferred term, there is an underlying theme of exploring how the process and products of scholarly communication may be better understood. The term bibliometrics, which is still widely used, can be traced back to the 1930s when Otlet introduced the French term bibliométrie (Otlet, 1934). Bibliometrics was later defined by Pritchard (1969) as "the application of mathematics and statistical methods to books and other media of communication." Around the same time, the term scientometrics (Naukometriya) was proposed by Nalimov and Mul'chenko (1969) as "the application of those quantitative methods which are dealing with the analysis of science viewed as an information process." Later, Nacke (1979) proposed the term informetrics (Informetrie) to encompass all quantitative aspects of the study of information, its production, dissemination, and use. More recently, the term webmetrics has been used to describe the application of metric approaches to information phenomena on the Internet, and more specifically the World Wide Web (Almind & Ingwersen, 1997). Space limitations prevent us from providing a detailed overview of these topics. Readers who are interested in finding out more about the history and scope of informetrics are encouraged to consult De Bellis (2009), Björneborn and Ingwersen (2004), as well as Egghe and Rousseau (1990).

Informetrics research has expanded beyond the evaluation of traditional units of measure such as authors and journals and now includes a broader array of units of measure and assessment. At the same time, the availability of larger, more detailed datasets has made it possible to study more granular levels of data in the production, dissemination, and use of the products of scientific communication. In the current era of data-driven research, informetrics plays a vital role in the evaluation of research entities. The focus of this research has expanded to include entities such as the datasets used in papers, genes, or drugs mentioned in papers as a focus for analysis (Ding et al., 2013). More broadly, the scientific community has been calling for scrutiny of the practice and reproducibility of research, particularly in the biomedical arena (Researching the Researchers, 2014). Techniques used in informetrics research can play a major role in this endeavor. Recently developed methods, as outlined in this book—such as data and text mining methods, network analysis, and advanced statistical techniques to reveal hidden relationships or patterns within large datasets—are quickly becoming valuable tools for the assessment of scholarly impact.

To date, there have been only a small number of monographs that have addressed informetrics-related topics. None provide a comprehensive treatment of recent developments or hands-on perspectives on how to apply these new techniques. This book fills that gap. The objective of this edited work is to provide an authoritative handbook of current topics, technologies, and methodological approaches that may be used for the study of scholarly impact. The chapters have been contributed by leading international researchers. Readers of this work should bring a basic familiarity with the field of scholarly communication and informetrics, as well as some understanding of statistical methods. However, the tools and techniques presented should also be accessible and usable by readers who are relatively new to the study of informetrics.

Each contributed chapter provides an introduction to the selected topic and outlines how the topic, technology, or methodological approach may be applied to informetrics-related research. The contributed chapters are grouped into four themes: Network Tools and Analysis, the Science System, Statistical and Text-based Methods, and Visualization. The book concludes with a chapter by Börner and Polley that brings together a number of the ideas presented in the earlier chapters.

A summary of each chapter's focus, methods outlined, software tools applied (where applicable), and data sources used (where applicable) appears below.

Network Tools and Analysis

Chapter 1
Title: Community detection and visualization of networks with the map equation framework.
Author(s): Ludvig Bohlin (Sweden), Daniel Edler (Sweden), Andrea Lancichinetti (Sweden), and Martin Rosvall (Sweden).
Topic(s): Networks.
Aspect(s): Community detection, visualization.
Method(s): Map equation.
Software tool(s) used: Infomap, MapEquation software package.
Data source: None.

Chapter 2
Title: Link prediction.
Author(s): Raf Guns (Belgium).
Topic(s): Networks.
Aspect(s): Link prediction.
Method(s): Data gathering–preprocessing–prediction–evaluation; recall–precision charts; using predictors such as common neighbors, cosine, degree product, SimRank, and the Katz predictor.
Software tool(s) used: linkpred; Pajek; VOSViewer; Anaconda Python.
Data source: Web of Science (Thomson Reuters)—co-authorship data of informetrics researchers.

Chapter 3
Title: Network analysis and indicators.
Author(s): Staša Milojević (USA).
Topic(s): Network analysis—network indicators.
Aspect(s): Bibliometric applications.
Method(s): Study of collaboration and citation links.
Software tool(s) used: Pajek; Sci2.
Data source(s): Web of Science (Thomson Reuters)—articles published in the journal *Scientometrics* over the period 2003–2012.

Chapter 4
Title: PageRank-related methods for analyzing citation networks.
Author(s): Ludo Waltman (The Netherlands) and Erjia Yan (USA).
Topic(s): Citation networks.
Aspect(s): Roles played by nodes in a citation network and their importance.
Method(s): Page-rank-related methods.
Software tool(s) used: Sci2; MATLAB; Pajek.
Data source: Web of Science (Thomson Reuters)—all publications in the journal subject category Information Science and Library Science that are of document type article, proceedings paper, or review and that appeared between 2004 and 2013.

The Science System

Chapter 5
Title: Systems Life Cycle and its relation with the Triple Helix.
Author(s): Robert K. Abercrombie (USA) and Andrew S. Loebl (USA).
Topic(s): Life cycle.
Aspect(s): Seen from a triple helix aspect.
Method(s): Technology Readiness Levels (TRLs).
Software tool(s) used: None.
Data source: From Lee et al. "Continuing Innovation in Information Technology."
    Washington, DC: The National Academies Press; plus diverse other sources.

Chapter 6
Title: Spatial scientometrics and scholarly impact: A review of recent studies, tools
    and methods.
Author(s): Koen Frenken (The Netherlands) and Jarno Hoekman (The
    Netherlands).
Topic(s): Spatial scientometrics.
Aspect(s): Scholarly impact, particularly, the spatial distribution of publication and
    citation output, and geographical effects of mobility and collaboration on cita-
    tion impact.
Method(s): Review.
Software tool(s) used: None.
Data source: Web of Science (Thomson Reuters): post 2008.

Chapter 7
Title: Researchers' publication patterns and their use for author disambiguation.
Author(s): Vincent Larivière and Benoit Macaluso (Canada).
Topic(s): Authors.
Aspect(s): Name disambiguation.
Method(s): Publication patterns.
Software tool(s) used: None.
Data source: List of distinct university-based researchers in Quebec; classification
    scheme used by the US National Science Foundation (NSF); Web of Science
    (Thomson Reuters); Google.

Chapter 8
Title: Knowledge integration and diffusion: Measures and mapping of diversity
    and coherence.
Author(s): Ismael Rafols (Spain and UK).
Topic(s): Knowledge integration and diffusion.
Aspect(s): Diversity and coherence.

Method(s): Presents a conceptual framework including cognitive distance (or proximity) between the categories that characterize the body of knowledge under study.

Software tool(s) used: Leydesdorff's overlay toolkit; Excel; Pajek; additional software available at http://www.sussex.ac.uk/Users/ir28/book/excelmaps.

Data source: Web of Science (Thomson Reuters)—citations of the research center ISSTI (University of Edinburgh) across different Web of Science categories.

Statistical and Text-Based Methods

Chapter 9

Title: Limited dependent variable models and probabilistic prediction in informetrics.

Author(s): Nick Deschacht (Belgium) and Tim C.E. Engels (Belgium).

Topic(s): Regression models.

Aspect(s): Studying the probability of being cited.

Method(s): logit model for binary choice; ordinal regression; models for multiple responses and for count data.

Software tool(s) used: Stata.

Data source: Web of Science—Social Sciences Citation Index (Thomson Reuters)—2,271 journal articles published between 2008 and 2011 in five library and information science journals.

Chapter 10

Title: Text mining with the Stanford CoreNLP.

Author(s): Min Song (South Korea) and Tamy Chambers (USA).

Topic(s): Text mining.

Aspect(s): For bibliometric analysis.

Method(s): Provides an overview of the architecture of text mining systems and their capabilities.

Software tool(s) used: Stanford CoreNLP.

Data source(s): Titles and abstracts of all articles published in the *Journal of the American Society for Information Science and Technology* (JASIST) in 2012.

Chapter 11

Title: Topic Modeling: Measuring scholarly impact using a topical lens.

Author(s): Min Song (South Korea) and Ying Ding (USA).

Topic(s): Topic modeling.

Aspect(s): Bibliometric applications.

Method(s): Latent Dirichlet Allocation (LDA).

Software tool(s) used: Stanford Topic Modeling Toolbox (TMT).

Data source(s): Web of Science (Thomson Reuters)—papers published in the *Journal of the American Society for Information Science (and Technology)* (JASIS(T)) between 1990 and 2013.

Chapter 12
Title: The substantive and practical significance of citation impact differences between institutions: Guidelines for the analysis of percentiles using effect sizes and confidence intervals.
Author(s): Richard Williams (USA) and Lutz Bornmann (Germany).
Topic(s): Analysis of percentiles.
Aspect(s): Difference in citation impact.
Method(s): Statistical analysis using effect sizes and confidence intervals.
Software tool(s) used: Stata.
Data source: InCites (Thomson Reuters)—citation data for publications produced by three research institutions in German-speaking countries from 2001 to 2002.

Visualization

Chapter 13
Title: Visualizing bibliometric networks.
Author(s): Nees Jan van Eck (The Netherlands) and Ludo Waltman (The Netherlands).
Topic(s): Bibliometric networks.
Aspect(s): Visualization.
Method(s): As included in the software tools; tutorials.
Software tool(s) used: VOSviewer; CitNetExplorer.
Data source: Web of Science (Thomson Reuters)—journals *Scientometrics* and *Journal of Informetrics* and journals in their citation neighborhood.

Chapter 14
Title: Replicable science of science studies.
Author(s): Katy Börner (USA) and David E. Polley (USA).
Topic(s): Science of Science.
Aspect(s): Data preprocessing, burst detection, visualization, geospatial, topical and network analysis, career trajectories.
Method(s): Use of freely available tools for the actions described under "aspects."
Software tool(s) used: Sci2 toolset.
Data source: Data downloaded from the Scholarly Database.

References

Almind, T. C., & Ingwersen, P. (1997). Informetric analyses on the World Wide Web: Methodological approaches to 'webometrics'. *Journal of Documentation, 53*(4), 404–426.

Björneborn, L., & Ingwersen, P. (2004). Toward a basic framework for webometrics. *Journal of the American Society for Information Science and Technology, 55*(14), 1216–1227.

De Bellis, N. (2009). *Bibliometrics and citation analysis*. Lanham, MD: Scarecrow Press.

Ding, Y., Song, M., Han, J., Yu, Q., Yan, E., Lin, L., & Chambers, T. (2013). Entitymetrics: Measuring the impact of entities. *PloS One, 8*(8), e71416.

Egghe, L., & Rousseau, R. (1990). *Introduction to informetrics: Quantitative methods in library, documentation and information science*. Elsevier Science Publishers. Retrieved from https://uhdspace.uhasselt.be/dspace/handle/1942/587.

Nacke, O. (1979). Informetrie: Ein neuer name für eine neue disziplin. *Nachrichten für Dokumentation, 30*(6), 219–226.

Nalimov, V. V., & Mul'chenko, Z. M. (1969). Наукометрия, Изучение развития науки как информационного процесса [*Naukometriya, the study of the development of science as an information process*]. Moscow: Nauka.

Otlet, P. (1934). *Traité de documentation: Le livre sur le livre*. Bruxelles, Éditions Mundaneum.

Pritchard, A. (1969). Statistical Bibliography or Bibliometrics? *Journal of Documentation, 25*(4), 348–349.

Researching the Researchers [Editorial]. (2014). *Nature Genetics, 46*(5), 417.

Bloomington, IN, USA
Antwerp, Belgium
Milwaukee, WI, USA

Ying Ding
Ronald Rousseau
Dietmar Wolfram

Contents

Part I
Network Tools and Analysis

Chapter 1
Community Detection and Visualization of Networks with the Map Equation Framework

Ludvig Bohlin, Daniel Edler, Andrea Lancichinetti, and Martin Rosvall

Abstract Large networks contain plentiful information about the organization of a system. The challenge is to extract useful information buried in the structure of myriad nodes and links. Therefore, powerful tools for simplifying and highlighting important structures in networks are essential for comprehending their organization. Such tools are called community-detection methods and they are designed to identify strongly intraconnected modules that often correspond to important functional units. Here we describe one such method, known as the map equation, and its accompanying algorithms for finding, evaluating, and visualizing the modular organization of networks. The map equation framework is very flexible and can identify two-level, multi-level, and overlapping organization in weighted, directed, and multiplex networks with its search algorithm Infomap. Because the map equation framework operates on the flow induced by the links of a network, it naturally captures flow of ideas and citation flow, and is therefore well-suited for analysis of bibliometric networks.

1.1 Introduction

Ever since Aristotle put the basis of natural taxonomy, classification and categorization have played a central role in philosophical and scientific investigation to organize knowledge. To keep pace with the large amount of information that we collect in the sciences, scholars have explored different ways to automatically categorize data ever since the dawn of computer and information science.

We now live in the era of Big Data and fortunately we have several tools for classifying data from many different sources, including point clouds, images, text documents (Blei, Ng, & Jordan, 2003; Kanungo et al., 2002; Ward, 1963), and networks. Networks of nodes and links are simple yet powerful representations of

L. Bohlin • D. Edler • A. Lancichinetti • M. Rosvall (✉)
Integrated Science Lab, Department of Physics, Umeå University, 901 87, Umeå, Sweden
e-mail: martin.rosvall@physics.umu.se

© Springer International Publishing Switzerland 2014
Y. Ding et al. (eds.), *Measuring Scholarly Impact*,
DOI 10.1007/978-3-319-10377-8_1

3

datasets of interactions from a great number of different sources (Barrat, Barthelemy, & Vespignani, 2008; Dorogovtsev & Mendes, 2003; Newman, 2010). Metabolic pathways, protein–protein interactions, gene regulation, food webs, the Internet, the World Wide Web, social interactions, and scientific collaboration are just a few examples of networked systems studied across the sciences.

In this chapter, we will focus on classification of nodes into communities and on visualization of the community structure. In network science, communities refer to groups of nodes that are densely connected internally. Community detection in networks is challenging, and many algorithms have been proposed in the last few years to tackle this difficult problem. We will briefly mention some of the possible approaches to community detection in the next section. The rest of this chapter is devoted to providing a theoretical background to the popular and efficient map equation framework and practical guidelines for how to use its accompanying search algorithm Infomap (Rosvall & Bergstrom, 2008).

The current implementation of Infomap is both fast and accurate. It can classify millions of nodes in minutes and performs very well on synthetic data with planted communities (Aldecoa & Marín, 2013; Lancichinetti & Fortunato, 2009). Furthermore, the map equation framework is also naturally flexible and can be straightforwardly generalized to analyze different kinds of network data. For instance, Infomap not only provides two-level, multi-level, and overlapping solutions for analyzing undirected, directed, unweighted, and weighted networks, but also for analyzing networks that contain higher-order data, such as memory networks (Rosvall, Esquivel, West, Lancichinetti, & Lambiotte, 2013) and multiplex networks.

This chapter is organized as follows. In Sect. 1.2, we provide some background on community detection in networks, in Sect. 1.3, we introduce the mathematics of the map equation and the Infomap algorithm, and, in Sect. 1.4, we explain how to run the software in the web applications and from the command line. We provide a collaboration network and a journal citation network as examples. For illustration, Fig. 1.1 shows a number of visualizations that can be created with the applications.

1.2 Overview of Methods

Most networks display a highly organized structure. For instance, many social systems show *homophily* in their network representations: nodes with similar properties tend to form highly connected groups called communities, clusters, or modules. For a limited number of systems, we might have some information about the classification of the nodes. For example, Wikipedia is a large network of articles connected with hyperlinks, and each article is required to belong to at least one category. In general, however, these communities are not known a priori, and, in the few fortunate cases for which some information about the classification is available, it is often informative to integrate it with the information contained in the network structure. Therefore, community detection is one of the most used techniques

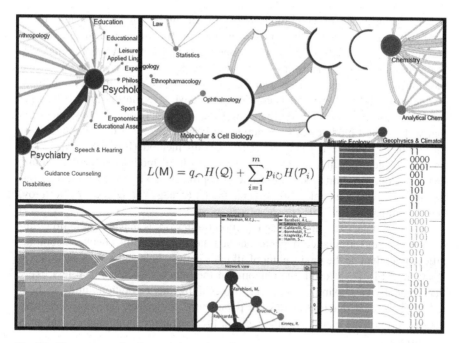

$$L(M) = q_\curvearrowright H(\mathcal{Q}) + \sum_{i=1}^{m} p_{i\circlearrowright} H(\mathcal{P}_i)$$

Fig. 1.1 The map equation framework consists of several tools for analyzing and visualizing large networks

among researchers when studying networks. Moreover, recommendation systems and network visualizations are just two of many highly useful applications of community detection.

Fortunato (2010) provided a comprehensive overview of community-detection methods. Here we just mention three of the main approaches:

Null models	Methods based on null models compare some measure of connectivity within groups of nodes with the expected value in a proper null model (Blondel, Guillaume, Lambiotte, & Lefebvre, 2008; Lancichinetti, Radicchi, Ramasco, & Fortunato, 2011; Newman & Girvan, 2004). Communities are identified as the sets of nodes for which the connectivity deviates the most from the null model. This is the approach of modularity (Newman & Girvan, 2004), which the commonly used Louvain method (Blondel et al., 2008) implements.
Block models	Methods based on block models identify blocks of nodes (Gopalan & Blei, 2013; Karrer & Newman, 2011; Peixoto, 2013) with common properties. Nodes assigned to the same block are statistically equivalent in terms of their connectivity to nodes within the block and to other blocks. The latent block structure is identified by maximizing the likelihood of observing the empirical data.
Flow models	Methods based on flows (Rosvall & Bergstrom, 2008; van Dongen, 2000) operate on the dynamics on the network rather than on its topological structure per se. The rationale is that the prime function of networks is to capture the flow between the components of the real systems they represent. Accordingly, communities consist of nodes among which flow persists for a long time once entered. As we explain in great detail in the next section, the map equation is a flow-based method.

The performance of community-detection methods can be evaluated using synthetic data generated with planted community structure, which the algorithms are supposed to detect from the network topology only (Lancichinetti, Fortunato, & Radicchi, 2008). By performing these benchmark tests, it has been found that modularity optimization suffers from a so called resolution limit (Fortunato & Barthelemy, 2007). A method is considered to have a resolution limit if the size of identified groups depends on the size of the whole network. A consequence is that well-defined modules can be merged in large networks. Several solutions to this problem have been proposed. The Louvain method (Blondel et al., 2008) provides a hierarchical tree of clusters based on local modularity maxima found during the optimization procedure. Another approach is to use so called resolution-free methods with a tunable resolution parameter (Traag, Van Dooren, & Nesterov, 2011; Waltman & Eck, 2012).

One advantage of using the map equation framework is that the resolution limit of its two-level formulation depends on the total weight of links between communities rather than on the total weight of all links in the whole network (Kawamoto & Rosvall, 2014). As a result, the outcome of the map equation is much less likely than modularity maximization to be affected by the resolution limit, without resorting to local maxima or tunable parameters. Moreover, for many networks the resolution limit vanishes completely in the multilevel formulation of the map equation.

Hierarchical block models (Peixoto, 2013) are also less likely to be affected by the resolution limit, and which method to pick ultimately depends on the particular system under study. For example, block models can also detect bipartite group structures, and are therefore well suited for analyzing food webs. On the other hand, the map equation framework is developed to capture how flow spreads across a system, and is therefore well suited for analyzing weighted and directed networks that represent the constraints on flow in the system. Since citation networks are inherently directed, the map equation framework is a natural choice for analyzing bibliometric networks.

Most of the algorithms mentioned above are implemented in open source software. Of particular relevance are three network libraries that provide several of these methods in a unified framework: NetworkX (http://networkx.github.io/), graph-tool (http://graph-tool.skewed.de/), and igraph (http://igraph.sourceforge. net/). Although igraph also has a basic implementation of Infomap, we recommend the most updated and feature-rich implementation introduced in the next section and available from www.mapequation.org.

1.3   The Map Equation Framework

Here we provide an overview of the map equation framework and the software for network analysis that we have developed on top of it and made available on www. mapequation.org. First we explain the flow-based and information-theoretic rationale behind the mathematics of the map equation, and then we outline the algorithm implemented in Infomap for minimizing the map equation over possible network partitions.

1.3.1 The Map Equation

The map equation is built on a flow-based and information-theoretic foundation and was first introduced in Rosvall and Bergstrom (2008). Specifically, the map equation takes advantage of the duality between finding community structure in networks and minimizing the description length of a random walker's movements on a network. That is, for a given modular partition of the network, there is an associated information cost for describing the movements of the random walker, or of empirical flow, if available. Some partitions give shorter and some give longer description lengths. The partition with the shortest description length is the one that best captures the community structure of the network with respect to the dynamics on the network.

The underlying code structure of the map equation is designed such that the description can be compressed if the network has regions in which the random walker tends to stay for a long time. Therefore, with a random walker as a proxy for real flow, minimizing the map equation over all possible network partitions reveals important aspects of network structure with respect to the dynamics on the network. That is, the map equation is a direct measure of how well a given network partition captures modular regularities in the network.

The flow-based foundation of the map equation is well-suited for bibliometric analysis, since a random walker on a citation network is a good model for how researchers navigate scholarly literature by reading articles and following citations. See Fig. 1.2 for an illustration of a random walker moving across a network. For example, with a network of citing articles, or citing articles aggregated at the journal level, the modules identified by the map equation would correspond to research fields. Similarly, with a network of collaborators obtained from coauthorships, the modules identified by the map equation would correspond to research groups.

To explain the machinery of the map equation, we first derive its general expression and then illustrate with examples from the interactive map equation demo available on www.mapequation.org/apps/MapDemo.html (Figs. 1.2, 1.3, and 1.4). However, we begin with a brief review of the foundations of the map equation: the mathematics of random walkers on networks and basic information theory. Readers not interested in the details of the map equation's theoretical foundation can skip to Sect. 1.3.1.4 for illustrative examples.

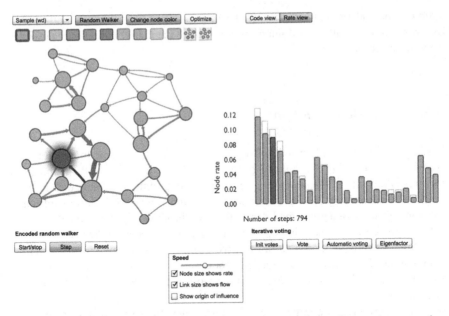

Fig. 1.2 The rate view of the map equation demo showing a random walker moving across the network

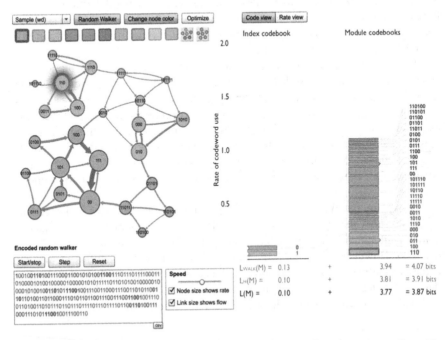

Fig. 1.3 The code view of the map equation demo showing the encoding of a random walker with a two-module solution

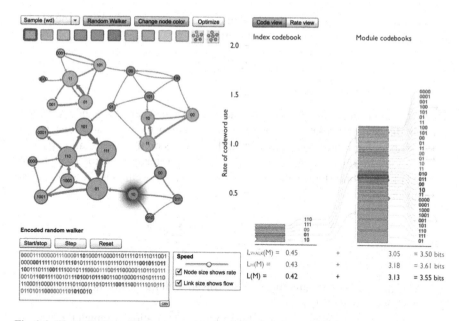

Fig. 1.4 The code view of the map equation demo showing the encoding of a random walker with the optimal five-module solution

1.3.1.1 Dynamics on Networks Modeled with Random Walkers

The map equation measures the per-step theoretical lower limit of a modular description of a random walker on a network. Instead of measuring the codelength of a long walk and dividing by the number of steps, it is more efficient to derive the codelength from the stationary distribution of the random walker on the nodes and links. In general, given a network with n nodes and weighted directed links $W_{\alpha \to \beta}$ between nodes $\alpha, \beta \in 1, 2, \ldots, n$, the conditional probability that the random walker steps from node α to node β is given by the relative link weight

$$p_{\alpha \to \beta} = \frac{W_{\alpha \to \beta}}{\sum_{\beta} W_{\alpha \to \beta}}. \tag{1.1}$$

In the Flash application available on www.mapequation.org/apps/MapDemo.html, the random walker moves from node to node following the directed links proportional to their weights. In the snapshot displayed in Fig. 1.2, the random walker has taken 794 steps and visited the nodes according to the frequency distribution to the right. The currently visited node is highlighted both in the network and in the histogram.

Assuming that the stationary distribution is given by p_α, it can in principle be derived from the recursive system of equations

$$p_\alpha = \sum_\beta p_\beta p_{\beta \rightarrow \alpha}. \tag{1.2}$$

However, to ensure a unique solution independent of where the random walker starts in directed networks, at a low rate τ the random walker instead *teleports* to a random node. For more robust results that depend less on the teleportation parameter τ, we most often use teleportation to a node proportional to the total weights of the links to the node (Lambiotte & Rosvall, 2012). The stationary distribution is then given by

$$p_\alpha = (1 - \tau) \sum_\beta p_\beta p_{\beta \rightarrow \alpha} + \tau \frac{\sum_\beta W_{\beta \rightarrow \alpha}}{\sum_{\alpha,\beta} W_{\beta \rightarrow \alpha}}. \tag{1.3}$$

This system of equations can efficiently be solved with the power-iteration method (Golub & Van Loan, 2012).

To make the results even more robust to the teleportation parameter τ, we use *unrecorded* teleportation steps and only record steps along links (Lambiotte & Rosvall, 2012). We capture these dynamics with an extra step without teleportation on a stationary solution similar to Eq. (1.3) but with teleportation to a node proportional to the total weights of the links *from* the node,

$$p_\alpha^* = (1 - \tau) \sum_\beta p_\beta^* p_{\beta \rightarrow \alpha} + \tau \frac{\sum_\beta W_{\alpha \rightarrow \beta}}{\sum_{\alpha,\beta} W_{\beta \rightarrow \alpha}}. \tag{1.4}$$

The unrecorded visit rates on links $q_{\beta \rightarrow \alpha}$ and nodes p_α can now be expressed:

$$q_{\beta \rightarrow \alpha} = p_\beta^* p_{\beta \rightarrow \alpha} \tag{1.5}$$

$$p_\alpha = \sum_\beta q_{\beta \rightarrow \alpha}. \tag{1.6}$$

This so called smart teleportation scheme ensures that the solution is independent of where the random walker starts in directed networks with minimal impact on the results from the teleportation parameter. A typical value of the teleportation rate is $\tau = 0.15$, but in practice the clustering results show only small changes for teleportation rates in the range $\tau \in (0.05, 0.95)$ (Lambiotte & Rosvall, 2012). For example, for undirected networks the results are completely independent of the teleportation rate and identical to results given by Eq. (1.2). For directed networks, a teleportation rate too close to 0 gives results that depend on how the random walker was initiated and should be avoided, but a teleportation value equal to 1 corresponds to using the link weights as the stationary distribution. Accordingly, the unrecorded teleportation scheme also makes it possible to describe the raw flow given by the links themselves without first inducing dynamics with a random walker. The Infomap code described in Sect. 1.3.2 can use any of these dynamics described above, but we recommend the unrecorded teleportation scheme proportional to link weights for most robust results.

The map equation is free from external resolution parameters. Instead, the resolution scale is set by the dynamics. The dynamics described above correspond to encoding one node visit per step of the random walker, but the code rate can be set both higher and lower (Schaub, Lambiotte, & Barahona, 2012). A higher code rate can be achieved by adding self-links and a lower code rate can be achieved by adding nonlocal links to the network (Schaub et al., 2012). A higher code rate gives smaller modules because the random walker becomes trapped in smaller regions for a longer time. The Infomap code allows an increase of the code rate from the natural value of encoding by one node visit per step of the random walker.

1.3.1.2   Basic Information Theory

While the map equation gives the theoretical lower limit of a modular description of a random walker on a network, the interactive map equation demo illustrates the description with real codewords. We use Huffman codes (Huffman, 1952), which are optimal in the sense that no binary codes can come closer to the theoretical limit. However, for identifying the optimal partition of the network, we are only interested in the compression rate and not the actual codewords. Accordingly, the Infomap algorithm only measures the theoretical limit given by the map equation.

Shannon's source coding theorem (Shannon, 1948) states that the per-step theoretical lower limit of describing a stream of n independent and identically distributed random variables is given by the entropy of the probability distribution. That is, given the probability distribution $P = \{p_i\}$ such that $\sum_i p_i = 1$, the lower limit of the per-step codelength is given by

$$L(P) = H(P) \equiv -\sum_i p_i \log p_i, \qquad (1.7)$$

with the logarithm taken in base 2 to measure the codelength in bits. In other words, no codebook with codewords for the events distributed according to P can use fewer bits on average.

Accordingly, the best compression of random walker dynamics on a network is given by the entropy rate (Shannon, 1948)

$$\sum_\alpha p_\alpha H(p_{\alpha \to \beta}), \qquad (1.8)$$

which corresponds to the average codelength of specifying the next node visit given current node position, averaged over all node positions. This coding scheme takes advantage of the independent and identically distributed next node visits given current node position, but cannot be used to take advantage of the modular structure of the network. Instead, the map equation uses the extra constraint that the only available information from one step to the next is the currently visited module, or that the random walk switches between modules, forcing independent and identically distributed events within and between modules. From this assumption naturally follows a modular description that is maximally compressed by the network partition that best represents the modular structure of the network with respect to the dynamics on the network.

1.3.1.3 The Mathematics of the Map Equation

Given a network partition, the map equation specifies the theoretical modular description length of how concisely we can describe the trajectory of a random walker guided by the possibly weighted, directed links of the network. We use M to denote a network partition of the network's n nodes into m modules, with each node α assigned to a module i. We then seek to minimize the description length $L(M)$ given by the map equation over possible network partitions M. Again, the network partition that gives the shortest description length best captures the community structure of the network with respect to the dynamics on the network.

The map equation can be expressed in closed form by invoking Shannon's source coding theorem in Eq. (1.7) for each of multiple codebooks, and by weighting them by their rate of use. Both the description length and the rate of use can be expressed in terms of the node-visit rates p_α and the module-transition rates $q_{i\curvearrowright}$ and $q_{i\curvearrowleft}$ at which the random walker enters and exits each module i, respectively:

$$q_{i\curvearrowright} = \sum_{\alpha \in j \neq i, \beta \in i} q_{\alpha \to \beta} \tag{1.9}$$

$$q_{i\curvearrowleft} = \sum_{\alpha \in i, \beta \in j \neq i} q_{\alpha \to \beta}. \tag{1.10}$$

To take advantage of the modular structure of the network, m *module codebooks* and one *index codebook* are used to describe the random walker's movements within and between modules, respectively. Module codebook i has one codeword for each node $\alpha \in i$ and one exit codeword. The codeword lengths are derived from the frequencies at which the random walker visits each of the nodes in the module, $p_{\alpha \in i}$, and exits the module, $q_{i\curvearrowleft}$. We use $p_{i\cup}$ to denote the sum of these frequencies, the total use of codewords in module i, and P^i to denote the normalized probability distribution. Similarly, the index codebook has codewords for module entries. The codeword lengths are derived from the set of frequencies at which the random walker enters each module, $q_{i\curvearrowright}$. We use $q_{i\curvearrowright}$ to denote the sum of these frequencies, the total use of codewords to move into modules, and Q to denote the normalized probability distribution. We want to express the average length of codewords from the index codebook and the module codebooks weighted by their rates of use. Therefore, the map equation is

$$L(M) = q_\curvearrowright H(Q) + \sum_{i=1}^{m} P_{i\cup} H(P_i).\qquad(1.11)$$

Below we explain the terms of the map equation in detail and in Figs. 1.3 and 1.4 we provide examples with Huffman codes for illustration.

$L(M)$	The per-step description length for module partition M. That is, for module partition M of n nodes into m modules, the lower bound of the average length of the code describing a step of the random walker. The bottom right bit counts in Figs. 1.3 and 1.4 show the description length for the given network partition.
$q_\curvearrowright = \sum_{i=1}^{m} q_{i\curvearrowright}$	The rate at which the index codebook is used. The per-step use rate of the index codebook is given by the total probability that the random walker enters any of the m modules. The total height of the blocks under *Index codebook* in Figs. 1.3 and 1.4 corresponds to this rate.
$H(Q) = -\sum_{i=1}^{m} (q_{i\curvearrowright}/q_\curvearrowright)\log(q_{i\curvearrowright}/q_\curvearrowright)$	The frequency-weighted average length of codewords in the index codebook. The entropy of the relative rates to use the module codebooks measures the smallest average codeword length that is theoretically possible. The heights of individual blocks under *Index codebook* in Figs. 1.3 and 1.4 correspond to the relative rates and the codeword lengths approximately correspond to the negative logarithm of the rates in base 2.
$p_{i\cup} = \sum_{\alpha \in i} p_\alpha + q_{i\curvearrowright}$	The rate at which the module codebook i is used, which is given by the total probability that any node in the module is visited, plus the probability that the random walker exits the module and the exit codeword is used. This rate corresponds to the total height of similarly colored blocks associated with the same module under *Module codebooks* in Figs. 1.3 and 1.4.
$H(P^i) = -(q_{i\curvearrowright}/p_{i\cup})\|\log(q_{i\curvearrowright}/p_{i\cup})$ $\quad\quad - \sum_{\alpha \in i}(p_\alpha/p_{i\cup})\log(p_\alpha/p_{i\cup})$	The frequency-weighted average length of codewords in module codebook i. The entropy of the relative rates at which the random walker exits module i and visits each node in module i measures the smallest average codeword length that is theoretically possible. The heights of individual blocks under *Module codebooks* in Figs. 1.3 and 1.4 correspond to the relative rates and the codeword lengths approximately correspond to the negative logarithm of the rates in base 2.

1.3.1.4 The Machinery of the Map Equation

The map equation demo available on www.mapequation.org/apps/MapDemo.html provides an interactive interface to help understand the machinery of the map equation. The demo has two modes accessible with the two buttons *Rate view* and *Code view*. The purpose of the rate view shown in Fig. 1.2 is to illustrate how we use the random walker to induce flow on the network. The purpose of the code view shown in Figs. 1.3 and 1.4 is to illustrate the duality between finding regularities in the network structure and compressing a description of the flow induced by the network structure.

In the Flash application available on www.mapequation.org/apps/MapDemo. html, the random walker currently visits the green node with codeword 110 (Fig. 1.3). The height of a block under *Index codebook* represents the rate at which the random walker enters the module. The bit sequence next to each block is the associated codeword. Similarly, the height of a block under *Module codebooks* represents the rate at which the random walker visits a node, or exits a module. The blocks representing exit rates have an arrow on their right side. The text field in the bottom left corner shows the encoding for the previous steps of the random walker, ending with the step on the node with codeword 110. Steps in the two modules are highlighted with green and blue, respectively, and the enter and exit codewords are boldfaced. L(M) in the bottom right corner shows the theoretical limit of the description length for the chosen two-module network partition given by Eq. (1.11). L_M(M) shows the limit of the Huffman coding given by the actual codebooks and L_{WALK}(M) shows the average per-step description length of the realized walk in the simulation.

In the rate view, click *Random walker* and *Start/stop* and a random walker begins traversing the network. Moving from one node to another, the random walker chooses which neighbor to move to next proportional to the weights of the links to the neighbors according to Eq. (1.1). As described in Sect. 1.3.1.1, to ensure an ergodic solution, i.e., that the average visit rates will reach a steady-state solution independent of where the random walker starts, the random walker sometimes moves to a random node irrespective of the link structure. In this implementation, the random walker *teleports* to a random node with a 15 % chance every step, or about every six steps.

The histogram on the right in the rate view shows the node-visit distribution. The colored bars show the average distribution so far in the simulation, and the bars with a gray border show the ergodic solution. The visit rates of the ergodic solution correspond to the eigenvector of the leading eigenvalue of the transition matrix given by the network. This solution also corresponds to the PageRank of the nodes (Brin & Page, 1998). After a long time the average visit rates of the random walker will approach the ergodic solution, but this walk-based method is very inefficient for obtaining the ergodic solution. In practice, since the map equation only takes the ergodic visit rates as input, we use the power-iteration method to derive the ergodic solution (Golub & Van Loan, 2012). The power-iteration method works by operating on the probability distribution of random walkers rather than on a specific random walker. By selecting *Init votes*, each node receives an equal share of this probability. By clicking *Vote*, the probability at each node is pushed to neighbors proportional to the link weights, and 15 % is distributed randomly. As can be seen by clicking a few times on *Vote*, the probability distribution quickly approaches the ergodic solution.

Compared to the two-module solution in Fig. 1.3, the index codebook is larger and used more often. Nevertheless, and thanks to the more efficient encoding of movements within modules with the smaller module codebooks, the per-step codelength is 0.32 bits shorter on average.

In the code view, each node is labeled with its codeword as shown in Figs. 1.3 and 1.4. Each event, i.e. that the random walker visits a node, enters a module, or exits a module, is also represented as a block in the stacks on the right. The stack under *Index codebook* shows module-enter events, and the stack under *Module codebooks* shows within-module events. Mouseover a node or a block in the map equation demo highlights the corresponding block or node. The height of a block represents the rate at which the corresponding event occurs, and the bit string to the right of each block is the codeword associated with the event. The codewords are Huffman codes (Huffman, 1952) derived from their frequency of use. Huffman codes are optimal for symbol-by-symbol encoding with binary codewords given a known probability distribution. As explained in Sect. 1.3.1.2, the average codelength of a Huffman code is bounded below by the entropy of the probability distribution (Shannon, 1948). In general, the average codelength of a Huffman code is somewhat longer than the theoretical limit given by the entropy, which has no constraints on using integer-length codewords. For example, the average codelengths with actual binary codewords shown in the lower right corners of Figs. 1.3 and 1.4 are about a percent longer than the theoretical limit.

In practice, for taking advantage of the duality between finding the community structure and minimizing the description length, we use the theoretical limit given by the map equation. That is, we show the codewords in the map equation demo only for pedagogical reasons. For example, note that frequently visited nodes are assigned short codewords and that infrequently visited nodes are assigned longer codewords, such that the description length will be short on average. Similarly, modules which the random walker enters frequently are assigned short codewords, and so on. The varying codeword lengths take advantage of the regularity that some events happen more frequently than others, but does not take advantage of the community structure of the network. Instead, it is the modular code structure with an index codebook and module codebooks that exploits the community structure.

The optimal network partition corresponds to a modular code structure that balances the cost of specifying movements within and between modules. Figure 1.3 shows a network partition with two modules. The lower bound of the average description length for specifying movements within modules is 3.77 bits per step and only 0.10 bit per step for specifying movements into modules. Describing movements into modules is cheap because a single bit is necessary to specify which of the two modules the random walker enters once it switches between modules, and it only switches between modules every ten steps on average. However, the movements within modules are relatively expensive, since each module contains many nodes, each one with a rather long codeword. The more events a codebook contains, the longer the codewords must be on average, because there is a limited number of short codewords. Figure 1.4 shows the optimal network partition with five modules. In this partition, the smaller modules allow for more efficient encoding of movements within modules. On average, specifying movements within modules requires 3.13 bits per step, 0.64 bit less in the two-module solution in Fig. 1.3. This compression gain comes at the cost of more expensive description of movements into the five modules, but the overall codelength is nevertheless smaller, 3.55 bits in the optimal solution compared to 3.87 bits in the two-module solution. To better understand the inner-workings of the map equation, it is helpful to change the partition of the network in the map equation demo and study how the code structure and associated codelengths change.

Here we have described the basic two-level map equation. One strength of the map equation framework is that it is generalizable to higher-order structures, for example to hierarchical structures (Rosvall & Bergstrom, 2011), overlapping structures (Esquivel & Rosvall, 2011), or higher-order Markov dynamics (Rosvall et al., 2013). For details about those methods, we refer to the cited papers.

1.3.2   Infomap

We use *Infomap* to refer to the search algorithm for minimizing the map equation over possible network partitions. Below we briefly describe the algorithms for identifying two-level and multilevel solutions.

1.3.2.1   Two-Level Algorithm

The core of the algorithm follows closely the Louvain method (Blondel et al., 2008): neighboring nodes are joined into modules, which subsequently are joined into supermodules, and so on. First, each node is assigned to its own module. Then, in random sequential order, each node is moved to the neighboring module that results in the largest decrease of the map equation. If no move results in a decrease of the map equation, the node stays in its original module. This procedure is repeated, each time in a new random sequential order, until no move generates a decrease of the map equation. Then, the network is rebuilt, with the modules of the last level forming the nodes at this level, and, exactly as at the previous level, the nodes are joined into modules. This hierarchical rebuilding of the network is repeated until the map equation cannot be reduced further.

With this algorithm, a fairly good clustering of the network can be found in a very short time. Let us call this the core algorithm and see how it can be improved. The nodes assigned to the same module are forced to move jointly when the network is rebuilt. As a result, what was an optimal move early in the algorithm might have the opposite effect later in the algorithm. Two or more modules that merge together and form one single module when the network is rebuilt can never be separated again in this algorithm. Therefore, the accuracy can be improved by breaking the modules of the final state of the core algorithm in either of the two following ways:

Submodule movements. First, each cluster is treated as a network on its own and the main algorithm is applied to this network. This procedure generates one or more submodules for each module. Then all submodules are moved back to their respective modules of the previous step. At this stage, with the same partition as in the previous step but with each submodule being freely movable between the modules, the main algorithm is reapplied.

Single-node movements. First, each node is reassigned to be the sole member of its own module, in order to allow for single-node movements. Then all nodes are moved back to their respective modules of the previous step. At this stage, with the same partition as in the previous step but with each single node being freely movable between the modules, the main algorithm is reapplied. In practice, we repeat the two extensions to the core algorithm in sequence and as long as the clustering is improved. Moreover, we apply the submodule movements recursively. That is, to find the submodules to be moved, the algorithm first splits the submodules into subsubmodules, subsubsubmodules, and so on until no further splits are possible. Finally, because the algorithm is stochastic and fast, we can restart the algorithm from scratch every time the clustering cannot be improved further and the algorithm stops. The implementation is straightforward and, by repeating the search more than once—100 times or more if possible—the final partition is less likely to correspond to a local minimum. For each iteration, we record the clustering if the description length is shorter than the shortest description length recorded before.

1.3.2.2   Multilevel Algorithm

We have generalized our search algorithm for the two-level map equation to recursively search for multilevel solutions. The recursive search operates on a module at any level; this can be all the nodes in the entire network, or a few nodes at the finest level. For a given module, the algorithm first generates submodules if this gives a shorter description length. If not, the recursive search does not go further down this branch. But if adding submodules gives a shorter description length, the algorithm tests if movements within the module can be further compressed by additional index codebooks. Further compression can be achieved both by adding one or more coarser codebooks to compress movements between submodules, or by adding one or more finer index codebooks to compress movements within submodules. To test for all combinations, the algorithm calls itself recursively, both operating on the network formed by the submodules and on the networks formed by the nodes within every submodule. In this way, the algorithm successively increases and decreases the depth of different branches of the multilevel code structure in its search for the optimal hierarchical partitioning. For every split of a module into submodules, we use the two-level search algorithm described above.

1.4   Step-by-Step Instructions to the MapEquation Software Package

Here we provide detailed instructions on how to analyze networks with the map equation framework and the software we have developed on top of it. Networks can be analyzed either by using the MapEquation web applications, or by downloading the Infomap source code and run the program locally from the command line. Both the MapEquation software package and the Infomap source code can be found on the website www.mapequation.org. The MapEquation software package provides applications for both analyzing and visualizing networks. The web application interface offers helpful tools for visualizing the data, but the interface can be slow for large networks and is not useful for clustering more than on the order of 10,000 nodes. The Infomap source code, on the other hand, is fast and more flexible. It can cluster networks with hundreds of millions of nodes and comes with additional options for network analysis. After using Infomap, many of the results can be imported and visualized in the web application.

This section is organized as follows. In Sect. 1.4.1, we provide instructions for how to analyze and visualize networks with the MapEquation web applications. In Sect. 1.4.2, we explain how to analyze networks with the Infomap source code from the command line. Finally, in Sect. 1.4.3, we specify all available input and output formats.

1.4.1   The MapEquation Web Applications

The MapEquation web applications are available on www.mapequation.org/apps. html. They include the *Map and Alluvial Generator* for visualizing two-level modular structures and change in those structures, and the *Hierarchical Network Navigator* for exploring hierarchical and modular organization of real-world networks. Below we provide step-by-step instructions for how to use these applications.

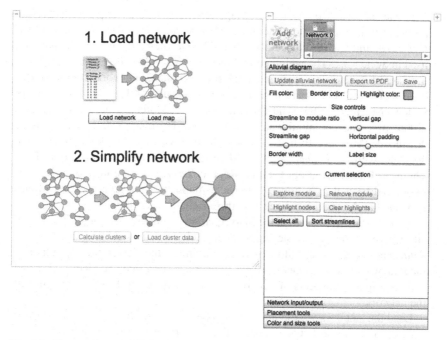

Fig. 1.5 The start frame of the Map and Alluvial Generator in the Flash application available on www.mapequation.org/apps/MapGenerator.html

1.4.1.1   The Map Generator

The Flash application on www.mapequation.org/apps/MapGenerator.html includes the Map Generator and the Alluvial Generator for analyzing and visualizing single networks and change in multiple networks, respectively. When you load your weighted or unweighted, directed or undirected network into the application, the Map Generator clusters the network based on the map equation and generates a map for you to customize and save as a pdf file. To simplify and summarize important structural changes in networks with alluvial diagrams, the Alluvial Generator makes it possible to load multiple networks with coinciding node names and partition them one by one with the Map Generator. Figure 1.5 shows the start frame of the Map and Alluvial Generator for loading networks.

To load, analyze, and view networks with the Map Generator, the following steps are essential:

1. Load the .net network file into the Map Generator by clicking the button *Load network* and choose between undirected and directed network. For very large networks, the load and clustering time can be long in the Flash-based Map Generator. If you are encountering problems because of large networks, you can run the clustering code offline and load the .map file into the application by clicking *Load map*.

If you just want to try out the Map Generator, a few sample networks are provided in the application, including *Modular demo network, Network scientists 2010,* and *Social science 2004.* The *Modular demo network* is the network used in Sect. 1.3.1. The weighted undirected network *Network scientists 2010* is the largest connected component of a coauthorship network compiled (for details of how weights are assigned, see Newman (2001)) from two network reviews Newman (2003)) and Boccaletti, Latora, Moreno, Chavez, and Hwang (2006) and one community detection review Fortunato (2010), and can be downloaded[1] in .net format. The weighted directed network *Social science 2004,* as well as *Science 1998–2004* provided under *Load map,* come from Thomson Reuters' Journal Citation Reports 2004. The data tally on a journal-by-journal basis the citations from articles published in a given year to articles published in the previous 5 years, with self-citations excluded.

2. Cluster the network based on the map equation by clicking *Calculate clusters* or alternatively provide a clustering in Pajek's .clu format by clicking *Load cluster data.* The Infomap algorithm tries to minimize the description length of a random walker's movement on the network as described in Sect. 1.3.2 and reveals important aspects of network structure with respect to the dynamics of the network.

   Load the .net network file into the Map Generator by clicking the button Load network and choose between undirected and directed network. When the network is loaded, click Calculate clusters to cluster the network based on the map equation and generate a map of the network.

3. Cluster the network based on the map equation by clicking *Calculate clusters* or alternatively provide a clustering in Pajek's .clu format by clicking *Load cluster data.* The Infomap algorithm tries to minimize the description length of a random walker's movement on the network as described in Sect. 1.3.2 and reveals important aspects of network structure with respect to the dynamics of the network.

4. The Map Generator displays the network as a map. Every module represents a cluster of nodes and the links between the modules represent the flow between the modules. The default name of a module is given by the node with the largest flow volume in the module. The size of a module is proportional to the average time a random walker spends on nodes in the module and the width of a link is proportional to the per step probability that a random walker moves between the

---

[1] Collaboration network of network scientists is available for download here: http://mapequation. org/downloads/netscicoauthor2010.net

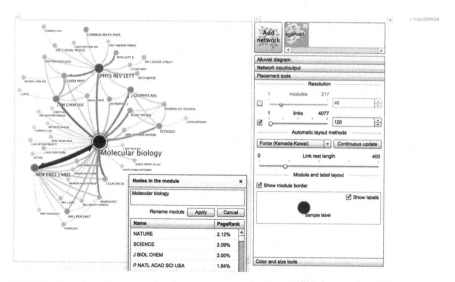

Fig. 1.6 The map and the control panel of the Map Generator available on www.mapequation. org/apps/MapGenerator.html

modules. The module's internal flow is also distinguished from the flow out of the module by a layer as shown for citation data below.

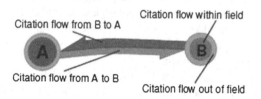

5. Customize the map by changing the position of the modules manually or automatically (see Placement tools in the control panel), by changing the names of the modules (double-click on a module to edit the name and list the nodes within the module together with their flow values), by changing the color of modules and links (see Color and size tools in the control panel), by moving the labels, etc. All adjustments can also be applied only to selected modules (shift-click selects a single module and shift-drag selects multiple modules).
6. Save the customized map in scalable vector graphics as a pdf file or as a .map file for later access in the Map Generator.

Figure 1.6 shows the Map Generator after following steps 1–5 above with the journal citation network Science 2004 provided in the application.

In this example, the network *Science 2004* was loaded and the scientific fields were identified and visualized by clicking *Calculate clusters*. The largest field has been renamed to *Molecular biology* and the open window lists the top journals assigned to the field together with their flow volume (PageRank).

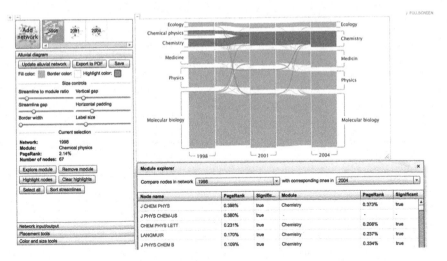

Fig. 1.7 The control panel and an alluvial diagram of the Alluvial Generator available on www.
mapequation.org/apps/MapGenerator.html

1.4.1.2 The Alluvial Generator

The Alluvial Generator is integrated with the Map Generator. It can load multiple
networks with coinciding node names and partition them one by one with the Map
Generator. The alluvial diagram can reveal organizational changes with streamlines
between modules of the loaded networks, as shown on the right in Fig. 1.7. The
figure shows three scientific journal networks for years 1998, 2001, and 2004
loaded with the button *Load map*.

In this example, the three journal citation networks *Science 1998*, *Science 2001*,
and *Science 2004* have been loaded as .map files. The alluvial diagram uses
streamlines to connect nodes assigned to modules in the different networks. Here
all journals assigned to the field Chemistry in 2004 have been highlighted in red,
and the streamlines trace back in which fields those journals were clustered in years
1998 and 2001. The Module explorer contains detailed information about individual journals.

To use the Alluvial Generator, the following steps are essential:

1. Follow the instructions for the Map Generator described in Sect. 1.4.1.1 above to
   load a first network.
2. Click *Add network* above the control panel to load additional networks. You can
   rearrange the order of loaded networks. Simply click a network thumbnail above
   the control panel and drag it to its preferred position.
3. The alluvial diagram is displayed to the right of the control panel. If you need
   more room, you can collapse the map by clicking the collapse button in the upper
   left corner of the map. It is easy to rearrange the modules and the streamlines,
   just click and drag to the new position. Highlighted nodes in a module can be

rearranged if you press the mouse button for 2 s. The modules are first named by the most important node in the module (highest PageRank), but all names can be selected and changed appropriately. To change the layout of the diagram, use the size controls under *Alluvial diagram* in the control panel.

4. To remove, highlight, or explore a module, just click the module and select one of the options under *Alluvial diagram* in the control panel. A double-click takes you directly to the *Module explorer*. Drag and select multiple modules to perform actions to multiple modules at the same time.

5. In the Module explorer, you can select and highlight individual or groups of nodes. The left column corresponds to the selected module(s) and the right column corresponds to the module assignment in the network marked in the drop-down list. Grayed out names belong to modules that are not included in the diagram. To include such a module, just double-click the grayed-out module name. A dash - means that the node does not exist in the network.

6. By clicking *FULLSCREEN* in the upper right corner you can use your entire screen. For security reasons, Flash does not allow for inputs from the keyboard in full screen mode and you cannot edit any text. Pressing *ESC* takes you back to normal mode.

To separate change from mere noise, we provide separate code that simultaneously identifies modules and performs significance analysis with bootstrap networks. The code outputs a file with extension .smap described in Sect. 1.4.3 below. To include significance information about the network clusters, download and run the code conf-infomap[2] on each network and load the resulting.smap files instead of the networks by clicking *Load map*.

1.4.1.3   The Hierarchical Network Navigator

We have developed the Hierarchical Network Navigator to make it easier to explore the hierarchical and modular organization of real-world networks. When you load your network, the network navigator first runs the Infomap algorithm to generate a hierarchical map. Then it loads the solution into the *Finder* and *Network view* for you to explore. The Finder simultaneously shows modules in multiple levels but no link structure, whereas the Network view shows the link structure within a single module. Figure 1.8 shows the Hierarchical Network Navigator loaded with the undirected network *Network scientists 2010* provided in the application.

In this example, we have loaded the collaboration network *Network scientists 2010* provided in the application and navigated three steps down to the finest level in the hierarchical organization to show the connections between the actual scientists in the module *Latora, V.,.* named after the researcher with strongest connections and largest flow value in the module. We use a period or multiple periods after

---

[2] Code for generating significance modules is available for download here: http://www.tp.umu.se/~rosvall/code.html

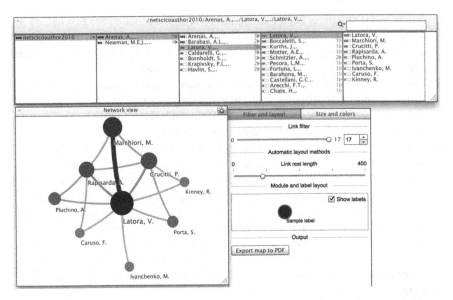

Fig. 1.8 The Finder and the Network view in the Hierarchical Network Navigator available on www.mapequation.org/apps/NetworkNavigator.html

a module name to indicate that there are one or more levels of nested modules within the module.

To use the Hierarchical Network Navigator, the following steps are essential:

1. Click on the button *Load network* and browse your network file matching the file formats.
2. Choose between Undirected/Directed links and Multilevel/Two-level clustering.
3. If you loaded a link list, you have to check Zero-based numbering if the node numbers in the file starts from 0, otherwise they are assumed to start from 1.
4. Click on *Calculate clusters*.
5. The Infomap algorithm will start to cluster your network and print its progress in the Network loader window. When it is finished, the Network loader window will automatically close and you can navigate your network in the Finder and the Network view. If Infomap exited with an error, please check the error message in the Network loader window.

Because the Hierarchical Network Navigator makes it possible to navigate very large networks, we have developed a streamable file format that supports fast navigation without first loading the entire network file and subsequently clustering the network in the Flash application with the integrated and therefore slower Infomap code as described in steps 1–5 above. The binary file with suffix .bftree can be generated with the fast Infomap source code described in Sect. 1.4.2. The file format includes the hierarchical structure in a breath first order, including the flow links of each subnetwork. In this way, only a small part of the file needs to be loaded to visualize the top structures, and the deeper structures can be loaded on demand. If the stand-alone Infomap source code is used, the following steps are essential:

1. Run Infomap on the network with the flag **–bftree** (see instructions in Sect. 1.4.2).
2. Click *Load map* in the Hierarchical Network Navigator and browse to your . bftree file generated by Infomap.
3. The Network loader window will automatically close and the Finder and Network view will open with the top level data visible.

Whichever method was used to generate the result, the following commands are available to navigate and customize the network:

Navigation	Use the keyboard arrows →, ↓, ←, and ↑ to navigate in the Finder. Use *click* and *alt + click* to navigate down and up, respectively, in the Network view. Use the search field in the Finder to directly navigate to nodes anywhere in the hierarchical structure.
Customization	Click the button in the top right in the Network view window to toggle the Control panel. Filter the number of nodes by setting the link limit. Change the size and color scales in the corresponding tab.
Explanation	The horizontal bars to the left of the node names in the Finder shows the flow volume of the nodes as a fraction of the whole network (color) and as a fraction of their parent nodes (size). The size is scaled logarithmically so that for each halving the flow is reduced by a factor of 10. Thus, the two vertical lines within the bar measures, from left to right, 1 and 10 % of the parent flow. The numbers after the modules in the Finder shows the maximum depth under that node in the tree. The same information is also encoded in the color of the arrow. The module names are automatically generated from their largest children, adding a new period for each level. Every module represents a cluster of nodes and the links between the modules represent the flow between the modules. The size of a module is proportional to the average time a random walker spends on nodes in the module and the width of a link is proportional to the per step probability that a random walker moves between the modules.

1.4.2   The Infomap Command Line Software

For large networks, the load and clustering time can be very long in the Flash-based Map Generator. To overcome this problem, the Infomap source code for clustering large networks can be executed from the command line. The code generates output files that can be loaded in the MapEquation web applications. Here we describe how to install and run Infomap from the command line with different options.

1.4.2.1 Installation

The latest Infomap code can be downloaded on www.mapequation.org/code.html.
The source code is also available at Bit-bucket.[3] Infomap is written in C++ and can
be compiled with gcc using the included Makefile. To extract the zipped archive
and compile the source code in a Unix-like environment, open a terminal and type:

```
cd [path/to/Infomap]
unzip Infomap-0.13.5.zip
cd Infomap-0.13.5
make
```

Substitute [path/to/Infomap] with the folder where Infomap was downloaded,
e.g., ~/Downloads. To be able to run on a Windows machine, you can install
MinGW/MSYS to get a minimalist development environment where the above
instructions will work. Please follow the instructions on MinGW—Getting Started[4]
or download the complete MinGW-MSYS Bundle.[5]

1.4.2.2 Running

To run Infomap on a network, use the following command:

```
./Infomap [options] network_data dest
```

The optional arguments can be put anywhere. Run ./Infomap –help or see
Sect. 1.4.2.3 for available options. The option network_data should point to a
valid network file and dest to a directory where Infomap should write the output
files. If no option is given, Infomap will assume an undirected network and try to
partition it hierarchically.

1.4.2.3 Options

For a complete list of options, run ./Infomap –help. Below we provide the most
important options to Infomap:

---

[3] https://bitbucket.org/mapequation/infomap

[4] http://www.mingw.org/wiki/GettingStarted

[5] http://sourceforge.net/projects/mingwbundle/files/latest/download

```
------- Input -------
-i<o>, --input-format=<o> # Specify input format ('pajek' or 'link-list') to override format
    possibly implied by file extension.
-z, --zero-based-numbering # Assume node numbers start from zero in the input file instead of
    one.
-k, --include-self-links # Include links with the same source and target node. (Ignored by
    default.)
-c<p>, --cluster-data=<p> # Provide an initial two-level solution (.clu format).

------- Output -------
--tree # the hierarchy in .tree format. (default true)
--map # the top two-level modular network in the .map format.
--clu # the top cluster indices for each node.
--node-ranks # the calculated flow for each node to a file.
--btree # the tree in a streamable binary format.
--bftree # the tree including horizontal flow links in a streamable binary format.

------- Dynamics -------
-u, --undirected # Assume undirected links. (default)
-d, --directed # Assume directed links.
-t, --undirdir # Two-mode dynamics: Assume undirected links for calculating flow, but directed
    when minimizing codelength.



-y<f>, --self-link-teleportation-probability=<f> # The probability of teleporting to itself.
    Effectively increasing the code rate, generating more and smaller modules.

------- Miscellaneous -------
-2, --two-level # Optimize a two-level partition of the network.
-s<n>, --seed=<n> # A seed to the random number generator.
-N<n>, --num-trials=<n> # The number of outer-most loops to run before picking the best
    solution.
-v, --verbose # Verbose output on the console. Add additional 'v' flags to increase verbosity
    up to -vvv.
-h, --help # Prints help message with full list of commands.
```

Argument types are defined by:

```
<n> # a positive integer
<f> # a floating point number
<p> # a path to a file or directory
<o> # a string that matches one of the listed options
```

1.4.2.4 Examples

We begin with a simple example with the network file ninetriangles.net provided in the root directory of the source code. First we create a new directory where we want the put the results, and feed the destination to Infomap together with the input network and the option to -N 10, which tells the code to pick the best result from ten attempts.

```
mkdir output
./Infomap ninetriangles.net output/ -N 10
```

Now Infomap will try to parse the file ninetriangles.net as an undirected network, try to partition it hierarchically, and write the best result out of ten attempts to the output directory as a .tree file described in Sect. 1.4.3. In the second example specified below, Infomap will treat the network as directed and try to find the optimal two-level partition with respect to the flow. With this command, Infomap will also create an output .map file that can be visualized with the Map Generator, for example.

```
./Infomap ninetriangles.net output/ -N 10 --directed --two-level --map
```

As mentioned earlier, for viewing large networks in the Hierarchical Network Navigator, we recommend options –btree or –bftree to generate streamable binary . btree and .bftree files, respectively.

For acyclic or time-directed networks, such as article-level citation networks, we recommend the option –undirdir, which makes Infomap use a random walk model with movements both along and against link directions but encoding only along link directions. With the standard random walk model, a disproportional amount of flow will reach the oldest articles.

1.4.3   Network Input and Output Formats

Here we specify all file formats that can be loaded as input or generated as output from the applications in the MapEquation software package.

1.4.3.1   Pajek's .net Format

Network data stored in Pajek's .net format can be loaded in all applications of the MapEquation software package. The Pajek format specifies both the nodes and the links in two different sections of the file. In the .net file, the network nodes have one unique identifier and a label. The definition of nodes starts with the line *Vertices N, where N is the number of nodes in the network. Labels or node names are quoted directly after the node identifier. The link section starts with the line *Edges L or *Arcs L (case insensitive), where L is the number of links. Weights can be given to nodes by adding a third column with positive numbers. Below we show an example network with six nodes and eight directed and weighted links.

```
# A network in Pajek's .net format
*Vertices 6
1 "Node 1" 1.0
2 "Node 2" 1.0
3 "Node 3" 1.0
4 "Node 4" 1.0
5 "Node 5" 1.0
6 "Node 6" 1.0
*Arcs 8
```

```
1 2 3.0
2 3 2.0
3 1 2.0
1 4 1.0
4 5 2.0
5 6 2.0
6 4 2.0
4 1 1.0
```

Pajek uses *Edges for undirected links and *Arcs for directed links. The MapEquation software package accepts both *Edges and *Arcs and the choice of load button in the user interface determines whether the algorithm treats the network as undirected or directed. Directed links have the form from to weight. That is, the first link in the list above goes from node 1 to node 2 and has weight 3.0. The link weight is optional and the default value is 1 (we aggregate the weights of links defined more than once). Node weights are optional and sets the relative proportion to which each node receives teleporting random walkers in the directed version of the code.

1.4.3.2   The Link List Format

The Hierarchical Network Navigator can, in addition to a Pajek .net file, also load a link list. A link list is a minimal format to describe a network by only specifying a set of links as shown below. Each line corresponds to the triad source target weight, which describes a weighted link between the nodes with specified numbers. The weight can be any nonnegative value. If omitted, the default link weight is 1. The nodes are assumed to start from 1 and the total number of nodes will be determined by the maximum node number.

```
# A network in link list format
1 2 1
1 3 1
2 3 2
3 5 0.5
. . .
```

1.4.3.3   Pajek's .clu Format

For a given network input file, it is also possible to specify the clustering of the nodes in all applications. The cluster information must be provided in Pajek's .clu format. In the web applications, the file can be loaded after the network by clicking *Load cluster data*. Infomap reads the cluster information with the option -c clusterfile.clu Pajek's .clu format is just a list of module assignments as shown below.

```
# A clustering in Pajek's .clu format
*Vertices 6
2
2
2
1
1
1
```

The cluster file above specifies that nodes 1–3 in the network belong to module 2 and that nodes 4–6 belong to module 1. Infomap generates a .clu file with the option –clu.

1.4.3.4   The .map and .smap Formats

Information contained in the network and the cluster file together can be loaded in the web applications as a single .map file, which also include link and node information aggregated at the module level. The .map file begins with information about the number of nodes, modules, and links in the network, followed by the modules, nodes, and the links between modules in the network as shown below.

```
# A network clustering file in the .map format
# modules: 2
# modulelinks: 2
# nodes: 6
# links: 8
# codelength: 2.51912
*Directed

*Modules 2
1 "Node 1,..." 0.5 0.0697722
2 "Node 4,..." 0.5 0.0697722
*Nodes 6
1:1 "Node 1" 0.209317
1:2 "Node 3" 0.147071
1:3 "Node 2" 0.143613
2:1 "Node 4" 0.209317
2:2 "Node 6" 0.147071
2:3 "Node 5" 0.143613
*Links 2
1 2 0.0697722
2 1 0.0697722
```

This .map file also contains the codelength and the flow volumes of the nodes, and was generated with Infomap. In the output from Infomap, the names under *Modules are by default derived from the node with the highest flow volume within the module and 0.5 0.0697722 represent, respectively, the aggregated flow volume of all nodes within the module and the per step exit flow from the module. The nodes are listed with their module assignments together with their flow volumes. Finally, all links between the modules are listed in order from high flow to low flow. Infomap generates a .map file with the option –map.

The .smap file below corresponds to the .map file above with additional significance information.

```
# A network significance clustering file in the .smap format
# modules: 2
# modulelinks: 2
# nodes: 6
# links: 8
# codelength: 2.51912
*Directed
*Modules 2
1 "Node 1,..." 0.5 0.0697722
2 "Node 4,..." 0.5 0.0697722
*Insignificants 1
2 < 1
*Nodes 6
1:1 "Node 1" 0.209317
1:2 "Node 3" 0.147071
1;3 "Node 2" 0.143613
2:1 "Node 4" 0.209317
2:2 "Node 6" 0.147071
2;3 "Node 5" 0.143613
*Links 2
1 2 0.0697722
2 1 0.0697722
```

The .smap file contains the necessary information to generate a significance map in the Alluvial Generator. Compared to the .map file above, this file also contains information about which modules that are not significantly standalone and which modules they most often are clustered together with. The notation 2 < 1 under *Insignificants 1 in the example above means that the significant nodes in module 2 are clustered together with the significant nodes in module 1 more often than the confidence level. In the module assignments, we use colons to denote significantly clustered nodes and semicolons to denote insignificantly clustered nodes. For example, the colon in 1:1 "Node 1" 0.209317 means that the node belongs to the by flow largest set of nodes that are clustered together more often than the confidence level number of bootstrap networks. This .smap file was generated with code described in Sect. 1.4.1.2. For more information, see Rosvall and Bergstrom (2010).

1.4.3.5 The .tree Format

The default output from Infomap is a .tree file that contains information about the identified hierarchical structure. The hierarchical structure in the .tree file below has three levels.

```
# A .tree file specifying a tree-level hierarchical structure
# Codelength = 3.48419 bits.
1:1:1 0.0384615 "7"
1:1:2 0.0384615 "8"
1:1:3 0.0384615 "9"
```

```
1:2:1 0.0384615 "4"
1:2:2 0.0384615 "5"
...
```

Each row begins with the multilevel module assignments of a node. The module assignments are colon-separated from coarse to fine level, and all modules within each level are sorted by the total PageRank of the nodes they contain. Further, the integer after the last comma is the rank within the finest-level module, the decimal number is the steady state population of random walkers, and finally is the node name within quotation marks. Infomap generates a .tree file by default or with the option –tree.

1.4.3.6 The .btree and .bftree Formats

To be able to navigate the network as soon as the hierarchical structure has been loaded into the Hierarchical Network Navigator, we use a customized streamable format that includes the tree structure in a breath first order (.btree and .bftree), including the flow links of each subnetwork (.bftree only). In this way, only a small part of the file has to be loaded to visualize the top structures, and the deeper structures can be loaded on demand. Infomap generates the .btree and .bftree files with the options –btree and –bftree, respectively.

Conclusions
This chapter is meant to serve as a guideline for how to use the map equation framework for simplifying and highlighting important structures in networks. We have described several applications developed for analysis and visualization of large networks. However, we haven't covered all features and new software is under development. Ultimately, for maximal usability and user-

(continued)

(continued)

friendly interface, we would like the web application to be as fast as the command line software. We are continuously investigating different solutions that become available with new web technology, and welcome all feedback. Therefore, we encourage you to contact us if you have any questions or comments. Please visit www.mapequation.org for the latest releases and updates, including an up-to-date version of this tutorial.

References

Aldecoa, R., & Marín, I. (2013). Exploring the limits of community detection strategies in complex networks. *Scientific Reports, 3*, 2216.

Barrat, A., Barthelemy, M., & Vespignani, A. (2008). *Dynamical processes on complex networks* (Vol. 574). Cambridge: Cambridge University Press.

Blei, D. M., Ng, A. Y., & Jordan, M. I. (2003). Latent Dirichlet allocation. *Journal of Machine Learning Research, 3*, 993–1022.

Blondel, V. D., Guillaume, J. L., Lambiotte, R., & Lefebvre, E. (2008). Fast unfolding of communities in large networks. *Journal of Statistical Mechanics: Theory and Experiment, 2008*(10), P10008.

Boccaletti, S., Latora, V., Moreno, Y., Chavez, M., & Hwang, D. U. (2006). Complex networks: Structure and dynamics. *Physics Reports, 424*(4), 175–308.

Brin, S., & Page, L. (1998). The anatomy of a large-scale hypertextual Web search engine. *Computer Networks and ISDN Systems, 30*(1), 107–117.

Dorogovtsev, S. N., & Mendes, J. F. F. (2003). *Evolution of networks: From biological Nets to the Internet and WWW*. Oxford: Oxford University Press.

Esquivel, A. V., & Rosvall, M. (2011). Compression of flow can reveal overlapping-module organization in networks. *Physical Review X, 1*(2), 021025.

Fortunato, S. (2010). Community detection in graphs. *Physics Reports, 486*(3), 75–174.

Fortunato, S., & Barthelemy, M. (2007). Resolution limit in community detection. *Proceedings of the National Academy of Sciences, 104*(1), 36–41.

Golub, G. H., & Van Loan, C. F. (2012). *Matrix computations* (Vol. 3). Baltimore, MD: JHU Press.

Gopalan, P. K., & Blei, D. M. (2013). Efficient discovery of overlapping communities in massive networks. *Proceedings of the National Academy of Sciences, 110*(36), 14534–14539.

Huffman, D. A. (1952). A method for the construction of minimum redundancy codes. *Proceedings of the IRE, 40*(9), 1098–1101.

Kanungo, T., Mount, D. M., Netanyahu, N. S., Piatko, C. D., Silverman, R., & Wu, A. Y. (2002). An efficient k-means clustering algorithm: Analysis and implementation. *IEEE Transactions on Pattern Analysis and Machine Intelligence, 24*(7), 881–892.

Karrer, B., & Newman, M. E. (2011). Stochastic blockmodels and community structure in networks. *Physical Review E, 83*(1), 016107.

Kawamoto, T., & Rosvall, M. (2014). The map equation and the resolution limit in community detection. *arXiv preprint arXiv:1402.4385*.

Lambiotte, R., & Rosvall, M. (2012). Ranking and clustering of nodes in networks with smart teleportation. *Physical Review E, 85*(5), 056107.

Lancichinetti, A., & Fortunato, S. (2009). Community detection algorithms: A comparative analysis. *Physical Review E, 80*(5), 056117.

Lancichinetti, A., Fortunato, S., & Radicchi, F. (2008). Benchmark graphs for testing community detection algorithms. *Physical Review E, 78*(4), 046110.

Lancichinetti, A., Radicchi, F., Ramasco, J. J., & Fortunato, S. (2011). Finding statistically significant communities in networks. *PLoS ONE, 6*(4), e18961.

Newman, M. E. (2001). Scientific collaboration networks. II. Shortest paths, weighted networks, and centrality. *Physical Review E, 64*(1), 016132.

Newman, M. E. (2003). The structure and function of complex networks. *SIAM Review, 45*(2), 167–256.

Newman, M. E. (2010). *Networks: An introduction.* New York, NY: Oxford University Press.

Newman, M. E., & Girvan, M. (2004). Finding and evaluating community structure in networks. *Physical Review E, 69*(2), 026113.

Peixoto, T. P. (2013). Hierarchical block structures and high-resolution model selection in large networks. *arXiv preprint arXiv:1310.4377.*

Rosvall, M., & Bergstrom, C. T. (2008). Maps of random walks on complex networks reveal community structure. *Proceedings of the National Academy of Sciences, 105*(4), 1118–1123.

Rosvall, M., & Bergstrom, C. T. (2010). Mapping change in large networks. *PLoS ONE, 5*(1), e8694.

Rosvall, M., & Bergstrom, C. T. (2011). Multilevel compression of random walks on networks reveals hierarchical organization in large integrated systems. *PLoS ONE, 6*(4), e18209.

Rosvall, M., Esquivel, A. V., West, J., Lancichinetti, A., & Lambiotte, R. (2013). Memory in network flows and its effects on community detection, ranking, and spreading. *arXiv preprint arXiv:1305.4807.*

Schaub, M. T., Lambiotte, R., & Barahona, M. (2012). Encoding dynamics for multiscale community detection: Markov time sweeping for the map equation. *Physical Review E, 86*(2), 026112.

Shannon, C. E. (1948). A mathematical theory of communication. *Bell System Technical Journal, 27*(3), 379–423.

Traag, V. A., Van Dooren, P., & Nesterov, Y. (2011). Narrow scope for resolution-limit-free community detection. *Physical Review E, 84*(1), 016114.

van Dongen, S. M. (2000). *Graph clustering by flow simulation.* Doctoral dissertation, University of Utrecht, the Netherlands.

Waltman, L., & Eck, N. J. (2012). A new methodology for constructing a publication-level classification system of science. *Journal of the American Society for Information Science and Technology, 63*(12), 2378–2392.

Ward, J. H., Jr. (1963). Hierarchical grouping to optimize an objective function. *Journal of the American Statistical Association, 58*(301), 236–244.

Chapter 2
Link Prediction

Raf Guns

Abstract Social and information networks evolve according to certain regularities. Hence, given a network structure, some potential links are more likely to occur than others. This leads to the question of link prediction: how can one predict which links will occur in a future snapshot of the network and/or which links are missing from an incomplete network?

This chapter provides a practical overview of link prediction. We present a general overview of the link prediction process and discuss its importance to applications like recommendation and anomaly detection, as well as its significance to theoretical issues. We then discuss the different steps to be taken when performing a link prediction process, including preprocessing, predictor choice, and evaluation. This is illustrated on a small-scale case study of researcher collaboration, using the freely available linkpred tool.

2.1 Introduction

The field of informetrics studies quantitative aspects of knowledge and information. This involves areas such as citation analysis, collaboration studies, web link studies, and bibliometric mapping. These areas often comprise relations between people, documents, social structures, and cognitive structures. Although such relations can be looked at from many different angles, one of the most promising approaches is the network perspective.

The study of networks is gaining increasing attention in fields of research as diverse as physics, biology, computer science, and sociology. In informetrics as well, network analysis is a central component. This is not a new phenomenon. It was for instance recognized early on that the mathematical study of graphs might also be beneficial to a better understanding of how documents influence each other (as reflected in citation relations). Important early landmarks include work by Pinski and Narin (1976), Price (1965), and Xhignesse and Osgood (1967).

R. Guns (✉)
Institute for Education and Information Sciences, IBW, University of Antwerp, Venusstraat 35, 2000, Antwerpen, Belgium
e-mail: raf.guns@uantwerpen.be

© Springer International Publishing Switzerland 2014
Y. Ding et al. (eds.), *Measuring Scholarly Impact*,
DOI 10.1007/978-3-319-10377-8_2

Later studies have expanded on these seminal works and broadened the scope to all kinds of informational phenomena. Indeed, many interactions studied in informetrics can be represented as networks, e.g. citation networks, collaboration networks, web link networks, and co-citation networks. In recent years, several studies employ measures and techniques borrowed from **social network analysis** (e.g., Otte & Rousseau, 2002).

While a lengthy overview of social network analysis would lead us too far, we briefly introduce the terminology used throughout this chapter. A **network** or **graph** $G = (V, E)$ consists of a set of **nodes** or **vertices** V and a set of **links** or **edges** E. Each edge $e = \{u, v\}$ connects the nodes u and v ($u, v \in V$) The set of nodes $N(v)$ connected to a given node v is called its **neighbourhood**. The number of nodes adjacent to v (the cardinality of the neighbourhood) is called its **degree** and denoted as $|N(v)|$.

Although networks are sometimes studied as if they are static, most social and information networks tend to be dynamic and subject to change. In a journal citation network, for instance, new journals emerge and old ones disappear, a journal may start citing a journal it had never cited before, and so on. This kind of change in a network is not entirely random; several mechanisms have been proposed that explain how networks evolve. We point out two important ones:

- **Assortativity** is the tendency of actors to connect to actors that are, in some way, similar to themselves ('birds of a feather flock together'). The criterion for similarity may be diverse: race, sex, age, interests, but also for instance node degree. In some networks **dissortativity** has been determined: the tendency to connect to others who are different from oneself ('opposites attract').
- **Preferential attachment** (Barabási & Albert, 1999) is the tendency to connect with successful actors, where success is usually measured by degree. Preferential attachment is a self-reinforcing 'rich-get-richer' mechanism, since every node that links to a high-degree node increases the latter's degree and thus its attractiveness to other nodes. This eventually evolves into a network where the degree distribution follows a power law (as is observed for many social networks). The mechanism is closely related to the Matthew effect (Merton, 1968) and Price's (1976) success-breeds-success principle.

Suppose that we have a snapshot of a network at some point in time G_t. Given the existence of mechanisms like assortativity and preferential attachment, some changes in G_t are more likely than others. For instance, the chance that two high-degree actors with similar social background will connect in the next snapshot G_{t+1} is probably higher than for two low-degree actors from different backgrounds.

Link prediction is a more formalized way of studying and evaluating this kind of intuitions. The link prediction problem can be formulated as follows (Liben-Nowell & Kleinberg, 2007): to what extent can one predict which links will occur in a network based on older or partial network data? We can distinguish between future link prediction and missing link prediction. **Future link prediction** (Guns, 2009, 2011; Guns & Rousseau, 2014; Huang, Li, & Chen, 2005; Spertus, Sahami, & Buyukkokten, 2005; Yan & Guns, 2014) involves predicting links in a future snapshot of the network based on a current one. **Missing link prediction** (Clauset, Moore, & Newman, 2008; Guimerà & Sales-Pardo, 2009; Kashima & Abe, 2006; Zhou, Lü, & Zhang, 2009) involves predicting all links based on an incomplete or damaged version of a network (missing certain links or containing spurious ones, e.g. because of sampling or measurement errors). Both types can be addressed using similar methods.

The remainder of this chapter is structured as follows. Section 2.2 reviews the link prediction process and its applications. In Sect. 2.3 we describe the data set that will be used as an example throughout the chapter. Section 2.4 introduces the linkpred tool and Sect. 2.5 shows how it can be used for link prediction. Finally, the last section provides the conclusions.

2.2   The Link Prediction Process and Its Applications

Figure 2.1 provides a schematic overview of the link prediction process. There are four major steps: data gathering, preprocessing, prediction, and evaluation. In some studies (e.g., Guns, 2009), an extra postprocessing step in between prediction and evaluation can be distinguished. Because postprocessing is rare, it is left out here.

Data gathering may seem like an obvious step. It is explicitly included, because the quality of input data has a profound effect on the quality of later predictions. In this step, we also make a distinction between the training and the test data, both of which are typically derived from the same data sources. We will refer to the network that predictions are based on as the **training network**. If one wants to compare the prediction results to a 'ground truth' network (e.g., a later snapshot of

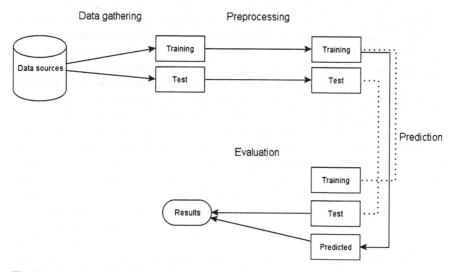

Fig. 2.1 Overview of the link prediction process

the same network), the latter network is called the **test network**. Note that the test network is only needed for evaluation purposes.

Preprocessing is very common but not mandatory. The main preprocessing step consists of filtering out certain nodes; the reasons and criteria for this are reviewed in Sect. 2.5.1. Preprocessing applies to both the training and test networks.

The prediction step operates on the training network. This step involves the choice of a **predictor**, a function or algorithm that calculates a likelihood score for each node pair (or for a subset of node pairs). Applying a predictor to the preprocessed training network yields a number of predictions. In practice, the prediction step results in a list of potential links with an associated likelihood score W. By ranking the potential links in decreasing order of W and choosing a threshold, one can obtain a predicted network. This step does not affect the training and test networks, which is reflected by the dotted lines in Fig. 2.1.

Evaluation involves comparing the test network with the predictions. There are several possible techniques and methods for this, most of which stem from information retrieval, data mining and related field (see Sect. 2.5.3).

At a high level, link prediction can be considered a statistical classification task. The items to be classified are node pairs, which should be classified into two groups: links and non-links. On the basis of empirical data (a training network, whose links we will refer to as attested links) one assigns a probability of linkage to each pair. These probability scores can then be used in different ways. We can generally distinguish between link prediction proper—that is, usage of probabilities for the actual purpose of prediction—and other applications, where these probabilities are exploited in different ways. Below, we list five possible applications of link prediction. The first two are mostly based on future link prediction, the next two build on missing link prediction, and the final one involves the importance of link prediction at a more theoretical level.

Let us first focus on link prediction proper. Typically, this happens when one is genuinely interested which links may arise in the future (or in a network which is related in a non-temporal way). In a scientometric context, this is interesting for policy makers, for whom a good understanding of likely future evolutions is crucial. Generally speaking, this use case is best served by a fairly small amount of very likely interactions or, in other words, by a focus on precision rather than recall. In the context of metadata generation, link prediction may help to alleviate sparsity of available associations (Rodriguez, Bollen, & Van de Sompel, 2009, p. 11).

A related application is **recommendation**: rather than actually trying to predict the future state of the network, one seeks to find likely but unattested links, often involving a specific node. For instance, given node a, one can create a (usually short) ranked list of candidate neighbours. These candidate neighbours are presented to a as recommendations. Interestingly, by doing so, one may influence the network's evolution. Recommendation is, for example, of great interest to smaller research groups that want to look into national or international collaboration: who are likely partners? There exists a rather large amount of literature on recommendation, recommender systems and collaborative filtering, where a and the items recommended to a are of different kinds, such as a researcher and papers she should read or cite and a library user and materials that might be of interest to him. Although most research into this kind of recommendation does not explicitly involve networks, it can be conceptualized as a link prediction problem within a two-mode network. Some studies explicitly study the role of link prediction in recommendation (Guns & Rousseau, 2014; Yan & Guns, 2014).

Thirdly, many networks that derive from actual data are in some ways incomplete. Link prediction can then be considered a tool to detect missing information. Conversely, erroneous data may sneak into a network and perhaps greatly affect research results. If the erroneous data cases the existence of a link that should not be present (similar to a completely random link inserted in the network), link prediction methods may single out the spurious link as highly unlikely.

Fourthly, detecting spurious links is related to a secondary application of link prediction: **anomaly detection**. Rattigan and Jensen (2005) suggest that anomaly detection offers a more fruitful line of inquiry than link prediction proper. The basic idea is that link prediction offers the tools to discover 'anomalous links', links that are unexpected and therefore interesting. For instance, an unexpected citation in a paper citation network may be a sign of interdisciplinarity (a paper building on methods or insights from other disciplines).

Finally, the main promise of link prediction at the theoretical level is as a practical way of testing and evaluating network formation and evolution models. Predictors normally derive from an explicit or implicit hypothesis of how and why links arise in a network. The performance of a predictor therefore also may help to test the validity of the underlying hypothesis. If a predictor performs markedly differently on different networks, this may point to variations in the factors that play a role in the evolution of these networks.

2.3 Data

We will use collaboration data to illustrate the process of link prediction. The data can be downloaded from https://raw.github.com/rafguns/linkpred/stable/examples/inf1990-2004.net and https://raw.github.com/rafguns/linkpred/stable/examples/inf2005-2009.net. The data represent a collaboration network between informetrics researchers, based on co-authorships. Thus, each node represents a researcher and each link represents a collaboration. A link's weight is the number of co-authorships of the two researchers. All data were downloaded from Thomson Reuters' Web of Science™.

The training network is in the file inf1990-2004.net (period 1990–2004) and the test network is in the file inf2005-2009.net (period 2005–2009). This data set is a subset of the data used by Guns, Liu, and Mahbuba (2011) and Guns (2011, 2012). Both files are in the Pajek format and can be read and visualized with several software packages, including Pajek and VOSviewer.

Table 2.1 summarizes basic descriptive statistics of these networks. It can be seen that the networks are very sparse and not well connected. Although the largest

Table 2.1 Descriptive
statistics of example data

	1990–2004	2005–2009
Number of nodes	632	634
Number of links	994	1,052
Density	0.0050	0.0052
Number of components	95	105
Size of largest component	238	173

component is significantly larger than the second largest component, it is not a real 'giant component' in either case. These properties may also affect the quality and feasibility of link prediction.

2.4   The Linkpred Tool

Linkpred is a tool that aims to make the most common link prediction actions available through a simple command-line interface. This means that linkpred is used by typing commands, rather than mouse input in a graphical user interface. Throughout the rest of this chapter, we assume that the operating system is Microsoft Windows. Because linkpred is written in the ubiquitous Python language (www.python.org), it can also be run under Apple's OS X and different flavours of Unix, including Linux.

An alternative software package that can be used for link prediction is LPmade (Lichtenwalter & Chawla, 2011), available from https://github.com/rlichtenwalter/LPmade.

2.4.1   Installation

The most straightforward way to get started is by installing the Anaconda Python distribution (https://store.continuum.io/cshop/anaconda/), which includes all required packages by default. One can then download and install linkpred as follows. Open a command line window and issue the following command:

```
> pip install https://github.com/rafguns/linkpred/archive/stable.
zip
```

This will display some output as linkpred is downloaded and installed. If no error messages are shown, linkpred is successfully installed. This can be verified by trying to run it, which should display the following output:

```
> linkpred
usage: linkpred training-file [test-file] [options]
linkpred: error: too few arguments
```

If an error message states that the command is unknown, linkpred can be run by referring to its exact location, for instance:

```
> C:\Anaconda\Scripts\linkpred
```

2.4.2 Basic Usage

Linkpred is a command-line tool. The command

```
> linkpred --help
```

should display basic usage information starting with the following line:

```
usage: linkpred training-file [test-file] [options]
```

We can see that linkpred expects a number of arguments: a training file, an optional test file, which are followed by options. The most important options are -p or --predictors, where a list of predictors is given, and -o or --output, where one can specify what kind of output is desired. A list of possible values is given in the output of linkpred --help.

The following examples give an idea how the tool can be used. This assumes that the current directory is the one where the data files inf1990-2004.net and inf2005-2009.net are located.

– To write all predictions based on the common neighbours predictor to a file:

```
> linkpred inf1990-2004.net -p CommonNeighbours -o cache-
predictions
```

The resulting file can be found alongside the training network and will have a name of the format *<training>*-*<predictor>*-predictions_*<timestamp>*.txt (e.g., inf1990-2004-CommonNeighbours-predictions_2013-10-25_14.47.txt).

– To apply the common neighbours and cosine predictors, and evaluate their performance with a recall-precision chart and a ROC chart (see Sect. 2.5.3 for details):

```
> linkpred inf1990-2004.net inf2005-2009.net -p
CommonNeighbours Cosine -o recall-precision roc
```

The resulting charts can be found alongside the training network and will have a name of the format *<training>-<chart-type>_<timestamp>*.pdf (e.g., inf1990-2004-ROC_2013-10-25_14.47.pdf). If a different file format than PDF is desired, this can be obtained by setting the -f (or --chart-filetype) option to another value (e.g., png or eps).

2.5 Link Prediction in Practice

In this paragraph, we discuss the practical steps to be taken when performing a link prediction study. We discuss preprocessing, the choice of predictor(s), the actual prediction, and evaluation, all with the linkpred program. Basic information on how to use linkpred as a module within Python is provided in the Appendix.

2.5.1 Preprocessing

Typically, some basic preprocessing operations need to be carried out before the actual prediction takes place.

Link prediction, by definition, is the act of predicting links, not nodes. However, as networks evolve, their node set evolves as well. Hence, one cannot assume that the training and the test network both contain exactly the same set of nodes. To allow for a fair comparison, we need to restrict our analysis to nodes that are common to the training and test networks. This is automatically done by linkpred. This way, it becomes possible—at least in theory—to do a perfect prediction with 100 % precision and 100 % recall.

It is notoriously difficult to predict anything about isolate nodes (nodes without any neighbours), since one has no (network-based) information about them. Similarly, low-degree nodes are sometimes discarded because one has too little information about them (e.g., Liben-Nowell & Kleinberg, 2007). However, doing so may lead to overestimating the precision of the predictions (Guns, 2012; Scripps, Tan, & Esfahanian, 2009). By default, linkpred removes isolate nodes and leaves all other nodes intact. This can be changed by setting min_degree to a different value than 1 (in a profile, see Sect. 2.5.4).

Collaboration networks are undirected, i.e. links can be traversed in both directions. Directed networks, on the other hand, have links with an inherent direction; citation networks are a prime example. Almost all link prediction studies deal exclusively with undirected networks. Hence, linkpred has been mainly tested on undirected networks and does not support directed networks. Shibata, Kajikawa, and Sakata (2012) study link prediction in an article citation network. However, they essentially treat the network as undirected and only predict the presence or absence of a link between two nodes. It is then assumed that the more recent article cites the older one. A similar approach is possible with linkpred: first, convert the directed network to an undirected one, and subsequently apply link prediction.

2.5.2 Predictor Choice

At the moment, linkpred implements 18 predictors, several of which have one or more parameters. Here, we will limit ourselves to discussing some of the most important ones. We distinguish between local and global predictors. Most predictors—both local and global—implemented in linkpred have both a weighted and an unweighted variant.

Local predictors are solely based on the neighbourhoods of the two nodes. Many networks have a natural tendency towards triadic closure: if two links $a - b$ and $b - c$ exist, there is a tendency to form the closure $a - c$. This property is closely related to assortativity and was empirically confirmed in collaboration networks by Newman (2001), who showed that the probability of two researchers collaborating increases with the number of co-authors they have in common: 'A pair of scientists who have five mutual previous collaborators, for instance, are about twice as likely to collaborate as a pair with only two, and about 200 times as likely as a pair with none'. It can be operationalized by the **common neighbours** predictor (Liben-Nowell & Kleinberg, 2007):

$$W(u,v) = |N(u) \cap N(v)| \tag{2.1}$$

Despite its simplicity common neighbours has been shown to perform quite well on a variety of social networks: triadic closure turns out to be a powerful mechanism indeed.

The common neighbours predictor is sensitive to the size of the neighbourhood. If both u and v have many neighbours, they are automatically more likely to have more neighbours in common. Therefore, several normalizations have been introduced in the literature, such as Dice's index (Zhou et al., 2009) or the **cosine** measure (Spertus et al., 2005). The latter, for instance, is defined as:

$$W(u,v) = \frac{|N(u) \cap N(v)|}{\sqrt{|N(u)| \cdot |N(v)|}} \tag{2.2}$$

Other implemented normalizations include Jaccard, maximum and minimum overlap, N measure, and association strength. These indicators are fairly well-known in the information science literature, since they are frequently used as similarity measures in information retrieval and mapping studies (e.g., Ahlgren, Jarneving, & Rousseau, 2003; Boyce, Meadow, & Kraft, 1994; Salton & McGill, 1983; Van Eck & Waltman, 2009). In most empirical studies, normalizations of common neighbours have an adverse effect on performance. In other words, two nodes that have many neighbours in common are automatically likely to link to each other, regardless of their total number of neighbours. One exception to this general rule are bipartite networks, such as author–paper networks; Guns (2011) shows that the cosine measure in particular forms a good predictor for this kind of networks.

The **Adamic/Adar** predictor starts from the hypothesis that a 'rare' (i.e., low-degree) neighbour is more likely to indicate a social connection than a high-degree one. Its original definition stems from Adamic and Adar (2003) and was adapted for link prediction purposes by Liben-Nowell and Kleinberg (2007):

$$W(u,v) = \sum_{z \in N(u) \cap N(v)} \frac{1}{\log|N(z)|} \tag{2.3}$$

A very similar predictor is **resource allocation**, which was introduced by Zhou et al. (2009):

$$W(u,v) = \sum_{z \in N(u) \cap N(v)} \frac{1}{|N(z)|} \tag{2.4}$$

Resource allocation is based on a hypothesis similar to that of Adamic/Adar, but yields a slightly different ranking. In practice this predictor often outperforms Adamic/Adar. Both Adamic/Adar and resource allocation tend to yield strong predictions and outperform common neighbours.

If preferential attachment is the mechanism underlying network evolution, it can be shown that the product of the degrees of nodes u and v is proportional to the probability of a link between u and v (Barabási et al., 2002). Hence, we define the **degree product** predictor (also known as the preferential attachment predictor):

$$W(u,v) = |N(u)| \cdot |N(v)| \tag{2.5}$$

Note that the assumptions underlying Eq. (2.5) are almost the opposite of those underlying Eq. (2.2) and similar normalizations: whereas the former rewards high degrees, the latter punishes them. This strongly suggests that if degree product performs well as a predictor, normalized forms of common neighbours will have poor results, and vice versa. Whereas some studies report poor performance for degree product (e.g., Liben-Nowell & Kleinberg, 2007), others have found degree product to be fairly strong predictor (e.g., Yan & Guns, 2014). The cohesion and density of the network appear to be influential factors.

Many networks have weighted links. In our case study, for instance, the link of a weight is equal to the number of co-authored papers. Taking link weights into account for link prediction seems like a logical step. It is therefore quite surprising that this subject has usually been ignored (see Murata & Moriyasu, 2007 and Lü & Zhou, 2010 for some exceptions). Guns (2012) argues that the vector interpretation of set-based similarity measures (Egghe & Michel, 2002) offers a strong theoretical

basis for weighted neighbour-based predictors. For instance, the weighted variant
of the cosine predictor becomes:

$$W(u,v) = \frac{\sum x_i \cdot y_i}{\sqrt{\sum x_i^2 \sum y_i^2}} \tag{2.6}$$

For each node i there is a corresponding vector element x_i or y_i. If node i is connected
to u, then $x_i = w_{u,i}$, the weight of the link between u and i (likewise for v and y_i). If
node i is unconnected to u (or v) the corresponding vector element is zero. This way,
linkpred implements weighted versions of all local predictors.

Now let us turn to the global predictors. Even if two nodes do not share any
common neighbours, they still may be related and form a link in a later stadium. A
straightforward measure of relatedness is the **graph distance** between two nodes:

$$W(u,v) = \frac{1}{d(u,v)} \tag{2.7}$$

where $d(u,v)$ denotes the length of the shortest path from node u to v. If link weights
are taken into account, graph distance is defined as:

$$W(u,v) = \left(\sum_{i=1}^{t} \frac{1}{w_i}\right)^{-1} \tag{2.8}$$

where w_i $(i = 1, \ldots, t)$ denotes the weight of the ith link in the shortest path (with
length t between u and v). Weighted graph distance is a much better predictor than
unweighted graph distance (Guns, 2012).

In 1953, Leo Katz proposed a centrality indicator that aims to overcome the
limitations of plain degree centrality (Katz, 1953). Let $\mathbf{A}$ denote the (full) adjacency
matrix of the network. The element a_{ij} is 1 if there is a link between nodes v_i and v_j
or 0 if no link is present. Each element $a_{ij}^{(k)}$ of A^k (the k-th power of $\mathbf{A}$) has a value
equal to the number of walks with length k from v_i to v_j (Wasserman & Faust, 1994,
p. 159). The **Katz** predictor is then defined as:

$$W(v_i, v_j) = \sum_{k=1}^{\infty} \beta^k a_{ij}^{(k)} \tag{2.9}$$

Generally, longer walks indicate a weaker association between the start and end
node. Katz (1953) therefore introduces a parameter $\beta(0 < \beta < 1)$, representing the
'probability of effectiveness of a single link'. Thus, each walk with length k has a
probability of effectiveness β^k. Its underlying hypothesis is that more and shorter
walks between two nodes indicate a stronger relatedness. Guns and Rousseau
(2014) show that the weighted variant of the Katz predictor can best be described

in the context of a multigraph, a network where one pair of nodes can be connected by multiple links. This predictor is among the most studied and best performing of all predictors implemented in linkpred.

PageRank is well-known thanks to its implementation in the Google search engine (Brin & Page, 1998). Assume the existence of a random walker (a 'random web surfer', Page, Brin, Motwani, & Winograd, 1999), who starts at a random node, randomly chooses one of its neighbours and navigates to that neighbour, again randomly chooses a neighbour and so on. Moreover, at every node, there is a small chance that the walker is 'teleported' to a random other node in the network. The chance of advancing to a neighbour is α ($0 < \alpha < 1$) and the chance of teleportation is $1 - \alpha$. One can determine the probability that the walker is at a given node. Some nodes are more important than others and will have a higher associated probability. This probability is equal to that node's PageRank. **Rooted PageRank** is a variant of PageRank where the random walker traverses the network in the same way, except that the teleportation is not randomized: the walker is always teleported back to the same root node. The associated rooted PageRank score can be interpreted as a predictor of other nodes' relatedness to the root node.

SimRank is a measure of the similarity between two nodes in a network, proposed by Jeh and Widom (2002). The SimRank thesis can be summarized as: *nodes that link to similar nodes are similar themselves*. To compute SimRank we start from the assumption that any node is maximally similar to itself: $sim(a, a) = 1$. The unweighted formula is (Antonellis, Molina, & Chang, 2008):

$$W(u, v) = \frac{c}{|N(u)| \cdot |N(v)|} \sum_{p \in N(u)} \sum_{q \in N(v)} W(p, q) \tag{2.10}$$

Here, c ($0 < c < 1$) is the 'decay factor', which determines how quickly similarities decrease. It is interesting to note that SimRank's authors explicitly acknowledge its bibliometric heritage: they regard SimRank as 'a generalization of co-citation where the similarity of citing documents is also considered, recursively. [...] This generalization is especially beneficial for nodes with few neighbours (e.g., documents rarely cited)' (Jeh & Widom, 2002). The performance of rooted PageRank and SimRank seems to vary from study to study; even more than for other predictors it is key to carefully tune the parameters. Furthermore, several studies use their unweighted versions, whereas the weighted ones perform better in most cases.

2.5.3 Prediction

Once a predictor has been chosen, it can be applied to the training network. In most cases, one wants to predict *new* links only, i.e. links that are not present in the training network. The default setting in linkpred is to predict only new links. It is, however, possible and in some cases perfectly appropriate to predict both current and new links. Anomaly detection, for instance, is based on likelihood scores that are given to current links. Current links whose likelihood score is below a certain threshold (or the bottom n) are potential anomalies. Predicting both current and new links can be enabled with the -a (or –all) option. For instance, the following command saves *all* predictions (according to SimRank) for the 2005–2009 network to a file:

```
> linkpred inf2005-2009.net -a -o cache-predictions -p SimRank
```

In the resulting list the least expected links can be considered potential anomalies, co-authorships that are unlikely given the network topology. Note that this is the opposite of 'normal' link prediction: instead of the most plausible links, we seek the least plausible ones. Without attempting to interpret the results, the six least expected co-authorships according to SimRank are the following (all of which had the same likelihood score):

- Fowler, JH–Aksnes, DW
- Liu, NC–Liu, L
- Origgi, G–Laudel, G
- Prabowo, R–Alexandrov, M
- Rey-Rocha, J–Martin-Sempere, MJ
- Shin, J–Lee, W

2.5.4 Evaluation and Interpretation

The prediction step results in a (usually large) number of predictions, each with a certain relatedness score W. It depends on one's intentions what should be done next. If no test file is supplied, the only possible output is saving the list of predictions, by setting output to cache-predictions. The list is saved as a tab-separated file with three columns: first node, second node, and relatedness score. This format can be easily imported into a spreadsheet like MS Excel or a statistical package like SPSS for further analysis.

If a test file is supplied, it becomes possible to evaluate the prediction results by comparing them with the test network. The following additional output values are possible:

- cache-evaluations: saves the evaluation data to a tab-separated file. Each line contains the following four columns: true positive number, false positive number, false negative number, and true negative number.
- recall-precision: yields a recall-precision chart (this is the default).
- roc: yields a ROC chart, which plots the false positive rate against the true positive rate (recall).

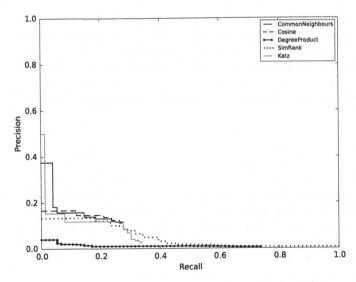

Fig. 2.2 Recall–precision chart for five predictors

- f-score: yields a chart showing the evolution of the F-score (the harmonic mean of recall and precision) as more predictions are made.
- fmax: yields a single-number indicator of performance, namely the highest F-score value.

Let us, as an example, try to compare the performance of the following predictors: common neighbours, cosine, degree product, SimRank, and Katz. We issue the following command:

```
> linkpred inf1990-2004.net inf2005-2009.net -p CommonNeighbours
Cosine DegreeProduct SimRank Katz
```

This will apply the chosen predictors to the training network with their default settings. Since no specific output has been specified, the default output (recall–precision) is chosen. Figure 2.2 displays the results. This chart illustrates the difficulty of comparing predictors. For instance, while Katz starts out as the best predictor (left side), it is overtaken by common neighbours and cosine for slightly higher recall levels. Degree product and SimRank appear to be weaker predictors in comparison, but achieve higher recall levels.

The ROC chart provides another view on the same data. Figure 2.3 was obtained with the command:

```
> linkpred inf1990-2004.net inf2005-2009.net -p CommonNeighbours
Cosine DegreeProduct SimRank Katz -o roc
```

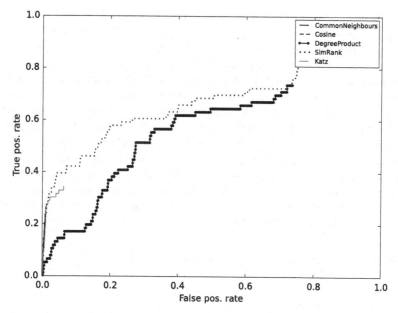

Fig. 2.3 ROC chart for five predictors

This chart should be interpreted as follows. If predictions were purely random-ized, they would follow the diagonal. A perfect prediction would follow a line from the bottom left to the top left corner, and from there to the top right corner. In other words, the higher, the better. Although this chart is based on exactly the same data as Fig. 2.2, it suggests a different interpretation: here, SimRank appears to be the best predictor. The main reason is that the largest part of Fig. 2.3 corresponds to recall values above (roughly) 0.3, which three of the five predictors do not achieve. Because most applications value precision over recall, we think recall–precision charts are usually to be preferred in link prediction evaluation.

2.5.5   Profiles

The linkpred tool supports the usage of **profiles**. A profile is a file that describes predictors and settings for a prediction run. Profiles have two advantages over setting predictors and settings through the command line:

1. One does not need to enter a potentially lengthy list of settings each time.
2. A profile allows for more fine-grained control over settings. For instance, predictor parameters can only be controlled through a profile.

Profiles can be written in the JSON and YAML formats.[1] We will use YAML here and demonstrate its use with a typical application: recommendation.

Suppose that we want to generate recommendations for collaboration between authors in informetrics. The main challenge here is to generate realistic recommendations that are at the same time not too obvious. As for preprocessing, it seems advisable that one does not exclude less prolific authors and sets min_degree to 1. The next step is prediction. Obviously we are only interested in unseen links, which is the default setting. Which predictors should be chosen? Since most local predictors yield rather obvious predictors, one will probably mainly use global predictors with good performance, such as rooted PageRank and Katz. The parameters of these predictors influence how far apart a predicted node pair can be; we choose to test two different values for each. Finally, since we are recommending, there is no (formal) evaluation step and we set output to cache-predictions. The corresponding profile looks like this:

```
output:
- cache-predictions
predictors:
- name: RootedPageRank
  displayname: Rooted PageRank (alpha = 0.5)
  parameters:
   alpha: 0.5
- name: RootedPageRank
  displayname: Rooted PageRank (alpha = 0.8)
  parameters:
   alpha: 0.8
- name: Katz
  displayname: Katz (beta = 0.01)
  parameters:
   beta: 0.01
- name: Katz
  displayname: Katz (beta = 0.001)
  parameters:
   beta: 0.001
```

Note that we can set predictor parameters and change a predictor's display name (the way it is displayed in chart legends, etc.) We save the profile as rootedpr.yaml and use it as follows:

```
> linkpred inf2005-2009.net --profile rootedpr.yaml
```

---

[1] See, e.g., http://en.wikipedia.org/wiki/JSON and http://en.wikipedia.org/wiki/YAML.

Table 2.2 Maximum
F-scores for rooted PageRank
with different α values

α	Fmax
0.1	0.1538
0.5	0.1923
0.9	0.1554

Because it is hard to assess which parameter values will yield the best predictions, it is often advisable to first test different settings on a related set of training and test networks. For instance, since we are recommending new links on the basis of the 2005–2009 data set, we could test which α value for rooted PageRank yields the best result for the 1990–2004 training network and the 2005–2009 test network. As indicated by the recall–precision chart (not shown) and the maximum F-scores (Table 2.2), a value around 0.5 appears to be optimal.

As a last example, we show how to compare the weighted and unweighted variants of a predictor, in this case Jaccard, with a recall–precision and ROC chart. The profile looks like this:

```
output:
- recall-precision
- roc
predictors:
- name: Jaccard
  displayname: Jaccard, unweighted
  parameters:
    weight: null
- name: Jaccard
  displayname: Jaccard, weighted
  parameters:
    weight: weight
```

Discussion and Conclusions

The use of network analysis techniques has become increasingly common in informetric studies over the last 10 years. Link prediction is a fairly recent set of techniques with both theoretical and practical applications. The techniques were illustrated on a case study of collaboration between researchers in the field of informetrics.

Generally speaking, it turns out that link prediction *is* possible, because link formation in social and information networks does not happen at random. Nonetheless, link prediction methods are also limited in several ways. First, these methods rely only on network topology and are unaware of social, cognitive and other circumstantial factors that may affect a network's evolution. To some extent, such factors are reflected in the network's topology but

(continued)

(continued)

the match is always imperfect. Second, the question which predictors are the best choice in a given concrete situation is difficult to answer. Some predictors (e.g., Katz) exhibit good performance in many studies and are therefore a good 'first choice' in the absence of more specific information. At the same time, it is typically a good idea to test different predictors on a comparable data set (e.g., Yan & Guns, 2014) in order to make a more informed decision on predictor choice. Third, there is a trade-off between prediction accuracy and non-triviality: good predictors often yield predictions that are predictable (!) and therefore less interesting.

As we have shown, the linkpred tool offers a simple but powerful way to perform link prediction studies. Its main limitations pertain to speed and network size: very large networks may be slow or even impossible to analyze with linkpred (also depending on which predictors are used). Nevertheless, the program has been applied to networks with several thousands of nodes and links without any problems. Moreover, it is always possible to export predictions (and evaluations) from linkpred for further analysis and processing in other software. Finally, we mention that linkpred is available as open source software (modified BSD license).

Appendix: Usage as a Python Module

Linkpred can be used both as a standalone tool and as a Python module. Here we provide basic instructions for usage as a Python module.

Once linkpred is properly installed, we can open a Python console and load the module from within Python.

```
> python
Python 2.7.3 (default, Apr 10 2012, 23:24:47) [MSC v.1500 64 bit
(AMD64)] on win32
Type "help", "copyright", "credits" or "license" for more information.
>>> import linkpred
```

The linkpred module is now loaded and can be used. First, let us open a network:

```
>>> G = linkpred.read_network("inf1990-2004.net")
11:49:00 - INFO - Reading file 'inf1990-2004.net'...
11:49:00 - INFO - Successfully read file.
```

We can now explore some properties of the network (stored in variable G), such as its number of nodes or links (see the NetworkX documentation at http://networkx.github.io/ for further information):

```
>>> len(G) # number of nodes
632
>>> G.size() # number of links
994
```

Predictors can be found in the linkpred.predictors submodule. Let us create a SimRank predictor for our network as an example. By setting only_new to True, we make sure that we only predict new links (i.e., links that are not present in the current network).

```
>>> simrank = linkpred.predictors.SimRank(G, only_new=True)
```

The line above only sets up the predictor, it does not actually apply it to the network. To do that, we invoke the predict method. Predictor parameters can be set here; we will set c to 0.5.

```
>>> simrank_results = simrank.predict(c=0.5)
```

Finally we take a look at the top five predictions and their scores.

```
>>> top = simrank_results.top(5)
>>> for authors, score in top.items():
...   print authors, score
...
Tomizawa, H - Fujigaki, Y 0.188686630053
Shirabe, M - Hayashi, T 0.143866427916
Garfield, E - Fuseler, EA 0.148097050146
Persson, O - Larsen, IM 0.138516589957
Vanleeuwen, TN - Noyons, ECM 0.185040358711
```

References

Adamic, L., & Adar, E. (2003). Friends and neighbors on the web. *Social Networks, 25*(3), 211–230.

Ahlgren, P., Jarneving, B., & Rousseau, R. (2003). Requirements for a cocitation similarity measure, with special reference to Pearson's correlation coefficient. *Journal of the American Society for Information Science and Technology, 54*(6), 550–560.

Antonellis, I., Molina, H. G., & Chang, C. C. (2008). Simrank++: Query rewriting through link analysis of the click graph. *Proceedings of the VLDB Endowment, 1*(1), 408–421.

Barabási, A.-L., & Albert, R. (1999). Emergence of scaling in random networks. *Science, 286* (5439), 509.

Barabási, A.-L., Jeong, H., Néda, Z., Ravasz, E., Schubert, A., & Vicsek, T. (2002). Evolution of the social network of scientific collaborations. *Physica A: Statistical Mechanics and its Applications, 311*(3–4), 590–614.

Boyce, B. R., Meadow, C. T., & Kraft, D. H. (1994). *Measurement in information science.* San Diego, CA: Academic.

Brin, S., & Page, L. (1998). The anatomy of a large-scale hypertextual web search engine. *Computer Networks and ISDN Systems, 30*(1–7), 107–117.

Clauset, A., Moore, C., & Newman, M. E. J. (2008). Hierarchical structure and the prediction of missing links in networks. *Nature, 453*(7191), 98–101.

de Solla Price, D. J. (1965). Networks of scientific papers. *Science, 149*(3683), 510–515.

de Solla Price, D. J. (1976). A general theory of bibliometric and other cumulative advantage processes. *Journal of the American Society for Information Science, 27*(5), 292–306.

Egghe, L., & Michel, C. (2002). Strong similarity measures for ordered sets of documents in information retrieval. *Information Processing & Management, 38*(6), 823–848.

Guimerà, R., & Sales-Pardo, M. (2009). Missing and spurious interactions and the reconstruction of complex networks. *Proceedings of the National Academy of Sciences, 106*(52), 22073–22078.

Guns, R. (2009). Generalizing link prediction: Collaboration at the University of Antwerp as a case study. *Proceedings of the American Society for Information Science & Technology, 46*(1), 1–15.

Guns, R. (2011). Bipartite networks for link prediction: Can they improve prediction performance? In E. Noyons, P. Ngulube, & J. Leta (Eds.), *Proceedings of the ISSI 2011 Conference* (pp. 249–260). Durban: ISSI, Leiden University, University of Zululand.

Guns, R. (2012). *Missing links: Predicting interactions based on a multi-relational network structure with applications in informetrics.* Doctoral dissertation, Antwerp University.

Guns, R., Liu, Y., & Mahbuba, D. (2011). Q-measures and betweenness centrality in a collaboration network: A case study of the field of informetrics. *Scientometrics, 87*(1), 133–147.

Guns, R., & Rousseau, R. (2014). Recommending research collaborations using link prediction and random forest classifiers. *Scientometrics.* doi:10.1007/s11192-013-1228-9.

Huang, Z., Li, X., & Chen, H. (2005). Link prediction approach to collaborative filtering. In *JCDL'05: Proceedings of the 5th ACM/IEEE-CS Joint Conference on Digital Libraries* (pp. 141–142). New York, NY: ACM Press.

Jeh, G., & Widom, J. (2002). SimRank: A measure of structural-context similarity. In *KDD'02: Proceedings of the Eighth ACM SIGKDD International Conference on Knowledge Discovery and Data Mining* (pp. 538–543). New York, NY: ACM.

Kashima, H., & Abe, N. (2006). A parameterized probabilistic model of network evolution for supervised link prediction. In *Proceedings of the 6th IEEE International Conference on Data Mining (ICDM2006)* (pp. 340–349). Washington, DC: IEEE Computer Society.

Katz, L. (1953). A new status index derived from sociometric analysis. *Psychometrika, 18*(1), 39–43.

Liben-Nowell, D., & Kleinberg, J. (2007). The link-prediction problem for social networks. *Journal of the American Society for Information Science and Technology, 58*(7), 1019–1031.

Lichtenwalter, R. N., & Chawla, N. V. (2011). LPmade: Link prediction made easy. *Journal of Machine Learning Research, 12*, 2489–2492.

Lü, L., & Zhou, T. (2010). Link prediction in weighted networks: The role of weak ties. *Europhysics Letters, 89*, 18001.

Merton, R.K. (1968). The Matthew effect in science. *Science, 159*(3810), 56–63.

Murata, T., & Moriyasu, S. (2007). Link prediction of social networks based on weighted proximity measures. In *Proceedings of the IEEE/WIC/ACM International Conference on Web Intelligence* (pp. 85–88). Washington, DC: IEEE Computer Society.

Newman, M. E. (2001). Clustering and preferential attachment in growing networks. *Physical Review E, 64*(2), 025102.

Otte, E., & Rousseau, R. (2002). Social network analysis: A powerful strategy, also for the information sciences. *Journal of Information Science, 28*(6), 441–453.

Page, L., Brin, S., Motwani, R., & Winograd, T. (1999). *The PageRank citation ranking: Bringing order to the web.* Stanford, CA: Stanford Digital Library Technologies Project.

Pinski, G., & Narin, F. (1976). Citation influence for journal aggregates of scientific publications: Theory with application to the literature of physics. *Information Processing & Management, 12*(5), 297–312.

Rattigan, M. J., & Jensen, D. (2005). The case for anomalous link discovery. *ACM SIGKDD Explorations Newsletter, 7*(2), 41–47.

Rodriguez, M. A., Bollen, J., & Van de Sompel, H. (2009). Automatic metadata generation using associative networks. *ACM Transactions on Information Systems, 27*(2), 1–20.

Salton, G., & McGill, M. J. (1983). *Introduction to modern information retrieval*. New York, NY: McGraw-Hill.

Scripps, J., Tan, P. N., & Esfahanian, A. H. (2009). Measuring the effects of preprocessing decisions and network forces in dynamic network analysis. In *Proceedings of the 15th ACM SIGKDD International Conference on Knowledge Discovery and Data Mining* (pp. 747–756). New York, NY: ACM.

Shibata, N., Kajikawa, Y., & Sakata, I. (2012). Link prediction in citation networks. *Journal of the American Society for Information Science and Technology, 63*(1), 78–85.

Spertus, E., Sahami, M., & Buyukkokten, O. (2005). Evaluating similarity measures: A large-scale study in the Orkut social network. In *Proceedings of the Eleventh ACM SIGKDD International Conference on Knowledge Discovery in Data Mining* (pp. 678–684). New York, NY: ACM.

Van Eck, N. J., & Waltman, L. (2009). How to normalize cooccurrence data? An analysis of some well-known similarity measures. *Journal of the American Society for Information Science and Technology, 60*(8), 1635–1651.

Wasserman, S., & Faust, K. (1994). *Social network analysis: Methods and applications*. Cambridge: University Press.

Xhignesse, L. V., & Osgood, C. E. (1967). Bibliographical citation characteristics of the psychological journal network in 1950 and in 1960. *American Psychologist, 22*(9), 778–791.

Yan, E., & Guns, R. (2014). Predicting and recommending collaborations: An author-, institution-, and country-level analysis. *Journal of Informetrics, 8*(2), 295–309.

Zhou, T., Lü, L., & Zhang, Y.-C. (2009). Predicting missing links via local information. *European Physical Journal B, 71*(4), 623–630.

Chapter 3
Network Analysis and Indicators

Staša Milojević

Abstract Networks have for a long time been used both as a metaphor and as a method for studying science. With the advent of very large data sets and the increase in the computational power, network analysis became more prevalent in the studies of science in general and the studies of science indicators in particular. For the purposes of this chapter science indicators are broadly defined as "measures of changes in aspects of science" (Elkana et al., Toward a metric of science: The advent of science indicators, John Wiley & Sons, New York, 1978). The chapter covers network science-based indicators related to both the social and the cognitive aspects of science. Particular emphasis is placed on different centrality measures. Articles published in the journal *Scientometrics* over a 10-year period (2003–2012) were used to show how the indicators can be computed in coauthorship and citation networks.

3.1 Introduction

Networks have become pervasive in modern life. We use them to describe and understand, among other phenomena, social connections, disease transmission, the human brain, and the Internet. The most commonly studied types of networks are: social (e.g., friendship, kinship, affiliation, collaboration); technological (e.g., the Internet, telephone network, power grids, transportation networks); biological (e.g., biochemical, neural, ecological); and information (e.g., document, citation, the World Wide Web). The "science of networks" includes perspectives from various fields, such as sociology, mathematics, physics, computer science, and biology. In particular, it is social network analysis and network science that lead the interdisciplinary effort to understand networks.

S. Milojević (✉)
Department of Information and Library Science, School of Informatics and Computing, Indiana University, Bloomington, IN, USA
e-mail: smilojev@indiana.edu

© Springer International Publishing Switzerland 2014
Y. Ding et al. (eds.), *Measuring Scholarly Impact*,
DOI 10.1007/978-3-319-10377-8_3

Social network analysis provides the methodology to analyze social relationships and patterns, and implications of those relationships (Wasserman & Faust, 1994). More broadly defined, network analysis provides a way to study any type of relationship among different entities (not only people). Sociologists and other social scientists made great advancements in the field of social network analysis over the past 60 years (Freeman, 2004; Wasserman & Faust, 1994). In the 1990s physicists and computer scientists started working in the area of what they called network science on the similar types of problems, but on a slightly different scales and aspects. While initially these two fields developed in parallel, there is recently more acknowledgement of each other's effort and cross-pollination of ideas (Carrington & Scott, 2011). Börner, Sanyal, and Vespignani (2007) provide an excellent overview of network science.

Network analysis/network science is today a mature field with its own introductory textbooks (Easley & Kleinberg, 2010; Newman, 2010; Scott, 2013; Wasserman & Faust, 1994). In addition, network analysis is now accessible to many researchers thanks to a number of free tools available for network analysis: Pajek, Network Workbench, NodeXL, Gephi, and The Science of Science (Sci2) Tool. Huisman and van Duijn (2011) provide an overview of the available software.

Social network methods developed as an integral part of the advances in social theory, empirical research, and formal mathematics and statistics. Wasserman and Faust (1994) identify graph theory, statistical and probability theory, and algebraic models, as the three major mathematical foundations of network methods. Graph theory is the branch of mathematics that studies graphs. A graph represents the structure of a network. In graph theory, itself a branch of set theory, a graph $G = (V,E)$ is a set of nodes or vertices (V) and a set of lines or edges (E) between specific pairs of nodes. In a graph, nodes represent actors, while lines represent ties between the actors. Graph theory provides both an appropriate representation of a social network and a set of concepts that can be used to study formal properties of social networks.

3.2   Networks and Bibliometrics

Networks have played a major role in studies of science (Ben-David & Collins, 1966; Crane, 1969, 1972; Mullins, 1972, 1973), both as a metaphor and as a method. Chubin (1976) suggested that "the imagery of a network has captured the fancy of most specialty analysts" (p. 463). Studies on the social production of knowledge (Babchuk, Keith, & Peters, 1999; Crane, 1972; Friedkin, 1998; Kuhn, 1996) showed the relationship between the social position of scientists and their beliefs and ideas. It is only recently, with the advancement of computer power and digital data availability, that large-scale network studies of entire disciplines became possible. A number of these studies were conducted by physicists who used coauthorship networks to study network dynamics (Barabási & Albert, 1999; Barabási et al., 2002; Farkas et al., 2002; Newman, 2001). Moody (2004) was among the first to use large-scale networks to study structure of a scientific discipline from the sociological standpoint. His study of the structure of social science collaboration network connects network topologies to empirical and theoretical findings from sociology of science.

One of the founders of the field of scientometrics, Derek de Solla Price, introduced the idea of studying documents through networks in 1965 paper titled "Networks of scientific papers" (De Solla Price, 1965). The study of networks has a long tradition in the field of scientometrics, but the connection with the networks literature in other fields was not fully established before the 2000s. More detailed description of these developments is outside the scope of this chapter. However, there are excellent articles already covering these topics. For example, Otte and Rousseau (2002) demonstrate the usage of social network analysis in information sciences, while Bar-Ilan (2008) provides a review of network studies of collaboration and White (2011) of scientific and scholarly networks.

The focus of this chapter is on network analysis used for science indicators. For the purposes of this chapter, science indicators are broadly defined as "measures of changes in aspects of science" (Elkana et al., 1978). However, indicators are more often viewed in a narrower sense as closely tied to research evaluation in general and evaluative bibliometrics in particular. In that sense indicators are developed and used to assess "the level of quality, importance, influence, or performance, of individual documents, people, journals, groups, domains (subject areas, fields, or disciplines), or nations" (Borgman & Furner, 2002, p. 6). As such, they are closely tied to science policy and rewards in science. Narin (1976) and Moed (2010) provide overviews of this subfield.

The remainder of the chapter will have the following structure. We will first introduce basic network terminology combined with the description of some network properties. We will then cover basics of network data. Following these introductions, we will use coauthorship, author-citation, and paper-citation networks for the articles published in journal *Scientometrics* in the 10-year period, 2003–2012, as examples. We will introduce a variety of centrality measures, describe how some of them have been used in scientometrics, and apply them to the three types of networks and discuss the results and implications of using those measures.

3.3 Basic Network Properties

Networks are a set of entities and relationships among these entities. Different fields have used different terms to describe these two basic network components. For example, in mathematics they are called vertices and edges/arcs; in computer science nodes and links and in sociology actors and ties/relations. The main focus of all network studies is to uncover the patterns within (social, informational, technological or biological) structures. In order to achieve this goal network studies focus on structural variables, i.e., the variables that measure the relationships between the entities.

Here are the definitions of the basic terms related to social network analysis.

Nodes are entities, such as discrete individuals and corporate or collective units, whose linkages we are concerned with. Sometimes we are interested in the pairs of actors and links between them, or *dyads*, or the triples of actors and links among them or *triads*.

Nodes are associated to one another by *links*. Links can have different attributes. They can be *directed*, i.e., they run in a particular direction from one node to another, or *undirected*, or reciprocal. In addition, the links can be either *dichotomous*, i.e., either present or absent, or *weighted*, i.e., have values that indicate either frequency or intensity of a relation. Other attributes associated with links can be rankings and types. Most network measures are defined only for undirected networks (Wasserman & Faust, 1994).

A *bipartite* or *two-mode* network is a special type of network in which there are two sets of entities and relations that exist between these two sets and not within them. An *affiliation* network is a type of two-mode network which has only one set of nodes and a set of events to which these nodes belong.

One of the basic properties of a node is a *degree*. A degree is the number of links connected to a node. In a directed network we differentiate *in-degree*, a number of incoming links, and *out-degree*, a number of outgoing links. In the case of weighted networks instead of a node degree one can calculate *node strength* as a sum of weights (Barrat, Barthélemy, Pastor-Satorras, & Vespignani, 2004; Newman, 2004a). Opsahl, Agneessens, and Skvoretz (2010) propose a measure that incorporates both node degree and degree strength.

The *density* of a network indicates what proportion of the connections that may exist between nodes is present. For the directed networks it is calculated as the ratio of the number of edges E to the number of possible edges $n(n-1)$, i.e., $D = E/(n(n-1))$ and for the undirected network as $D = 2E/(n(n-1))$.

The most common way to study connectedness in the network is through *components*. A network component is a maximally connected subgraph, i.e., "a subgraph in which there is a path between all pairs of nodes in the subgraph (all pairs of nodes in a component are reachable) and (since it is maximal) there is no path between a node in the component and any node not in the component." (Wasserman & Faust, 1994, p. 109). In undirected networks one talks simply about *connected components*. In directed networks there are *strongly connected components* in which each node within a component can reach every other node in the component by following the directed links and *weakly connected components* in which each node within a component can reach every other node in the component by disregarding the direction of the link. When the largest connected component encompasses a significant fraction of the network (usually over 50 %) it is called the *giant component*.

Finally, there are two measures that indicate how separated the entities in the network are. These are *geodesic distance*, or the *shortest path*, which is the shortest sequence of links connecting two nodes and the *diameter*, the longest shortest path in the network.

3.4 Network Data

The basic data collection techniques for network data are: questionnaires (roster vs. free-recall; free vs. fixed choice; and rating vs. complete ranking); interviews; observations; archival records; and experiments (Wasserman & Faust, 1994). Archival records in this context have a very broad meaning and include any type of record of interaction. Most bibliometric studies use archival records to collect data. Furthermore, they mostly focus on two types of data: bibliographic records (from one of the citation indexes, such as Thomson Reuters' Web of Science, Elsevier Scopus or Google Scholar) or patents. The usage of archival records has enabled studies of very large networks over time, which were either not possible, or would be much more labor-intensive using other techniques of data gathering.

There are three common ways to represent networks: *graphs*, *adjacency matrices*, and *adjacency lists* (Fig. 3.1). Adjacency matrices are used to present one-mode networks in which rows and columns index nodes. In an unweighted adjacency matrix links are represented as dichotomous values such that

$$A_{ij} = 1 \text{ if node } i \text{ has a link to node } j$$
$$= 0 \text{ if node } i \text{ does not have a link to node } j$$

$A_{ij} = A_{ji}$ if the network is undirected, or if i and j share a reciprocated link.

In a weighted adjacency matrix A_{ij} take values of weights, defined to be greater than zero.

Two-mode networks are represented by rectangular matrixes called *incidence matrices*.

Adjacency lists are much easier to work with than adjacency matrices, especially for large and/or sparse networks, which is why they are the most widely used network representation in different software tools.

One of the most widely used network formats today is .net, introduced in Pajek. However, Pajek in itself does not provide a way to automatically extract networks from bibliographic data retrieved from Web of Science or Scopus (although separate software tool, *WoS2Pajek*, has been developed to convert Web of Science

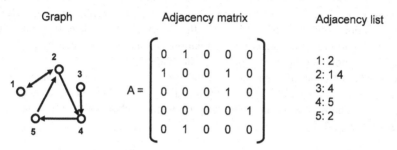

Fig. 3.1 A five-node network presented as a graph, adjacency matrix and adjacency list

to Pajek). Sci2 has the capability to extract network downloaded from these databases, when the records have been saved using particular formats. For example, for the Web of Science data the download format needs to be .isi, which is obtained by using "Save to Other Reference Software" option and then renaming the file to . isi. For the detailed description of data formats and types of networks that can be extracted please check Sci2 documentation at https://sci2.cns.iu.edu/user/documen tation.php.

3.5   Scientometrics Through Networks

For the remainder of the chapter we will focus on unweighted coauthorship, paper-citation and author-citation networks extracted from papers published in the journal *Scientometrics. Scientometrics* is the major specialized journal for the field of scientometrics. It began publication in 1978. The bibliographic records for the examples were downloaded from the Web of Science for the period 2003–2013. The data set includes 1,644 journal articles.

For the analyses described in this chapter we primarily used Pajek (version 3.08), while Sci2 (version 1.0 alpha) was used only for network extraction, computation of PageRank, and network visualizations. While many of the tasks could also have been performed using Sci2, the advantage of Pajek is the availability of a book (de Nooy, Mrvar, & Batagelj, 2011), which not only introduces major social network analysis concepts, but provides step-by-step instructions on how to obtain those measures.

Once the records have been downloaded from the Web of Science in batches of 500 they were combined into a single file, which was then loaded into Sci2 using File>Load… command (Fig. 3.2). After the data have been uploaded the details regarding them will be shown in the Data Manager area.

The next step is the extraction of networks. All of the networks can be extracted using Data Preparation suite of commands in Sci2. For the collaboration network we need to extract the coauthor network (Fig. 3.3). In the pop-up window choose "isi" file format. We will also use Extract Paper Citation Network to create the citation network. Please note that the link direction of the extracted paper citation network using Sci2 is different from what one would typically use. Namely, in paper citation networks created by Sci2 the direction of a link is from the cited paper to the citing paper, showing the relationship "paper B is cited by paper A," and not the more common one "paper A cites paper B."

Once the network has been created it will be displayed in the Data Management area. To save the network, select "Extracted Co-authorship network" and right-click. Then choose option Save. In a pop-up window choose the output data type Pajek .net (Fig. 3.4).

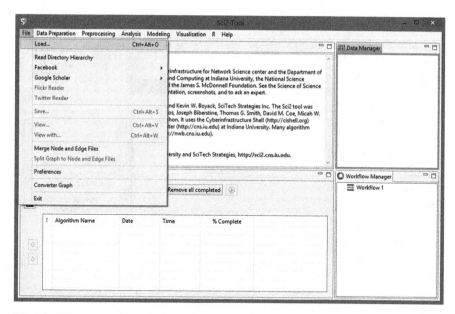

Fig. 3.2 Using Sci2 to load the data

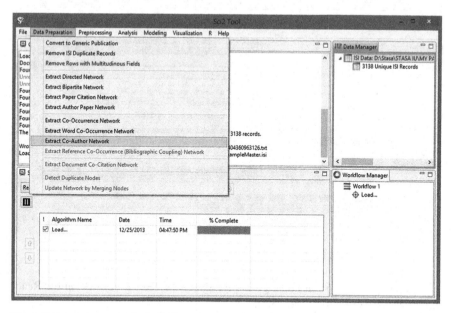

Fig. 3.3 Extracting networks in Sci2

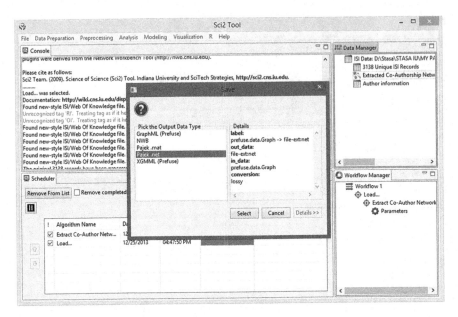

Fig. 3.4 Saving extracted networks in Pajek.net format

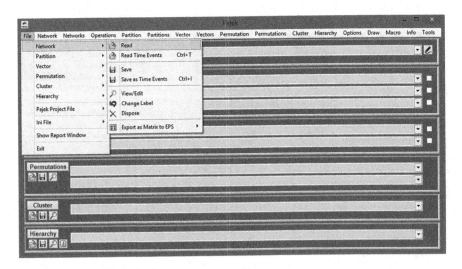

Fig. 3.5 Opening files in Pajek

To open the network in Pajek go to File>Network>Read, or click on the Read icon in the Networks row (Fig. 3.5).

Once the network has been successfully opened it will be shown in the Networks area. One can open multiple networks in Pajek. Use the drop down menu to choose the appropriate network. In order to get the basic network properties once the network is opened, go to the report screen for networks by choosing

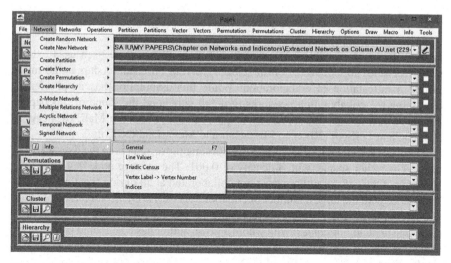

Fig. 3.6 Obtaining a report for the properties of a network in Pajek

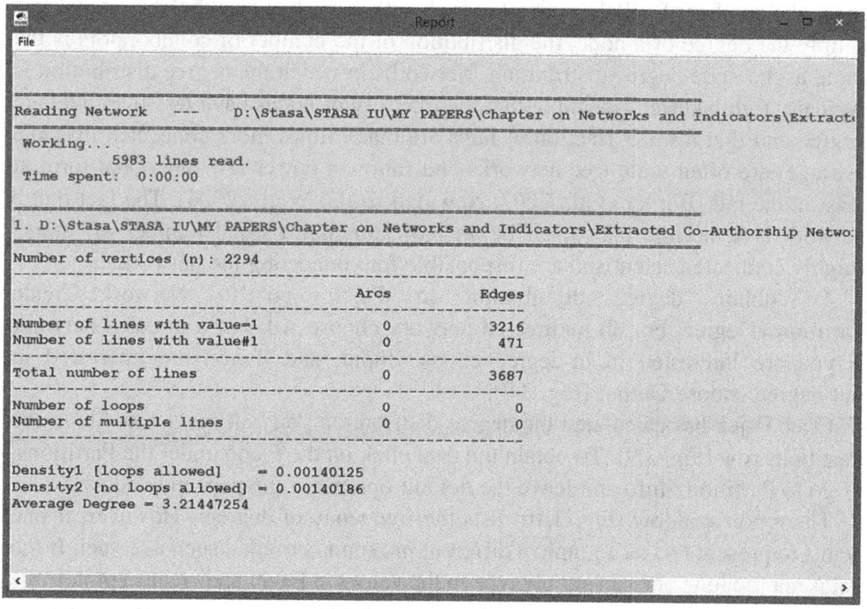

Fig. 3.7 Pajek report window showing basic network properties

Network>Info>General and in the pop-up window keep the default value (Fig. 3.6).

The report (Fig. 3.7) provides information on the number of nodes and lines, network density as well as the average node degree.

3.6   Collaboration Networks

We first focus on coauthorship networks. These networks have been used extensively for studying collaboration in science, relying on coauthorship as the most visible manifestation of collaboration. While this assumption has its problems, discussing the collaboration networks in detail is beyond the scope of this paper. Recent review articles that focus on coauthorship networks are Mali, Kronegger, Doreian, and Ferligoj (2012) and Biancani and McFarland (2013).

There are 2,294 authors and 3,687 links among those authors in the data set. The network is rather sparse, with a density of 0.001. The average node degree is 3.2. However, 7.8 % of the authors have not collaborated at all during this time period. The author with the largest number of collaborators (40) is Glänzel. Please note that we use the term *collaborators* to represent the totality of authors (or, more precisely, distinct author *names*) with whom a given author has coauthored papers over some time period, while the term *coauthors* is used to mean authors that appear on a single paper.

The distribution of the number of collaborators is an important indicator of the structure of an author network and of the processes that produce that structure. Since the number of collaborators that each author has in a coauthorship network is simply the degree of a node, the distribution of the number of collaborators is the same as the node degree distribution. Networks in which the degree distribution is typically right-skewed (meaning that the majority of nodes have less-than-average degree and that a small fraction of hubs are many times more connected than the average) are often scale-free networks and follow a power-law functional form at least in the tail (Börner et al., 2007; Newman, 2003; Watts, 2004). The fact that a network is scale-free may itself be an indicator that a small number of "stars" (highly connected scientists) are responsible for connecting the network.

To obtain degree distribution in Pajek, go to Network>Create Partition>Degree. For an undirected network choose All. For a directed network if you are interested in in-degree choose Input, and if you are interested in out-degree choose Output (Fig. 3.8).

Once Pajek has calculated the degree distribution you will see a new file in the Partitions row (Fig. 3.9). To obtain the data click on the *i* icon under the Partitions, or go to Partition>Info and leave the default options in the pop-up window.

The report window (Fig. 3.10) lists the frequency of degrees. However, if one wants to present this as a graph, a different program is needed such as Excel. If the list is not too long one can simply type in the values in Excel and create a plot. If the list is long, or one is more comfortable doing it that way, one should save the values by clicking the second icon (Save) under Partitions and choosing Save partition as Pajek partition in .clu format option. Once you have imported the file in Excel you will get only a single column with degree values. This is a list of degrees, rather than their frequency. One then needs to calculate the frequencies and plot them.

The node degree distribution (Fig. 3.11) in our example is similar to the large number of collaborator distributions identified in the literature (Barabási et al.,

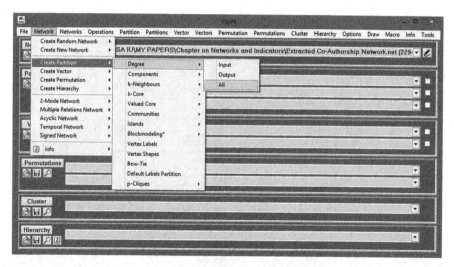

Fig. 3.8 Calculating degree distribution for undirected network in Pajek

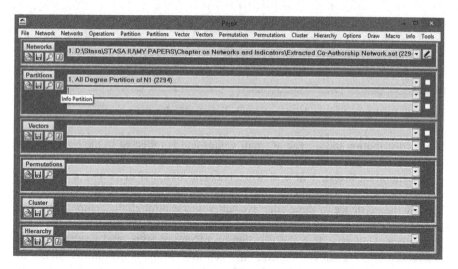

Fig. 3.9 The results after calculating the degree distribution

2002; Moody, 2004; Newman, 2004b) in that it is right-skewed. However, it also has a so-called hook for the authors with small number of collaborators, a feature identified by Wagner and Leydesdorff (2005) and further studied in Milojević (2010). This feature is important and the recent model of scientific research team formation and evolution (Milojević, 2014) has identified the likely underlying reason for its existence. It follows from two modes of knowledge production: one with relatively small core teams following Poisson distribution and another with extended teams which due to preferential attachment can become very large. The

```
┌─────────────────────────────────────────────────────────────────────────────┐
│ ≛                                   Report                              ─ □ ×  │
├─────────────────────────────────────────────────────────────────────────────┤
│ File                                                                          │
│ ═══════════════════════════════════════════════════════════════════════════  │
│ 1. All Degree Partition of N1 (2294)                                          │
│ ═══════════════════════════════════════════════════════════════════════════  │
│ Dimension: 2294                                                               │
│ The lowest value:  0                                                          │
│ The highest value: 40                                                         │
│                                                                               │
│ Frequency distribution of cluster values:                                     │
│                                                                               │
│   Cluster      Freq      Freq%   CumFreq   CumFreq%  Representative            │
│ ─────────────────────────────────────────────────────────────────────────    │
│         0       179     7.8030       179     7.8030  Velema, Ta               │
│         1       482    21.0113       661    28.8143  Shi, Y                   │
│         2       576    25.1090      1237    53.9233  Schifanella, C           │
│         3       388    16.9137      1625    70.8370  Wang, My                 │
│         4       236    10.2877      1861    81.1247  Amara, N                 │
│         5       123     5.3618      1984    86.4865  Landry, R                │
│         6        85     3.7053      2069    90.1918  Yu, Dr                   │
│         7        57     2.4847      2126    92.6765  Garcia-zorita, C         │
│         8        55     2.3976      2181    95.0741  Yang, Ly                 │
│         9        16     0.6975      2197    95.7716  Abramo, G                │
│        10        18     0.7847      2215    96.5562  Guan, Jc                 │
│        11        17     0.7411      2232    97.2973  Daniel, Hd               │
│        12         9     0.3923      2241    97.6896  Yu, G                    │
│        13         6     0.2616      2247    97.9512  Ma, Rm                   │
│        14         5     0.2180      2252    98.1691  Gorraiz, J               │
│        15         3     0.1308      2255    98.2999  Qiu, Jp                  │
│        16         4     0.1744      2259    98.4743  Chen, Dz                 │
│        17         4     0.1744      2263    98.6486  Huang, Mh                │
│        21         2     0.0872      2265    98.7358  Park, Hw                 │
│        22        23     1.0026      2288    99.7384  Wu, Ys                   │
│        23         1     0.0436      2289    99.7820  Debackere, K             │
│        25         1     0.0436      2290    99.8256  Ho, Ys                   │
│        26         2     0.0872      2292    99.9128  De Moya-anegon, F        │
│        36         1     0.0436      2293    99.9564  Rousseau, R              │
│        40         1     0.0436      2294   100.0000  Glanzel, W               │
│ ─────────────────────────────────────────────────────────────────────────    │
│       Sum      2294   100.0000                                                 │
└─────────────────────────────────────────────────────────────────────────────┘
```

Fig. 3.10 Pajek report for the frequency distribution of node degrees

former distribution has a peak at small values and is reflected in the hook. These processes affect the network topology.

An important property of a network is connectivity. If the network is disconnected, then some pairs of nodes cannot reach one another. One of the basic ways to identify the existence of these cohesive subgroups is through the components. The size of the giant component is expressed as the percentage of all nodes in the giant component. Understanding the distribution of components, and particularly the emergence of the giant component, has been used in the collaboration studies to study the emergence of a discipline. Namely, the existence of a robust giant component in a network has been interpreted as an indicator of field formation. Increases in the size of the largest connected component signify a transition from a relatively unorganized group of researchers into a scientific field.

To identify connected components in Pajek, go to Network>Create partition>Components>Strong (Fig. 3.12). As discussed above in Sect. 3.3, for a directed network there is a difference between the strongly connected and the weakly connected components. The choice will depend on the questions one wants to answer.

There are 586 connected components in the *Scientometrics* collaboration network. The largest connected component contains 647 authors, i.e., 28.2 % of all the

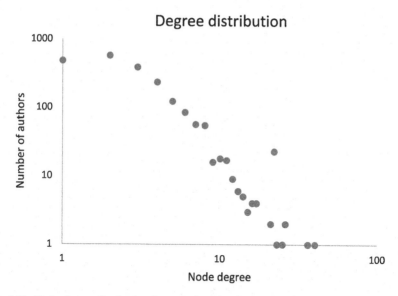

Fig. 3.11 Node degree distribution for authors who have published in journal *Scientometrics* 2003–2012

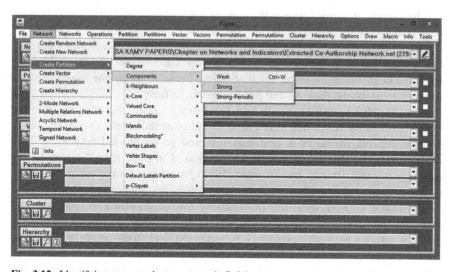

Fig. 3.12 Identifying connected components in Pajek

authors in the data set (Fig. 3.13). The next three largest components are much smaller and have 28, 20 and 20 authors each, respectively. To visualize the network in Sci2 go to Visualization>Networks>Guess, and then in the Layout choose option GEM. For more information on network visualization consult Chap. 13.

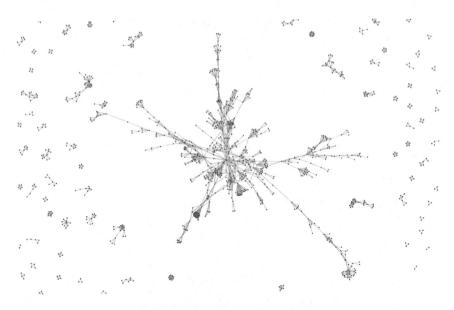

Fig. 3.13 The largest connected component of the collaboration network

The average distance between the authors within the largest connected component is 6.4 and the diameter is 17. To calculate the distances in Pajek go to Network>Create Vector>Distribution of Distances. The above results may suggest that researchers who publish in *Scientometrics* are not part of an established scientific field. However, to say more about the field formation one would need to evaluate longitudinal data and not just a snapshot in time.

After establishing these basic network properties let us turn our attention to some of the network measures that have potential utility as indicators. One of the most popular network measures is centrality (or prestige in directed networks). The measures of centrality try to answer the question of who the most "important" or central nodes in a network are. Freeman (1979) provides a comprehensive analysis of the literature on centrality.

The major network measures of centrality can be divided into local: degree centrality, and the ones relative to the rest of the network (i.e., based on node's location within a network): closeness, betweenness, and eigenvector (or Bonacich power) centrality. These measures have originally been developed for unweighted networks. While centrality measures for weighted networks have also been developed (Barrat et al., 2004; Newman, 2004a; Opsahl et al., 2010) the application and interpretation of those measures is not as straightforward and we will not explore those measures in this chapter. Many centrality and prestige measures have been used in scientometrics. For example, Leydesdorff explored the usefulness of betweenness centrality to measure the interdisciplinarity of journals (Leydesdorff, 2007; Leydesdorff & Rafols, 2011). He has found the normalized betweenness centrality to be a good indicator of interdisciplinarity. He provides centrality measures for 7,379 journals included in the *Science Citation Index and Social Science Citation Index 2004* (http://www.leydesdorff.net/jcr04/centrality/index.htm). In another study of journal indicators Leydesdorff (2009) has compared more traditional journal indicators (e.g., impact factor) with centrality measures and PageRank, and found that PageRank (which we will discuss later on) is not an impact indicator. Finally, Abbasi, Hossain, and Leydesdorff (2012) studied the evolution of collaboration networks and found that the authors with high betweenness centrality attract more collaborators than those with high degree and closeness centrality.

To compute degree, eigenvector, betweenness, and closeness centralities in Pajek go to Network>Create Vector>Centrality and choose Degree>All, Hubs-Authorities, Betweenness, and Closeness>All, respectively (Fig. 3.14). Hubs-Authorities in the directed networks computes hubs and authorities, but in undirected networks hubs and authorities are identical and they are the same as eigenvector centrality measures.

To view the results in the Report window click on the *i* icon in the Vectors area of the Pajek. One can view all the names and the values attached to them by clicking the magnifying glass icon ("View/Edit Vector") in the same area.

One of the simplest local measures of centrality is the degree centrality. This measure is tied to the idea of social capital. According to this measure of centrality, the most prominent nodes are the ones that have the most ties to other nodes in the network. These nodes can then use those connections to influence others, or can gain access to information quicker. According to degree centrality, the five most central authors in *Scientometrics* network are: W. Glänzel, R. Rousseau, F. De Moya-Anegon, B. Klingsporn, and Ys Ho (Table 3.1). A very interesting author in this group is B. Klingsporn. He illustrates the sensitivity of this measure to the existence of very large coauthorship teams. Namely, he has published only two papers in *Scientometrics* in the 10-year time period we have examined, one of those papers with 21 coauthors, which is the second largest number of coauthors per paper in the entire dataset. In the fields where the extensive authorship lists are becoming a norm alternative ways of creating collaboration networks may be warranted and measures such as degree centrality need to be carefully examined.

However, not all nodes are equal in terms of influence or power. Bonacich (1987) was the first to propose the extension of the simple degree centrality that takes into account node inequality. Namely, he proposed a measure that is based on the idea that a node's importance in a network may increase by having connections to other nodes that are themselves important. This measure is known as eigenvector centrality and is calculated by giving each node a score proportional to the sum of the scores of its neighbors. Thus, eigenvector centrality can be large either because a node has many neighbors or because it has important neighbors (or both). This measure works best for undirected networks. According to the eigenvector centrality, the five most central authors in *Scientometrics* network are: W. Glänzel, B. Thijs, A. Schubert, K. Debackere, and B. Schlemmer (Table 3.1). We see that the list of the most central authors has changed significantly in comparison to the one based on degree centrality (except for Glänzel who has the highest centrality in

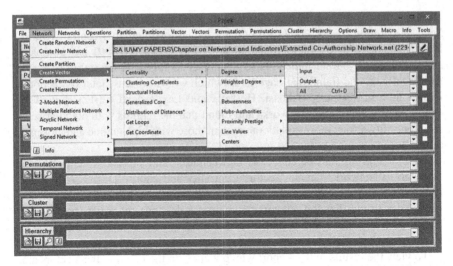

Fig. 3.14 Computing degree, eigenvector, betweenness and closeness centralities in Pajek

Table 3.1 The values and ranks (in parenthesis) for 14 authors that were top five authors in the *Scientometrics* dataset based on at least one of four measures of centrality: degree centrality, eigenvector centrality, betweenness centrality and closeness centrality

Name	Degree centrality	Eigenvector centrality	Betweenness centrality	Closeness centrality
Glänzel, W	40 (1)	0.63 (1)	0.05 (1)	0.08 (1)
Rousseau, R	36 (2)	0.07 (16)	0.02 (3)	0.07 (5)
De Moya-Anegon F	26 (3)	0.03 (26)	0.008 (12)	0.07 (6)
Klingsporn, B	26 (4)	0.004 (89)	0.005 (21)	0.05 (174)
Ho, Ys	25 (5)	0 (2096)	0.005 (22)	0.03 (613)
Thijs, B	11 (63)	0.49 (2)	0.001 (65)	0.07 (10)
Schubert, A	16 (36)	0.32 (3)	0.003 (38)	0.06 (27)
Debackere, K	23 (6)	0.31 (4)	0.006 (16)	0.07 (4)
Schlemmer, B	3 (670)	0.15 (5)	0 (382)	0.06 (33)
Meyer, M	14 (43)	0.07 (14)	0.02 (2)	0.07 (2)
Leydesdorff, L	12 (54)	0.02 (46)	0.02 (4)	0.06 (42)
Rafols, I	2 (1058)	0.004 (83)	0.01 (5)	0.06 (45)

The authors are not listed in any particular order

both). Apart from Glänzel and Debackere (who has degree centrality of 23, which makes him the sixth most central author in terms of degree centrality), other most central authors based on eigenvector centrality have relatively low degree centrality. The most obvious benefits of having a few connections, but to very influential individuals, can be seen in the case of Schlemmer who has degree centrality of 3 (which makes him 670th most central author according to the degree centrality) and yet is the fifth most central author in terms of eigenvector centrality. His position has been warranted by coauthoring all five of his papers with the most central person in the collaboration network, Glänzel. Interestingly, 21 out of 22 papers by Thijs (the second most important author in terms of eigenvector centrality) were coauthored with Glänzel.

Degree centrality and its variations have proven useful in the studies of power, status, and influence. However, these centrality measures may not be the best ones for all contexts. For example, degree centrality is insufficient to describe the ability to broker between groups, or likelihood that information originating anywhere in the network will reach a particular node. A better centrality measure for those instances is betweenness. Nodes with high betweenness centrality may serve both as enablers of communication between otherwise disjointed communities, but also as gatekeepers since they have the power to control the information that passes between others. According to the betweenness centrality, the five most central authors in the *Scientometrics* network are: W. Glänzel, M. Meyer, R. Rousseau, L. Leydesdorff, and I. Rafols (Table 3.1). The most interesting author in this group is I. Rafols who has published only three papers in this dataset with only two coauthors (M. Meyer and A. Porter). Nevertheless through these collaborations he is the only link between 57 researchers and the most central part of the graph around Glänzel. A look at the graph (Fig. 3.15) makes it very obvious that the most central authors according to this measure indeed serve as bridges.

Finally, there are contexts in which it is not very important to be in direct contact with many nodes, nor to broker between different communities, but to be close to the center. The measure that is used to identify individuals that can relatively quickly reach others in the network is called closeness centrality. Closeness centrality is based on the length of the average shortest path between a node and all other nodes in a network. Nodes with high closeness centrality, i.e., those that are fairly close to others in the network, can benefit from an opportunity to have their opinions heard more quickly in the community. According to the closeness centrality, the five most central authors in *Scientometrics* network are: W. Glänzel, M. Meyer, H. Kretschmer, K. Debackere, and R. Rousseau.

Collaboration networks have been studied with the aim to better understand the social structure of science, in general, and communication (and the potential for communication) within scientific communities, in particular. Thus, the different centrality measures we discussed have the potential to identify individuals who either play, or have a potential to play, important roles in conveying information through a particular community, share ideas, and promote interdisciplinarity and cross-pollination. Two authors who we have identified as the most central for the research community around the journal *Scientometrics* at the beginning of the twenty-first century are W. Glänzel and R. Rousseau. Glänzel has been identified to be the most central researcher according to all four measures of centrality (as well as the most productive author within the data set). The central role of this researcher for the community that publishes in this journal in a way has been confirmed by his recent appointment as the new editor of the journal *Scientometrics*. Another most central author, R. Rousseau, is among the top five authors on all centrality measures except eigenvector centrality (as well as the

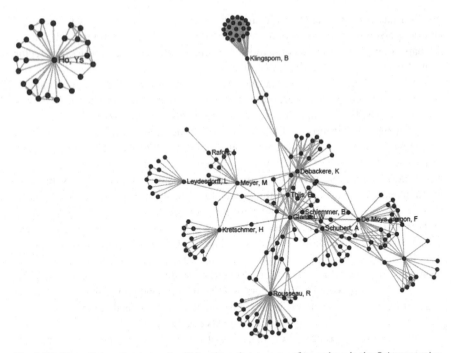

Fig. 3.15 The collaboration network of 14 authors that were top-five authors in the *Scientometrics* dataset based on at least one of four measures of centrality and their collaborators

second most productive author within the data set). His lower position (16th most central author) on eigenvector centrality may be the combined result of fewer repeated collaborations compared to other most active authors and his collaborations with larger number of less collaborative individuals. The important role that this researcher plays within the scientometrics community has an outside confirmation in the fact that he has been the president of the International Society for Scientometrics and Informetrics since 2007.

3.7   Citation Networks

The most common type of scientific and scholarly network is a citation network. There are different types of citation networks: author citation networks, which aggregate the data to the level of an author's oeuvre, paper citation networks, which focus on individual papers as units of analysis, and journal citation networks which focus on citation patterns among individual journals. Author citation networks have been used for evaluative purposes, i.e., to assess the impact of individual authors on a scientific field, rank them accordingly, and, finally, reward them. Despite the extensive usage of bibliometrics for evaluative purposes, there are still unresolved problems, and scientometrics researchers have long cautioned against the usage of any single metrics. Paper citation networks have been used primarily to understand the intellectual aspects of science in descriptive bibliometrics. They have been used to understand the knowledge base and the research front. This line of research follows the assumption proposed by Small (1978) that referenced papers stand in for particular ideas. The literature on citation networks is extensive and its review is outside the scope of this chapter. For a recent overview of this literature see Radicchi, Fortunato, and Vespignani (2012).

Typically, citation networks are acyclic directed networks, because in the majority of cases (except, for example, in special issues of journals) the cited reference cannot also cite the citing reference. Such networks require different measures from the ones we covered for undirected collaboration networks. Node degree can be computed for citation networks as well. However, instead of a single node degree, we can now compute two measures: in-degree and out-degree. In the case of a citation network the in-degree measure corresponds to the number of citations received by an author, paper, or a journal. Out-degree of a node is defined as the number of links emanating from that node. In citation networks out-degree corresponds to a number of references per paper. In the case of an author this measure corresponds to all the literature this author has referenced and in the case of a journal it corresponds to all the literature referenced in that journal. Let us examine the in-degree distribution of paper and author citation networks created using the *Scientometrics* dataset described above. Note that this analysis is focused on examining the knowledge base of research published in *Scientometrics* in the most recent 10-year period, and does not include the citations received from other sources.

As is obvious in Fig. 3.16 the in-degree distribution of the *Scientometrics* paper citation network approximately follows a power law distribution. This means that the contribution to the knowledge base is not evenly distributed, but that there are clear "hubs", or papers that attract disproportionately high attention. The paper that is referenced the most in our dataset is Hirsch's 2005 paper on h-index, with 202 citations. The second most cited paper is Katz & Martin's 1997 paper "What is research collaboration?" with 70 citations. These two papers are followed by de Solla Price's 1963 book *Big Science, Little Science* with 64 citations. King's 2004 *Nature* paper "The scientific impact of nations" with 63 citations is the fourth highest used document. Finally, the fifth most-cited document is Glänzel's 2001 "National characteristics in international scientific co-authorship relations" paper.

The in-degree distribution for author citation networks also approximately follows a power law (Fig. 3.17). The most central authors according to this measure are: W. Glänzel, E. Garfield, L. Leydesdorff, L. Egghe, and H. Moed.

These results are in agreement with recent findings on the cognitive focus of *Scientometrics* and its historical roots using different sources of evidence. Namely, among the most central papers two deal with different international trends. This is interesting because Milojević and Leydesdorff (2013) have found, by examining the title words of articles published in *Scientometrics* (2007–2011), that this journal, more than any other venue that publishes scientometrics research, focuses

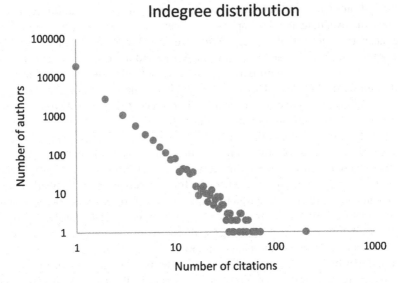

Fig. 3.16 In-degree distribution for *Scientometrics* paper citation network

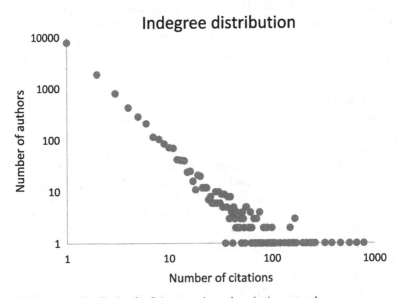

Fig. 3.17 In-degree distribution for *Scientometrics* author citation network

on geographic and international trends. They have also found a strong focus on collaboration, evaluation, and assessment, which correspond to remaining three most central papers. Some of the most central authors, on the other hand, correspond to the intellectual foundation of the field. Namely, in the study of historiography of iMetrics, Leydesdorff, Bornmann, Marx, and Milojević (2014) have found that *Scientometrics* has been intellectually shaped in the early 1960s by the work of historian of science, Derek de Solla Price and the creator of "citation indexing", Eugene Garfield. de Solla Price's seminal book *Big Science, Little Science* is among the top most cited works and Garfield is present in the list of most central authors.

According to the betweenness centrality the five most central papers are: Vanclay, J. (2012) "Impact factor: outdated artefact or stepping-stone to journal certification" *Scientometrics*; Boyack, K., Klavans, R., & Börner (2005) "Mapping the backbone of science" *Scientometrics*; Leydesdorff, (2003) "The mutual information of university-industry-government relations: An indicator of the triple helix dynamics" *Scientometrics*; Batista, P.D., Campiteli, M.G., & Kinouchi, O. (2006) "Is it possible to compare researchers with different scientific interests?" *Scientometrics*; Weingart, P. (2005) "Impact of bibliometrics upon the science system: Inadvertent consequences?" *Scientometrics*. The centrality of these papers is the result of their being on citation paths among many different papers. In a way they serve as boundary objects among different communities.

The most central authors in the author citation network based on the betweenness centrality are: W. Glänzel, L. Leydesdorff, L. Bornmann, L. Egghe, and M. Meyer. It is interesting that three of these five authors, Glänzel, Leydesdorff, and Meyer, have also been identified as the most central authors based on the betweenness centrality in collaboration network. The central position of these authors may be the result of multiple factors: high productivity, work on diverse topics that are of interest to different communities, or work on a topic that is of interest to multiple communities.

Finally, let us look at the PageRank of papers and authors. PageRank has been introduced by Brin and Page (1998) as an algorithm to rank the results of Web queries. At its core it has the same underlying assumption as eigenvector centrality that not all nodes are equal and therefore not all endorsements are equal. The main difference, which is particularly important in the Web environment, is that this measure takes into account the out-degree. Namely, if a page is endorsed by an influential node together with 100 other nodes, it should carry less weight than the same node being endorsed by the same influential node together with only five other nodes. PageRank also depends on a parameter called the damping factor. The most commonly used values for the parameter d are 0.85, 0.5, and 0.15. The damping factor of 0.85 stresses network topology, 0.5 stresses the short path of two, and 0.15 the random citation (Ding, 2011). Bollen, Rodriquez, and Van de Sompel (2006) used weighted PageRank to determine the prestige of journals. A number of authors have used PageRank, weighted PageRank and a number of its extensions to evaluate or rank papers and authors (Ding, 2011; Ding, Yan, Frazho, & Caverlee, 2009; Radicchi, Fortunato, Markines, & Vespignani, 2009; Yan & Ding, 2011; Życzkowski, 2010). For a more extensive discussion of PageRank please check Chap. 4.

To compute the PageRank one can use Sci2. Go to Analysis>Networks> Weighted & Directed>PageRank (Fig. 3.18). In the pop-up window choose the

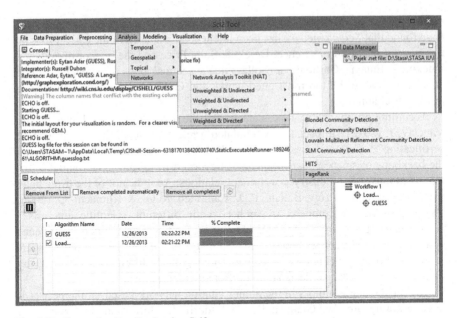

Fig. 3.18 Computing PageRank using Sci2

weight attribute ('Treat all edges as weight one' is the default, and is used here since our paper citation network is unweighted) and the desired damping value (0.85 is the default, and is used here).

Once the PageRank is obtained, one can inspect the top nodes: go to Preprocessing>Networks>Extract Top Nodes and choose the number of top nodes to be displayed. One can open the results by right-clicking on the results in the Data Manager (Fig. 3.19) and choosing the viewer application from the drop down menu in the pop-up window.

As an illustration, we list the top papers for the unweighted PageRank for the paper citation network with the damping value of 0.85: Hirsch's 2005 *PNAS* paper on h-index; Egghe's 2006 *Scientometrics* paper "Theory and practice of the g-index"; Bornmann & Daniel 2005 *Scientometrics* paper "Does the h-index for ranking of scientists really work?"; Ball 2005 *Nature* paper "Index aims for fair ranking of scientists"; and King 2004 *Nature* paper "The scientific impact of nations". Two of these five papers are also among the five top papers as measured using in-degree centrality. It is interesting that all of these papers are on impact and ranking. The five most central authors based on weighted PageRank applied to author citation network are: W. Glänzel, T. Braun, L. Egghe, E. Garfield, and L. Leydesdorff. Four of these authors are also most central as measured using in-degree centrality.

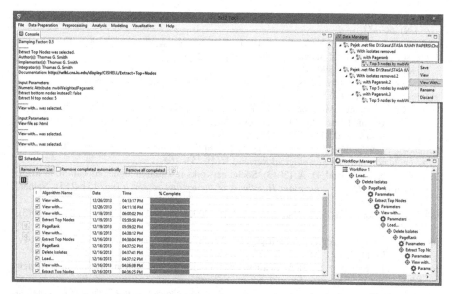

Fig. 3.19 Viewing the results of PageRank computation

Conclusions

In this chapter we have focused on showing how different centrality measures can be used to study science both as a social and an intellectual endeavor. The focus of these measures, as presented in this chapter, has not been on ranking and evaluating individual researchers or papers, but to show how these measures can add to a scientometrician's toolbox to study processes and trends in science and the roles that different researchers and papers play in these developments. While there are other network measures, related to community detection, diffusion of ideas, and the network dynamics, which would also be potentially useful to scientometrics, we focused on centrality measures for a number of reasons: they can be applied to both directed and the undirected networks, they are relatively simple to compute, and despite their usefulness they have so far not been used widely (with the exception of PageRank) to enhance our understanding of scientific and scholarly networks.

References

Abbasi, A., Hossain, L., & Leydesdorff, L. (2012). Betweenness centrality as a driver of prefer-ential attachment in the evolution of research collaboration networks. *Journal of Informetrics*, 6(3), 403–412.

Babchuk, N., Keith, B., & Peters, G. (1999). Collaboration in sociology and other scientific disciplines: A comparative trend analysis of scholarship in the social, physical, and mathe-matical sciences. *The American Sociologist*, 30(3), 5–21.

Barabási, A. L., & Albert, R. (1999). Emergence of scaling in random networks. *Science, 286* (5439), 509–512.

Barabási, A. L., Jeong, H., Néda, Z., Ravasz, E., Schubert, A., & Vicsek, T. (2002). Evolution of the social network of scientific collaborations. *Physica A: Statistical Mechanics and its Applications, 311*(3), 590–614.

Bar-Ilan, J. (2008). Informetrics at the beginning of the 21st century—A review. *Journal of Informetrics, 2*(1), 1–52.

Barrat, A., Barthélémy, M., Pastor-Satorras, R., & Vespignani, A. (2004). The architecture of complex weighted networks. *Proceedings of the National Academy of Sciences of the United States of America, 101*(11), 3747–3752.

Ben-David, J., & Collins, R. (1966). Social factors in the origins of a new science: The case of psychology. *American Sociological Review, 31*(4), 451–465.

Biancani, S., & McFarland, D. A. (2013). Social networks research in higher education. In M. B. Paulsen (Ed.), *Higher education: Handbook of theory and research* (pp. 151–215). Netherlands: Springer.

Bollen, J., Rodriquez, M. A., & Van de Sompel, H. (2006). Journal status. *Scientometrics, 69*(3), 669–687.

Bonacich, P. (1987). Power and centrality: A family of measures. *American Journal of Sociology, 92*(5), 1170–1182.

Borgman, C. L., & Furner, J. (2002). Scholarly communication and bibliometrics. *Annual Review of Information Science and Technology, 36*, 3–72.

Börner, K., Sanyal, S., & Vespignani, A. (2007). Network science. *Annual Review of Information Science and Technology, 41*(1), 537–607.

Brin, S., & Page, L. (1998). The anatomy of a large-scale hypertextual Web search engine. *Computer Networks and ISDN Systems, 30*(1), 107–117.

Carrington, P. J., & Scott, J. (2011). Introduction. In J. Scott & P. J. Carrington (Eds.), *The SAGE handbook of social network analysis* (pp. 1–8). Los Angeles, CA: Sage.

Chubin, D. E. (1976). The conceptualization of scientific specialties. *Sociological Quarterly, 17*, 448–476.

Crane, D. (1969). Social structure in a group of scientists: A test of the "invisible college" hypothesis. *American Sociological Review, 34*, 335–352.

Crane, D. (1972). *Invisible colleges: Diffusion of knowledge in scientific communities.* Chicago, IL: University of Chicago Press.

de Nooy, W., Mrvar, A., & Batagelj, V. (2011). *Exploratory social network analysis with Pajek.* Cambridge: Cambridge University Press.

De Solla Price, D. J. (1965). Networks of scientific papers. *Science, 149*, 510–515.

Ding, Y. (2011). Applying weighted PageRank to author citation networks. *Journal of the American Society for Information Science and Technology, 62*(2), 236–245.

Ding, Y., Yan, E., Frazho, A., & Caverlee, J. (2009). PageRank for ranking authors in co-citation networks. *Journal of the American Society for Information Science and Technology, 60*(11), 2229–2243.

Easley, D., & Kleinberg, J. (2010). *Networks, crowds, and markets: Reasoning about a highly connected world.* Cambridge: Cambridge University Press.

Elkana, Y., Lederberg, J., Merton, Y., Thackray, A., & Zuckerman, H. (1978). *Toward a metric of science: The advent of science indicators.* New York, NY: John Wiley & Sons.

Farkas, I., Derényi, I., Jeong, H., Neda, Z., Oltvai, Z. N., Ravasz, E., ... & Vicsek, T. (2002). Networks in life: Scaling properties and eigenvalue spectra. *Physica A: Statistical Mechanics and its Applications, 314*(1), 25–34.

Freeman, L. C. (1979). Centrality in social networks conceptual clarification. *Social Networks, 1* (3), 215–239.

Freeman, L. C. (2004). *The development of social network analysis: A study in the sociology of science.* Vancouver: Empirical Press.

Friedkin, N. E. (1998). *A structural theory of social influence*. Cambridge: Cambridge University Press.

Huisman, M., & van Duijn, M. A. (2011). A reader's guide to SNA software. In J. Scott & P. J. Carrington (Eds.), *The SAGE handbook of social network analysis* (pp. 578–600). Los Angeles, CA: Sage.

Kuhn, T. S. (1996). *The structure of scientific revolutions*. Chicago, IL: University of Chicago Press.

Leydesdorff, L. (2007). Betweenness centrality as an indicator of the interdisciplinarity of scientific journals. *Journal of the American Society for Information Science and Technology, 58*(9), 1303–1319.

Leydesdorff, L. (2009). How are new citation-based journal indicators adding to the bibliometric toolbox? *Journal of the American Society for Information Science and Technology, 60*(7), 1327–1336.

Leydesdorff, L., Bornmann, L., Marx, W., & Milojević, S. (2014). Referenced publication years spectroscopy applied to iMetrics: Scientometrics, journal of informetrics, and a relevant subset of JASIST. *Journal of Informetrics, 8*(1), 162–174.

Leydesdorff, L., & Rafols, I. (2011). Indicators of the interdisciplinarity of journals: Diversity, centrality, and citations. *Journal of Informetrics, 5*(1), 87–100.

Mali, F., Kronegger, L., Doreian, P., & Ferligoj, A. (2012). Dynamic scientific co-authorship networks. In A. Scharnhorst, K. Börner, & P. van den Besselaar (Eds.), *Models of science dynamics* (pp. 195–232). Berlin: Springer.

Milojević, S. (2010). Modes of collaboration in modern science: Beyond power laws and preferential attachment. *Journal of the American Society for Information Science and Technology, 61* (7), 1410–1423.

Milojević, S. (2014). Principles of scientific research team formation and evolution. *Proceedings of the National Academy of Sciences, 111*(11), 3984–3989.

Milojević, S., & Leydesdorff, L. (2013). Information metrics (iMetrics): A research specialty with a socio-cognitive identity? *Scientometrics, 95*(1), 141–157.

Moed, H. F. (2010). *Citation analysis in research evaluation* (Vol. 9). Dordrecht: Springer.

Moody, J. (2004). The structure of a social science collaboration network: Disciplinary cohesion from 1963 to 1999. *American Sociological Review, 69*(2), 213–238.

Mullins, N. C. (1972). The development of a scientific specialty: The phage group and the origins of molecular biology. *Minerva, 10*(1), 51–82.

Mullins, N. C. (1973). *Theories and theory groups in contemporary American sociology*. New York, NY: Harper & Row.

Narin, F. (1976). *Evaluative bibliometrics: The use of publication and citation analysis in the evaluation of scientific activity*. Washington, DC: Computer Horizons.

Newman, M. E. (2001). Clustering and preferential attachment in growing networks. *Physical Review E, 64*(2), 025102.

Newman, M. E. (2003). The structure and function of complex networks. *SIAM Review, 45*(2), 167–256.

Newman, M. E. (2004a). Analysis of weighted networks. *Physical Review E, 70*(5), 056131.

Newman, M. E. (2004b). Who is the best connected scientist? A study of scientific coauthorship networks. In M. E. Newman (Ed.), *Complex networks* (pp. 337–370). Berlin: Springer.

Newman, M. (2010). *Networks: An introduction*. New York, NY: Oxford University Press.

Opsahl, T., Agneessens, F., & Skvoretz, J. (2010). Node centrality in weighted networks: Generalizing degree and shortest paths. *Social Networks, 32*(3), 245–251.

Otte, E., & Rousseau, R. (2002). Social network analysis: A powerful strategy, also for the information sciences. *Journal of Information Science, 28*(6), 441–453.

Radicchi, F., Fortunato, S., Markines, B., & Vespignani, A. (2009). Diffusion of scientific credits and the ranking of scientists. *Physical Review E, 80*(5), 056103.

Radicchi, F., Fortunato, S., & Vespignani, A. (2012). Citation networks. In A. Scharnhorst, K. Börner, & P. Van den Besselaar (Eds.), *Models of science dynamics: Encounters between complexity theory and information sciences* (pp. 233–257). Berlin: Springer.

Scott, J. (2013). *Social network analysis.* Los Angeles, CA: Sage.

Small, H. G. (1978). Cited documents as concept symbols. *Social Studies of Science, 8*(3), 327–340.

Wagner, C. S., & Leydesdorff, L. (2005). Network structure, self-organization, and the growth of international collaboration in science. *Research Policy, 34*(10), 1608–1618.

Wasserman, S., & Faust, K. (1994). *Social network analysis: Methods and applications.* Cambridge: Cambridge University Press.

Watts, D. J. (2004). The "new" science of networks. *Annual Review of Sociology, 30*, 243–270.

White, H. D. (2011). Scientific and scholarly networks. In J. Scott & P. J. Carrington (Eds.), *The SAGE handbook of social network analysis* (pp. 271–285). Los Angeles, CA: Sage.

Yan, E., & Ding, Y. (2011). Discovering author impact: A PageRank perspective. *Information Processing & Management, 47*(1), 125–134.

Życzkowski, K. (2010). Citation graph, weighted impact factors and performance indices. *Scientometrics, 85*(1), 301–315.

Chapter 4
PageRank-Related Methods for Analyzing Citation Networks

Ludo Waltman and Erjia Yan

Abstract A central question in citation analysis is how the most important or most prominent nodes in a citation network can be identified. Many different approaches have been proposed to address this question. In this chapter, we focus on approaches that assess the importance of a node in a citation network based not just on the local structure of the network but instead on the network's global structure. For instance, rather than just counting the number of citations a journal has received, these approaches also take into account from which journals the citations originate and how often these citing journals have been cited themselves. The methods that we study are closely related to the well-known PageRank method for ranking web pages. We therefore start by discussing the PageRank method, and we then review the work that has been done in the field of citation analysis on similar types of methods. In the second part of the chapter, we provide a tutorial in which we demonstrate how PageRank calculations can be performed for citation networks constructed based on data from the Web of Science database. The Sci2 tool is used to construct citation networks, and MATLAB is used to perform PageRank calculations.

4.1 Introduction

How can we identify the most important or most prominent nodes in a citation network? For instance, in a citation network of journals, how can we determine the most important journals? Or in a citation network of individual publications, how can we infer each publication's importance? These are central questions in citation analysis, a research area within the fields of bibliometrics and scientometrics that is concerned with the study of citations in scientific literature. Mostly, researchers try

L. Waltman (✉)
Centre for Science and Technology Studies, Leiden University, Leiden, The Netherlands
e-mail: waltmanlr@cwts.leidenuniv.nl

E. Yan
College of Computing and Informatics, Drexel University, Philadelphia, PA, USA
e-mail: erjia.yan@drexel.edu

© Springer International Publishing Switzerland 2014
Y. Ding et al. (eds.), *Measuring Scholarly Impact*,
DOI 10.1007/978-3-319-10377-8_4

to answer the above questions using relatively simple counting and aggregation methods. Without doubt, the most famous examples of such methods are the journal impact factor (Garfield, 1972) and the h-index (Hirsch, 2005). Impact factors, h-indices, and other similar methods take into account only the local structure of a citation network. For instance, to calculate the impact factor of a journal, one needs to know the number of citations the journal has received, but it is not important to know from which journals these citations originate and how many times each of the citing journals has been cited itself. A citation originating from Nature or Science has the same weight as a citation originating from some obscure journal.

In this chapter, we consider methods that take into account the global structure of a citation network rather than only the local structure. The main idea of these methods is to give more weight to citations originating from highly cited nodes in a citation network than to citations originating from lowly cited nodes. For instance, being cited by a prestigious highly cited journal is considered more valuable than being cited by an unknown lowly cited journal. Typically, methods that take into account the global structure of a citation network are closely related to the well-known PageRank method (Brin & Page, 1998; Page, Brin, Motwani, & Winograd, 1999). The PageRank method is employed by the Google search engine to rank web pages. Using the PageRank method, the ranking of a web page increases if there are many other web pages that link to it and especially if these linking web pages also have a high ranking themselves. The PageRank idea can be translated relatively easily from hyperlink networks of web pages to citation networks of publications, journals, or authors. In fact, in citation analysis, the basic idea of PageRank was already proposed more than 20 years before PageRank was introduced as a method for ranking web pages (Pinski & Narin, 1976). However, the introduction of PageRank as a method for ranking web pages has led to a renewed interest in the use of PageRank-related methods in citation analysis. Especially in recent years, a significant amount of work has been dedicated to this topic.

Our aim in this chapter is to explain the PageRank idea, to provide an overview of the literature on PageRank-related methods in citation analysis, and to demonstrate in a step-by-step manner how PageRank can be applied to citation networks using two well-known software tools. We start by explaining the PageRank idea in Sect. 4.2, followed by a review of the literature in Sect. 4.3. Next, in Sect. 4.4, we provide a tutorial on the combined use of the Sci2 tool and MATLAB for performing PageRank calculations. Finally, we conclude the chapter in Sect. 4.5.

4.2 PageRank

PageRank was introduced as a method for ranking web pages by Brin and Page (1998) and Page et al. (1999). In this section, we offer a brief explanation of the PageRank method. We refer to Langville and Meyer (2006) for an extensive discussion of the PageRank method and the underlying mathematics.

4.2.1 Definition of PageRank

The PageRank value of a web page i, denoted by p_i, is given by

$$p_i = \alpha \sum_{j \in B_i} \frac{p_j}{m_j} + (1 - \alpha)\frac{1}{n}, \qquad (4.1)$$

where α denotes a so-called damping factor parameter, B_i denotes the set of all web pages that link to web page i, m_j denotes the number of web pages to which web page j links, and n denotes the total number of web pages to be ranked. The damping factor parameter α can be set to a value between 0 and 1. Following Brin and Page (1998) and Page et al. (1999), the typical value is 0.85.

Based on Eq. (4.1), the following observations can be made:

- The larger the number of web pages that link to web page i, the higher the PageRank value of web page i.
- The higher the PageRank values of the web pages that link to web page i, the higher the PageRank value of web page i.
- For those web pages that link to web page i, the smaller the number of other web pages to which these web pages link, the higher the PageRank value of web page i.
- The closer the damping factor parameter α is set to 1, the stronger the above effects.

So the PageRank idea is that for a web page to be considered important there must be a large number of other web pages linking to it, these linking web pages must be sufficiently important themselves, and the number of other web pages to which they link must not be too large. Somewhat informally, the PageRank method can also be interpreted in terms of a voting system. In this interpretation, a link from one web page to another represents a vote of the former web page for the latter one. The more important a web page, the higher the weight of its vote. A web page may vote for multiple web pages, in which case the weight of its vote is divided equally over each of the web pages. The importance of a web page is determined by total weight of all votes it receives.

A difficulty for the PageRank method is that some web pages do not link to any other web page. Web pages without outgoing links are referred to as dangling nodes in the PageRank literature. The most common way to deal with dangling nodes is to create artificial links from each dangling node to all other web pages. We refer to Langville and Meyer (2006) for a more extensive discussion of the dangling nodes issue.[1]

---

[1] For an empirical analysis of the dangling nodes issue in the context of citation networks, see Yan and Ding (2011a).

4.2.2 Calculation of PageRank

The PageRank definition given in Eq. (4.1) is recursive. To determine the PageRank value of a web page, one needs to know the PageRank values of the web pages that link to this web page. To determine the PageRank values of these linking web pages, one needs to know the PageRank values of the web pages that link to these linking web pages, and so on. In the end, one needs to have PageRank values $p_1, p_2, \ldots, p_n$ such that for all $i = 1, 2, \ldots, n$ the equality in Eq. (4.1) is satisfied. Each of these PageRank values will be between 0 and 1, and the sum of the values will be equal to 1. In practice, PageRank values are typically calculated using the power method. The power method is an iterative method. It starts by assigning the same PageRank value to all web pages. Hence, using $n_i^{(k)}$ to denote the PageRank value of web page i in iteration k, it starts with $p_1^{(0)} = p_2^{(0)} = \ldots = p_n^{(0)} = 1/n$. The PageRank values are then updated iteratively according to

$$p_i^{(k+1)} = \alpha \sum_{j \in B_i} \frac{p_j^{(k)}}{m_j} + (1 - \alpha)\frac{1}{n}. \qquad (4.2)$$

The power method continues iterating until the PageRank values have converged, that is, until the PageRank values in two successive iterations are very close to each other.

4.2.3 Damping Factor Parameter

It is important to understand the role of the damping factor parameter α in the PageRank method. If this parameter is set to 1, which means that there is no damping effect, it can in general not be guaranteed that the PageRank method works properly. More specifically, without a damping effect, the PageRank values $r_1, r_2, \ldots, r_n$ need not be uniquely defined and the power method need not converge. By setting the damping factor parameter α to a value below 1, PageRank values are guaranteed to be uniquely defined and the power method is guaranteed to converge. Nevertheless, values just below 1 for the damping factor parameter α are generally not recommended, since they may cause PageRank values to be extremely sensitive to small changes in the network of links between one's web pages. For a more detailed discussion of the role of the damping factor parameter α in the PageRank method, we refer to Langville and Meyer (2006).[2]

---

[2] In the context of citation networks, in addition to the typical value of 0.85 for the damping factor parameter, the value of 0.5 is also sometimes used. See Chen et al. (2007) for some further discussion.

4.2.4   Interpretation of PageRank in Terms of Random Surfing

Finally, we mention an alternative interpretation of the PageRank method. This interpretation is based on the idea of a random surfer (Brin & Page, 1998; Page et al., 1999). A random surfer is someone who is surfing on the Web by randomly following hyperlinks. The random surfer randomly chooses a hyperlink on the web page on which he or she currently finds himself or herself and then moves to the web page to which the hyperlink points. On this new web page, the random surfer again randomly chooses a hyperlink, and he or she then again follows this link. In this way, the random surfer keeps moving from one web page to another in a kind of random walk. Suppose now that occasionally the random surfer does not follow a randomly chosen hyperlink on his or her current web page but that instead the random surfer is "teleported" to a new web page chosen completely at random, with each web page being equally likely to be selected. In this situation, there turns out to be a close relationship between the frequency with which a web page is visited by the random surfer and the PageRank value of the web page. More specifically, if each time the random surfer moves to a new web page the probability of being "teleported" equals 1 minus the damping factor parameter α of the PageRank method, then in the long run the proportion of time the random surfer spends on a web page is equal to the PageRank value of the web page. Hence, PageRank values turn out to have a convenient interpretation in terms of random surfing behavior. The PageRank value of a web page can simply be seen as the proportion of time a random surfer spends on the web page.

4.3   Literature Review

In this section, we provide an overview of the literature on PageRank-related methods in citation analysis. We first discuss predecessors of the PageRank method. These predecessors have important elements in common with the PageRank method, even though they were developed before the introduction of this method in 1998. We then discuss methods that were developed more recently and that have been inspired by the PageRank method. We first focus on PageRank-inspired methods for analyzing journal citation networks, and we then consider methods for analyzing author and publication citation networks.

4.3.1 Predecessors of PageRank

When the PageRank method was introduced in 1998, the basic idea of the method was not new. An interesting overview of the roots of the PageRank method is given by Franceschet (2011). It turns out that before the introduction of the PageRank method closely related ideas had already been studied in the fields of citation analysis (Pinski & Narin, 1976), sociometry (Katz, 1953), and econometrics (Leontief, 1941). We now discuss in more detail the early developments in citation analysis.

In 1976, Gabriel Pinski and Francis Narin published their seminal work on citation-based influence measures for journals (Pinski & Narin, 1976). Their main idea is that to measure the influence of a journal one should not only count the citations received by the journal, like the journal impact factor does, but also weigh each citation based on the influence of the citing journal. In other words, citations from highly influential journals should be given more weight than citations from less influential journals. Clearly, this idea is closely related to the PageRank idea of the importance of a web page being dependent not only on the number of other web pages that link to it but also on the importance of these linking web pages. However, there are a number of differences as well. First, the PageRank method, as it was originally proposed, is based on binary relations, that is, a web page does or does not link to another web page. Relations between journals, on the other hand, are of a weighted nature, with weights being determined by the number of citations given from one journal to another. Second, in the case of journals, one normally needs to correct for the fact that some journals have more publications than others. There is no need for a similar type of correction in the PageRank method. And third, in the methodology proposed by Pinski and Narin, there is no parameter similar to the damping factor parameter in the PageRank method. In other words, the methodology of Pinski and Narin can best be compared with the PageRank method with the damping factor parameter set to a value of 1.

Early work building on the ideas of Pinski and Narin (1976) was done by Geller (1978) and Todorov (1984). Geller pointed out the relationship between the methodology developed by Pinski and Narin and the mathematical literature on Markov chains. Other work was done by Doreian (1985, 1987), who independently developed a methodology similar to the one proposed by Pinski and Narin.

Some interesting work has also been published outside the citation analysis literature, in particular in the fields of economics and management. In the economic literature, a methodology related to the one developed by Pinski and Narin (1976) was introduced by Liebowitz and Palmer (1984). More recently, an axiomatic characterization of this methodology was presented by Palacios-Huerta and Volij (2004). In the management literature, a similar type of methodology was proposed by Salancik (1986).

4.3.2   PageRank-Inspired Methods for Analyzing Journal Citation Networks

Applications of PageRank-related methods to journal citation networks have been quite popular. As indicated above, all or almost all work in citation analysis on predecessors of the PageRank method is related to journal citation networks. After the introduction of the PageRank method in 1998, a number of new PageRank-inspired methods for analyzing journal citation networks have been introduced. We now discuss the most important work that has been done.

The first PageRank-inspired method for analyzing journal citation networks was proposed by Bollen, Rodriguez, and Van de Sompel (2006). These authors referred to their method as a weighted PageRank method. This is because, unlike link relations between web pages, citation relations between journals are of a weighted nature. The weighted PageRank method of Bollen et al. does not correct for the fact that some journals have more publications than others. So journals with more publications will generally have higher weighted PageRank values. This is a difference with the method of Pinski and Narin (1976), in which a correction for journal size can be made. Another difference between the two methods is that the weighted PageRank method includes a damping factor parameter while the method of Pinski and Narin does not.

A second PageRank-inspired method for analyzing journal citation networks is the Eigenfactor method (Bergstrom, 2007; West, Bergstrom, & Bergstrom, 2010a). This method in fact provides two values for each journal in a journal citation network: An Eigenfactor value and an article influence value. The Eigenfactor value of a journal depends on the size of the journal, just like the above-discussed weighted PageRank value, while the article influence value of a journal has been corrected for size. In this respect, article influence values are similar to journal impact factors. A difference between the Eigenfactor method and the above-discussed weighted PageRank method is that in the weighted PageRank method each journal is equally likely to be selected in the case of "teleportation" while in the Eigenfactor method the probability with which a journal is selected is proportional to the number of publications of the journal. A special property of the Eigenfactor method is that journal self citations are not counted. On the one hand this makes the Eigenfactor method more robust to manipulation, but on the other hand it also introduces a disadvantage for larger journals compared with smaller ones. Like journal impact factors, Eigenfactor values and article influence values are reported in Thomson Reuters' Journal Citation Reports. In addition, Eigenfactor values and article influence values are freely available at www.eigenfactor.org, but the most recent values are missing on this website. A number of papers have appeared in which the Eigenfactor method is discussed and analyzed (Davis, 2008; Franceschet, 2010a, 2010b, 2010c; West, Bergstrom, & Bergstrom, 2010b). The relationship between the Eigenfactor method and the method of Pinski and Narin (1976) was discussed by Waltman and Van Eck (2010).

A third PageRank-inspired method for analyzing journal citation networks is the SCImago Journal Rank (SJR) method. The SJR method was introduced by González-Pereira, Guerrero-Bote, and Moya-Anegón (2010). Recently, the method was revised (Guerrero-Bote & Moya-Anegón, 2012). Like article influence values, SJR values have been corrected for journal size. The SJR method is of a more complex nature than the above-discussed weighted PageRank and Eigenfactor methods. The SJR method for instance includes two free parameters instead of only one. In addition, in the case of the revised SJR method, the weight of a citation depends not only on the "prestige" of the citing journal but also on the "thematic closeness" of the citing journal to the cited journal. SJR values are available in Elsevier's Scopus database, but they can also be obtained freely at www.scimagojr. com.

4.3.3 PageRank-Inspired Methods for Analyzing Author Citation Networks

Perhaps the best-known PageRank-inspired method for analyzing author citation networks is the SARA (science author rank algorithm) method proposed by Radicchi, Fortunato, Markines, and Vespignani (2009). A very similar method, referred to as the author-level Eigenfactor method, was introduced by West, Jensen, Dandrea, Gordon, and Bergstrom (2013), apparently independently from the work by Radicchi et al. The two methods have a lot in common with the above-discussed methods for analyzing journal citation networks, but with journals replaced by authors. Replacing journals by authors is not entirely trivial. This is because a publication can be associated with more than one author, while it can be related to only one journal. Both the SARA and the author-level Eigenfactor method deal with this by fractionalizing citations based on the number of authors of the citing and the cited publication. For instance, if a publication with two authors cites a publication with three authors, then for each of the $2 \times 3 = 6$ combinations of a citing and a cited author this counts as a citation with a weight of 1/6. The SARA method and the author-level Eigenfactor method are very similar, but a difference between the two methods is that author self citations are not counted in the author-level Eigenfactor method while they are counted in the SARA method.

In addition to the SARA method and the author-level Eigenfactor method, a number of other proposals for PageRank-inspired methods for analyzing author citation networks can be found in the literature. The interested reader is referred to Ding (2011), Fiala (2012), Fiala, Rousselot, and Ježek (2008), Yan and Ding (2011b), Zhou, Lü, and Li (2012), and Życzkowski (2010).

4.3.4   PageRank-Inspired Methods for Analyzing Publication Citation Networks

The first application of the PageRank method to publication citation networks was reported by Chen, Xie, Maslov, and Redner (2007). These authors employed the standard PageRank method, although instead of the commonly used value of 0.85 for the damping factor parameter they used a value of 0.5.

A fundamental difference between publication citation networks on the one hand and journal and author citation networks on the other hand is that publication citation networks contain no or almost no cycles (e.g., if publication A cites publication B, then publication B usually will not cite publication A), while in journal and author citation networks cycles are very common (e.g., if journal A cites journal B, then in many cases journal B will also cite journal A). Because of the acyclic nature of publication citation networks, applying the standard PageRank method to these networks may not give satisfactory results. In particular, there will be a tendency for older publications to have higher PageRank values than more recent publications. A modification of the PageRank method that aims to correct for this was proposed by Walker, Xie, Yan, and Maslov (2007). The proposed method, referred to as the CiteRank method, has an additional parameter that can be used to give more weight to recent publications compared with older ones.

After the appearance of the work by Chen et al. (2007) and Walker et al. (2007), a number of other studies have been reported in which PageRank-related methods are applied to publication citation networks. Ma, Guan, and Zhao (2008) followed Chen et al. by performing an analysis using the standard PageRank method with a value of 0.5 for the damping factor parameter. However, most studies proposed improvements to the standard PageRank method or alternatives to it. These studies have been reported by Gualdi, Medo, and Zhang (2011), Li and Willett (2009), Su et al. (2011), and Wu, He, and Pei (2010).

We end this section by mentioning the possibility of developing more complex PageRank-inspired methods that, for instance, combine ideas from journal-level, author-level, and publication-level methods into a hybrid approach. For an example of such an approach, we refer to Yan, Ding, and Sugimoto (2011).

4.4   Tutorial

We now provide a step-by-step demonstration of the combined use of two well-known software tools, the Sci2 tool and MATLAB, for applying PageRank to citation networks. The Sci2 tool is freely available at https://sci2.cns.iu.edu/. MATLAB is not freely available. Readers who do not have access to MATLAB may consider the use of Octave, which is a freely available software tool that is very similar to MATLAB. Octave can be obtained from www.gnu.org/software/octave/.

In the tutorial presented in this section, we focus on the application of PageRank to journal citation networks. PageRank can be applied to other types of citation networks, in particular author citation networks and publication citation networks, in a very similar way, but we do not discuss this in the tutorial.

The tutorial is organized into three parts. We first briefly discuss how to download bibliographic data from the Web of Science database, we then consider the construction of a journal citation network using the Sci2 tool, and finally we explain the use of MATLAB for performing PageRank calculations.

4.4.1 Downloading Bibliographic Data from the Web of Science Database

To download the relevant bibliographic data from Thomson Reuters' Web of Science database, the following steps need to be taken:

1. In your web browser, go to www.webofscience.com.
2. Select the **Web of Science™ Core Collection** option.
3. Enter your search query and press the **Search** button. If needed, use the **Advanced Search** functionality.
4. Select the **Save to Other File Formats** option.
5. Select the **Records … to …** option, and indicate the number of the first and the last record that you want to download. In the **Record Content** drop-down box, select the **Full Record and Cited References** option. In the **File Format** drop-down box, select the **Plain Text** option. Press the **Send** button, and save the resulting text file at an appropriate location. Notice that at most 500 records can be downloaded at a time. If you need to download more than 500 records, you need to do so in batches that each includes at most 500 records.

Figure 4.1 provides screenshots that illustrate steps 1–5.

For the purpose of this tutorial, we have downloaded bibliographic data of all publications in the journal subject category Information Science & Library Science that are of the document type article, proceedings paper, or review and that appeared between 2004 and 2013. The data was downloaded on November 27, 2013. The number of publications is 29,303, distributed over 103 journals.

4.4.2 Constructing a Journal Citation Network Using the Sci2 Tool

We use version 0.5.2 alpha of the Sci2 tool. In addition, two plug-ins are used: The Database plug-in and Web of Science plug-in.[3] We do not use the most recent version of the Sci2 tool, version 1.1 beta. This is because we need the Database

---

[3] See http://wiki.cns.iu.edu/display/SCI2TUTORIAL/3.2+Additional+Plugins

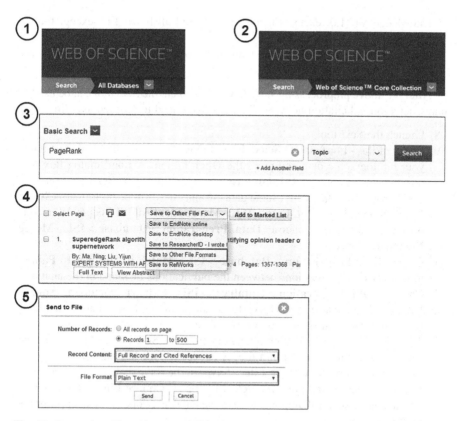

Fig. 4.1 Screenshots illustrating steps 1–5 in the tutorial (downloading Web of Science data)

plug-in and this plug-in is supported only by older versions of the Sci2 tool. We further note that in order to process the bibliographic data of our 29,303 publications we needed to allocate additional memory to the Sci2 tool. We increased the amount of memory allocated to the tool by 1 GB.[4]

Before you can load your bibliographic data into the Sci2 tool, you first need to take some preprocessing steps:

6. Assuming that you are dealing with more than 500 publications, you need to merge the various text files downloaded from the Web of Science database into a single file. This can be done using a text editor, but with many thousands of publications this can be quite tedious. There may then be better solutions available. For instance, on Windows systems, a command such as `copy *.txt merged_data.txt` can be entered in the Command Prompt tool. In the resulting file, make sure to remove all lines 'FN Thomson Reuters Web of

---

[4] See http://wiki.cns.iu.edu/display/SCI2TUTORIAL/3.4+Memory+Allocation

Knowledge VR 1.0' except for the first one and all lines "EF" except for the last one.

7. Change the extension of the text file that contains your bibliographic data from .txt into .isi.

After the above preprocessing steps have been taken, the Sci2 tool can be used to construct a journal citation network. To do so, you need to proceed as follows:

8. Launch the Sci2 tool.
9. Choose **File** > **Load**. In the **Select Files** dialog box, select the file that contains your bibliographic data and press the **Open** button. A **Load** dialog box will now appear. Choose the **ISI database** option and press the **Select** button.
10. To improve the data quality, some data cleaning needs to be performed. Choose **Data Preparation** > **Database** > **ISI** > **Merge Identical ISI People**. When this operation is finished, choose **Data Preparation** > **Database** > **ISI** > **Merge Document Sources**.
11. Choose **Data Preparation** > **Database** > **ISI** > **Match References to Papers** to identify citation relations between the publications included in the analysis.
12. Choose **Data Preparation** > **Database** > **ISI** > **Extract Document Source Citation Network (Core Only)** to obtain a network of citation relations between the journals included in the analysis.
13. To transfer your journal citation network from the Sci2 tool to MATLAB, the network needs to be saved in a file. A Pajek network file can for instance be used for this purpose. Choose **File** > **Save**. In the **Save** dialog box, choose the **Pajek .net** option and press the **Select** button. A **Choose File** dialog box will now appear. Use this dialog box to save the Pajek network file at an appropriate location.

Figure 4.2 provides screenshots that illustrate steps 8–13.

The Pajek network file that we have obtained based on our 29,303 information science and library science publications can be downloaded from www.ludowaltman.nl/pagerank/LIS_journals.net.

4.4.3   Performing PageRank Calculations Using MATLAB

PageRank calculations can be performed using the Sci2 tool. However, there are some difficulties. Because of a bug in version 0.5.2 of the Sci2tool, it is not possible in this version to perform PageRank calculations based on weighted networks. This is possible in more recent versions of the Sci2 tool, but these versions do not support the Database plug-in that is needed to create journal citation networks. Given these difficulties, in this tutorial we choose to use MATLAB to perform PageRank calculations.

First, you need to load your journal citation network into MATLAB. This can be done as follows:

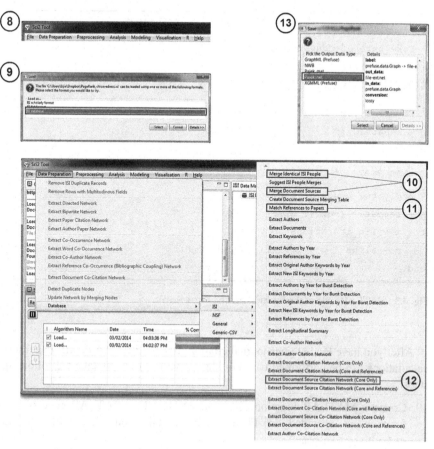

Fig. 4.2 Screenshots illustrating steps 8–13 in the tutorial (constructing a journal citation network using the Sci2 tool)

14. Open the Pajek network file created in step 13 in a text editor.
15. Search for the line '*Arcs' and copy all lines below this line to the clipboard.
16. Launch MATLAB.
17. Paste the contents of the clipboard into the MATLAB workspace. Make sure that you obtain a single matrix rather than multiple column vectors. Label the matrix `cit`.
18. In the MATLAB command window, enter the following commands:

```
n = max(max(cit(:, 1:2)));
C = sparse(cit(:, 2), cit(:, 1), cit(:, 3), n, n);
```

These commands create a journal citation matrix `C`. Element `C(i, j)` of this matrix represents the number of citations from journal `i` to journal `j`.

```
function p = calc_PageRank(C, alpha, n_iterations)

% Take care of dangling nodes.
m = sum(C, 2);
C(m == 0, :) = 1;

% Create a row-normalized matrix.
n = length(C);
m = sum(C, 2);
C = spdiags(1 ./ m, 0, n, n) * C;

% Apply the power method.
p = repmat(1 / n, [1 n]);
for i = 1:n_iterations
    p = alpha * p * C + (1 - alpha) / n;
end
```

Fig. 4.3 MATLAB code of the function `calc_PageRank`

After you have loaded your journal citation network into MATLAB, PageRank calculations can be performed based on the network. To do so, take the following steps:

19. Create a MATLAB function `calc_PageRank`. The MATLAB code of this function is provided in Fig. 4.3. Save this code in a file named `calc_PageRank.m`.
20. Set the current folder in MATLAB to the folder in which the file `calc_PageRank.m` has been saved.
21. In the MATLAB command window, enter the following command:
    `p = calc_PageRank(C, 0.85, 100);`
    The PageRank calculations will now be performed. The damping factor parameter is set to a value of 0.85, and the number of iterations of the power method is set to 100. Performing 100 iterations of the power method is usually sufficient to obtain accurate PageRank values. When the PageRank calculations are finished, the result of the calculations is available in the vector p in the MATLAB workspace. Element `p(i)` of this vector represents the PageRank value of journal i.

Figure 4.4 provides screenshots that illustrate steps 14–21.

Using the Pajek network file created in step 13, the PageRank values calculated in MATLAB can be linked to journal titles. For our 103 information science and library science journals, the journal titles and the corresponding PageRank values

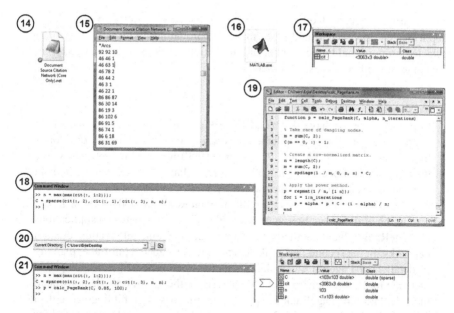

Fig. 4.4 Screenshots illustrating steps 14–21 in the tutorial (performing PageRank calculations using MATLAB)

Table 4.1 The ten information science and library science journals with the highest PageRank values. The PageRank values reported in the right column have been multiplied by 100

Journal of the American Society for Information Science and Technology	8.49
MIS Quarterly	8.28
Scientometrics	7.75
Information Systems Research	4.24
Information & Management	3.59
Journal of the American Medical Informatics Association	3.34
Information Processing & Management	3.06
Journal of Management Information Systems	3.02
Journal of Academic Librarianship	2.30
Journal of Informetrics	1.95

are available in a text file that can be downloaded from www.ludowaltman.nl/ pagerank/LIS_journals.txt. The ten journals with the highest PageRank values are also listed in Table 4.1. We emphasize that the PageRank values that we have calculated are sensitive to the size of a journal. Other things equal, journals that have more publications will also have higher PageRank values. To correct for journal size, the PageRank value of a journal needs to be divided by the number of publications of the journal. We also note that in the MATLAB code in Fig. 4.3

each journal is equally likely to be selected in the case of "teleportation." A more sophisticated approach would be to make the probability with which a journal is selected proportional to the number of publications of the journal, as is done in the Eigenfactor method discussed in Sect. 4.3.

Conclusion

In this chapter, we have discussed the use of PageRank-related methods in citation analysis. We first explained the original PageRank method developed by Brin and Page (1998) and Page et al. (1999) for ranking web pages. We then provided an overview of the literature on PageRank-related methods in citation analysis, from the early work by Pinski and Narin (1976) to recent work inspired by the introduction of the PageRank method in 1998. In our discussion of the recent literature, we made a distinction between applications of PageRank-related methods to journal, author, and publication citation networks. In the second part of the chapter, we provided a tutorial explaining in detail how PageRank calculations can be performed by combining two software tools: The Sci2 tool and MATLAB. In the tutorial, we demonstrated the calculation of PageRank values for journals based on bibliographic data from the Web of Science database. Similar calculations can be performed for authors and individual publications.

References

Bergstrom, C. T. (2007). Eigenfactor: Measuring the value and prestige of scholarly journals. *College and Research Libraries News, 68*(5), 314–316.

Bollen, J., Rodriguez, M. A., & Van de Sompel, H. (2006). Journal status. *Scientometrics, 69*(3), 669–687.

Brin, S., & Page, L. (1998). The anatomy of a large-scale hypertextual Web search engine. *Computer Networks and ISDN Systems, 30*(1–7), 107–117.

Chen, P., Xie, H., Maslov, S., & Redner, S. (2007). Finding scientific gems with Google's PageRank algorithm. *Journal of Informetrics, 1*(1), 8–15.

Davis, P. M. (2008). Eigenfactor: Does the principle of repeated improvement result in better estimates than raw citation counts? *Journal of the American Society for Information Science and Technology, 59*(13), 2186–2188.

Ding, Y. (2011). Applying weighted PageRank to author citation networks. *Journal of the American Society for Information Science and Technology, 62*(2), 236–245.

Doreian, P. (1985). A measure of standing of journals in stratified networks. *Scientometrics, 8* (5–6), 341–363.

Doreian, P. (1987). A revised measure of standing of journals in stratified networks. *Scientometrics, 11*(1–2), 71–80.

Fiala, D. (2012). Time-aware PageRank for bibliographic networks. *Journal of Informetrics, 6*(3), 370–388.

Fiala, D., Rousselot, F., & Ježek, K. (2008). PageRank for bibliographic networks. *Scientometrics, 76*(1), 135–158.

Franceschet, M. (2010a). The difference between popularity and prestige in the sciences and in the social sciences: A bibliometric analysis. *Journal of Informetrics, 4*(1), 55–63.

Franceschet, M. (2010b). Journal influence factors. *Journal of Informetrics, 4*(3), 239–248.
Franceschet, M. (2010c). Ten good reasons to use the Eigenfactor metrics. *Information Processing and Management, 46*(5), 555–558.
Franceschet, M. (2011). PageRank: Standing on the shoulders of giants. *Communications of the ACM, 54*(6), 92–101.
Garfield, E. (1972). Citation analysis as a tool in journal evaluation. *Science, 178*, 471–479.
Geller, N. L. (1978). On the citation influence methodology of Pinski and Narin. *Information Processing and Management, 14*(2), 93–95.
González-Pereira, B., Guerrero-Bote, V. P., & Moya-Anegón, F. (2010). A new approach to the metric of journals' scientific prestige: The SJR indicator. *Journal of Informetrics, 4*(3), 379–391.
Gualdi, S., Medo, M., & Zhang, Y. C. (2011). Influence, originality and similarity in directed acyclic graphs. *EPL, 96*(1), 18004.
Guerrero-Bote, V. P., & Moya-Anegón, F. (2012). A further step forward in measuring journals' scientific prestige: The SJR2 indicator. *Journal of Informetrics, 6*(4), 674–688.
Hirsch, J. E. (2005). An index to quantify an individual's scientific research output. *Proceedings of the National Academy of Sciences, 102*(46), 16569–16572.
Katz, L. (1953). A new status index derived from sociometric analysis. *Psychometrika, 18*(1), 39–43.
Langville, A. N., & Meyer, C. D. (2006). *Google's PageRank and beyond: The science of search engine rankings.* Princeton, NJ: Princeton University Press.
Leontief, W. W. (1941). *The structure of American economy, 1919–1929.* Cambridge, MA: Harvard University Press.
Li, J., & Willett, P. (2009). ArticleRank: A PageRank-based alternative to numbers of citations for analysing citation networks. *Aslib Proceedings, 61*(6), 605–618.
Liebowitz, S. J., & Palmer, J. P. (1984). Assessing the relative impacts of economics journals. *Journal of Economic Literature, 22*(1), 77–88.
Ma, N., Guan, J., & Zhao, Y. (2008). Bringing PageRank to the citation analysis. *Information Processing and Management, 44*(2), 800–810.
Page, L., Brin, S., Motwani, R., & Winograd, T. (1999). The PageRank citation ranking: Bringing order to the Web (Technical Report). *Stanford InfoLab.*
Palacios-Huerta, I., & Volij, O. (2004). The measurement of intellectual influence. *Econometrica, 72*(3), 963–977.
Pinski, G., & Narin, F. (1976). Citation influence for journal aggregates of scientific publications: Theory, with application to the literature of physics. *Information Processing and Management, 12*(5), 297–312.
Radicchi, F., Fortunato, S., Markines, B., & Vespignani, A. (2009). Diffusion of scientific credits and the ranking of scientists. *Physical Review E, 80*(5), 056103.
Salancik, G. R. (1986). An index of subgroup influence in dependency networks. *Administrative Science Quarterly, 31*(2), 194–211.
Su, C., Pan, Y. T., Zhen, Y. N., Ma, Z., Yuan, J. P., Guo, H., …, & Wu, Y. S. (2011). PrestigeRank: A new evaluation method for papers and journals. *Journal of Informetrics, 5*(1), 1–13.
Todorov, R. (1984). Determination of influence weights for scientific journals. *Scientometrics, 6*(2), 127–137.
Walker, D., Xie, H., Yan, K. H., & Maslov, S. (2007). Ranking scientific publications using a model of network traffic. *Journal of Statistical Mechanics: Theory and Experiment, 2007*, P06010.
Waltman, L., & Van Eck, N. J. (2010). The relation between Eigenfactor, audience factor, and influence weight. *Journal of the American Society for Information Science and Technology, 61*(7), 1476–1486.
West, J. D., Bergstrom, T. C., & Bergstrom, C. T. (2010a). The Eigenfactor metrics: A network approach to assessing scholarly journals. *College and Research Libraries, 71*(3), 236–244.

West, J. D., Bergstrom, T. C., & Bergstrom, C. T. (2010b). Big Macs and Eigenfactor scores: Don't let correlation coefficients fool you. *Journal of the American Society for Information Science and Technology, 61*(9), 1800–1807.

West, J. D., Jensen, M. C., Dandrea, R. J., Gordon, G. J., & Bergstrom, C. T. (2013). Author-level Eigenfactor metrics: Evaluating the influence of authors, institutions, and countries within the social science research network community. *Journal of the American Society for Information Science and Technology, 64*(4), 787–801.

Wu, H., He, J., & Pei, Y. J. (2010). Scientific impact at the topic level: A case study in computational linguistics. *Journal of the American Society for Information Science and Technology, 61*(11), 2274–2287.

Yan, E., & Ding, Y. (2011a). The effects of dangling nodes on citation networks. In *Proceedings of the 13th International Conference on Scientometrics and Informetrics*, July 4–8, 2011, Durban, South Africa.

Yan, E., & Ding, Y. (2011). Discovering author impact: A PageRank perspective. *Information Processing and Management, 47*(1), 125–134.

Yan, E., Ding, Y., & Sugimoto, C. R. (2011). P-Rank: An indicator measuring prestige in heterogeneous scholarly networks. *Journal of the American Society for Information Science and Technology, 62*(3), 467–477.

Zhou, Y. B., Lü, L., & Li, M. (2012). Quantifying the influence of scientists and their publications: Distinguishing between prestige and popularity. *New Journal of Physics, 14*, 033033.

Życzkowski, K. (2010). Citation graph, weighted impact factors and performance indices. *Scientometrics, 85*(1), 301–315.

Part II
The Science System

Chapter 5
Systems Life Cycle and Its Relation with the Triple Helix

Robert K. Abercrombie and Andrew S. Loebl

Abstract This chapter examines the life cycle of complex systems in light of the dynamic interconnections among the university, industry, and government sectors. Each sector is motivated in its resource allocation by principles discussed elsewhere in this book and yet remains complementary establishing enduring and fundamental relationships. Industry and government depend upon an educated workforce; universities depend upon industry to spark the R&D which is needed and to sponsor some basic research and much applied research. Government depends upon industry to address operational needs and provide finished products while universities offer government (along with industry) problem solving and problem solving environments. The life cycle of complex systems in this chapter will be examined in this context, providing historical examples. Current examples will then be examined within this multidimensional context with respect to the phases of program and project life cycle management from requirements definition through retirement and closeout of systems. During the explanation of these examples, the advances in research techniques to collect, analyze, and process the data will be examined.

5.1 Introduction and Motivation

The concept of a "life cycle" is inherent to any notion of project or program development. It may be viewed as an outgrowth of systems theory or management theory but is more simply a framework which provides a sequence of activities for designers, developers, managers, etc. to follow. In doing so, it allows for people of different interests, training, disciplines and backgrounds to focus on common

R.K. Abercrombie (✉)
Computational Sciences & Engineering Division, Oak Ridge National Laboratory,
1 Bethel Valley Road, Oak Ridge, TN 37831, USA
e-mail: abercrombier@ornl.gov

A.S. Loebl
ScalarWave, LLC, P. O. Box 30614, Knoxville, TN 37930, USA
e-mail: loeblas@comcast.net

© Springer International Publishing Switzerland 2014
Y. Ding et al. (eds.), *Measuring Scholarly Impact*,
DOI 10.1007/978-3-319-10377-8_5

principles and allow each to understand the roles and responsibilities of each other. One by-product of this process is the formation of teams of people who contribute to a common vision through a commonly understood method of accomplishment. A life cycle consists of a set of steps or phases in which each phase of the life cycle uses the results of the previous phase. In any application of the concept, slight variations in activities and phases occur. According to Taylor (Taylor, 2004), the project life cycle encompasses all the activities of the project, while the systems development life cycle focuses on realizing the product requirements. These specified requirements of the user must be satisfied by an implemented system.

Generically this framework is recognizable as consisting of important phases that are essential to the development or the project and consist of planning, analysis, design, and implementation, followed by project completion and close out for the building, use, and termination of the system. A number of species of this framework have emerged. Models, for example, of the software development life cycle have been developed to address the context of development and application particular to the user's need. Commonly they include: waterfall, fountain, spiral, build and fix, rapid prototyping, incremental, and synchronize and stabilize ("Information Resources Management, The Department of Justice Systems Development Life Cycle Guidance Document," 2003). The oldest of these, and the best known, is the waterfall model (Mohapatra, 2010; Schach, 1999): a sequence of stages in which the output of each stage becomes the input for the next (Kay, 2002). While the statements in Kay's article are over a decade old, it is enlightening to know that very little has changed. The advances that have occurred in terms of both software and hardware since 2002 is amazing, but the process in designing and implementing an efficient and scalable system has not changed much. The waterfall model and basic components of the concept are still taught in colleges today. While Kay focuses on the development of primarily software the concepts and steps used for the much broader issue of a comprehensive IT system are the same (Drogo, 2012).

This is not to say that the waterfall model is inclusive. Many users and developers fault this line of reasoning because of the very nature of its past success. In most instances users and developers are not always consistent about uses and systems requirements (Boehm, 1988). The waterfall approach relies on a progressive sequence which includes a complete understanding of requirements and agreement upon requirements between user and developer. With IT systems complexity growing and expectations for software to be more useful and complete, it is clear that understanding requirements and uses are often evolutionary and not sequential (Mohapatra, 2010).

Another dimension to the life cycle of complex systems deals with the perspective of readiness and system use as a metric. Systems engineers, policy makers, and also users of technology need a quantifiable metric for measuring the readiness of a system. Such a metric, the Technology Readiness Level (TRL), has been defined. It is a measure used to assess the maturity of evolving technologies (devices, materials, components, software, work processes, etc.) during its research, development, and implementation life cycles, and in some cases during early operations ("Technology Readiness Level," 2013). While this TRL model has evolved and been extended into concepts of Integration Readiness Level (IRL) and System Readiness Level (SRL) and the mathematical properties have been articulated (McConkie, Mazzuchi, Sarkani, & Marchette, 2013), the newer models are still being validated for general application (Harper, Van-Nostrnad, Pennock, & Algosos, 2010; Kujaswki, 2010). We, therefore, believe that the TRL model is sufficient for discussions as they apply to the interaction of the complex systems life cycle and its relation with the Triple Helix.

The concept of the Triple Helix of university-industry-government relations has evolved into a model for studying knowledge-based economies. A series of workshops, conferences, and special issues of journals have developed under this title since 1996 (Leydesdorff & Etzkowitz, 1996; Park, 2014). The Triple Helix model is an evolution of the interplay between the relationship of the university, industry and government (Etzkowitz & Leydesdorff, 2000). Initially, the interplay was described as an "elastic" model (e.g., the relationship that existed in the former Soviet Union and Eastern European countries), which was then modified to describe the "laissez-fare" model (e.g., the relationship that existed around 2000 in Sweden and the United States of America [USA]), and finally the Triple Helix model (e.g., the current relationship described in 2000 whose objective is to an innovation arrangement to foster innovation and is still applicable today) (Etzkowitz & Leydesdorff, 2000). The Triple Helix Association provides a current definition of the "Triple Helix" model with respect to innovation ("Concept: Theoretical Framework," 2014).

5.2 Background Work Related to This Study

As discussed initially, each entity had been perceived to work solely, often completely exclusive of one another. Under the Triple Helix model, the industrial sector operates as the locus of production; the government as the source of contractual relations that guarantee stable interactions and exchange; and the university as a source of new knowledge and technology, the generative principle of knowledge-based economies ("Concept: Theoretical Framework," 2014). The model of the Triple Helix of industry, academia and government has been invoked in many successful examples during the twentieth and thus far in the twenty-first centuries.

The Triple Helix model is being applied in both regional areas and nations. Its previous successes have been documented in Amsterdam, Mexico, Brazil, Canada, Sweden, and the USA (Etzkowitz & Leydesdorff, 2000; Etzkowitz & Ranga, 2010), in Algeria, India, and Malaysia, United Arab Emirates, Thai dessert industry, Tunisian pharmaceutical industry, Norwegian solar photovoltaic industry ("VIII Triple Helix International Conference on University, Industry and Government Linkages Book of Abstracts," 2010; "XI Triple Helix International Conference Program—Listing of Papers," 2013), in technical facilities of the Czech Republic (Kostalova & Tetrevova, 2013), and most recently with South Korea (Park, 2014).

The Triple Helix model is further being applied with success in business-like settings. In the Triple Helix IX International Conferences, the theme was "Silicon Valley: Global Model or Unique Anomaly" (Smallwood, 2011). Further discussed are successful science parks which embody the Triple Helix model of industry, academic and government collaboration. One example is the Research Triangle Park in North Carolina. Another version of a "science park" is a corporate-sponsored research lab or an incubation lab on university campuses. Examples include the Disney Research Lab at Carnegie Mellon University and the Stanford Research Institute (Smallwood, 2011) and virtual "science parks," which are really just open collaboration practices. The University of California has also been demonstrated this theme via its Industry-University Cooperative Research Program (IUCRUP), a matching grant program created to catalyze innovation (Coccia, 2014b). Examples of private company collaborations include security research projects of CA Technologies (a private company) with Dalhousie University and the US Department of Homeland Security (Smallwood, 2011), or a project on insider threat with Royal Melbourne Institute of Technology (RMIT) University and the Australian Research Council ("CA Labs and Royal Melbourne Institute of Technology win Australian Government Grant for Research into Insider Threat," 2011).

5.3   Hypothesis to Test

The consistent theme in the previous examples, however, seems to be a sparking mechanism to which these three sectors can respond and agree to collaborate (Kostalova & Tetrevova, 2013). The success of these endeavors is the correct amount of management involvement (Coccia, 2014b). In the previous examples, there was also little to drive an empirical understanding of sector-specific contributions made nor had there been a need to understand the execution of the Triple Helix, parsimoniously. Metrics of performance and contributions were not collected or sought. At other extremes there appears to be a number of examples of initiatives that were sought to engage the Triple Helix that did not accomplish that goal very effectively (Tetřevová & Kostalova, 2012). Recently, some of the failures addressed, using Wales as an example (Pugh, 2013), were described as a lack of cooperation. In the Wales decade long study, one of the findings indicates that the business entity (industry) did not seem to be interacting with the established governmental programs (Pugh, 2013). From a Czech study (Tetřevová & Kostalova, 2012), similarly, a lack of mutual cooperation for the period from 2007 to 2013, identified the following risks: information disarrangement; numerous application documentation process demands; non-transparent process of application evaluation; changing rules; administrative demands; absence of preventive checks; obligation to run a tender process; change management; complicated check mechanism; formal orientation of the checking process; strict and rigid financial management of the projects; and sustainability of project outputs. The authors (Tetřevová & Kostalova, 2012) advocate that it will be necessary to mitigate the above risks and problems both on the national and the European levels, in order for the Triple Helix model to reach its full potential. Also, all entities (university-industry-government) in the partnership must share in the mitigation of these risks and problems, for the Triple Helix model to be successful (Coccia, 2014a). The following sections of this chapter will investigate if the application of the TRL model is sufficient to mitigate risk and problems identified with the context of the interaction of the complex systems life cycle and its relation with the Triple Helix.

5.4 Measurable States During the Life Cycle of a Technology

Previous works have illustrated a model that tracks the life cycle or emergence of an identified technology from initial discovery (via original scientific and conference literature), through critical discoveries (via original scientific, conference literature and patents), transitioning through Technology Readiness Levels (TRLs) and ultimately on to commercial application (Abercrombie, Schlicher, & Sheldon, 2014; Abercrombie, Udoeyop, & Schlicher, 2012). During the period of innovation and technology transfer during the life cycle of a complex system as evidenced by the impact of scholarly works, patents and online web news sources can be identified. As trends develop, currency of citations, collaboration indicators, and online news patterns can be identified. The combinations of four distinct searchable online networked sources (i.e., scholarly publications and citation, worldwide patents, news archives, and online mapping networks) can be assembled to become one collective network (a data set for analysis of relations).

The TRL is a tool which helps define the maturity of a system (devices, materials, components, software, work processes, etc.) during its development and in some cases during early operations ("Defense Acquisition Guidebook," 2012). This model can serve to provide almost metric-level determination of the specific state of the system and can be employed as a helpful standard and shorthand for evaluating and classifying technology maturity (in the context of the life cycle). However, this standard of definition must be applied with expert, experienced, professional judgment. Expanding this model to extensively address the impact of TRLs leads to a generalized metric (Abercrombie, Schlicher, & Sheldon, 2013) that can be related to the Triple Helix models.

5.5 Step-by-Step Use of a Tool to Generate Results

The purpose of this section is to address the step-by-step use of a process to illustrate the relationships among multiple disparate sources of information as a way to systematically explain the life cycle emergence of technologies from innovation through to commercial application. Two examples will be explained to illustrate the process. The logical sequence of milestones derived from our analysis of a previously documented data set and technology (i.e., Simple Network Management Protocol [SNMP]) includes the initial discovery (evident via original scientific and conference literature), the subsequent critical discoveries (evident via original scientific, conference literature and patents), and the transitioning through the various TRLs ultimately to commercial application (Abercrombie et al., 2012). In another example, we investigate the combinations of five distinct sources (i.e., university R&D, industry R&D, life cycle product documentation, and two levels of annual market revenue [$1B and $10B]) (Abercrombie et al., 2013). The data set from the second example was assembled by the United States National Research Council of the National Academies, initially in 2003 (Innovation in Information Technology, 2003) and then updated in 2010, being published in 2012 (Lee et al., 2012). These established relationships become the basis from which to analyze the temporal flow of activity (searchable events) for the two cases we investigated.

We previously articulated the emergence of one particular innovation into a foundational technology enabling other innovations. The specific innovation we investigated is the well-known network management protocol SNMP Version 1, 2, and 3 (Note: each major version made earlier versions "obsolete"), and their impact as a standard operations and maintenance Internet protocol (Frye, Levi, Routhier, & Lucent, 2003). We selected the SNMP to both illustrate the process and test the TRL model hypothesis. SNMP is a standard operations and maintenance Internet protocol (Case, Fedor, Schoffstall, & Davin, 1990). SNMP-based management produces management solutions for systems, applications, complex devices, and environmental control systems, as well as supporting Web services. SNMPv3, the most recent standard approved by the Internet Engineering Task Force (IETF), adds secure capabilities (including encryption). SNMP is used in network management systems to monitor network-attached devices (hubs, routers, bridges, etc.) for conditions that warrant administrative attention. SNMP is a component of the Internet Protocol Suite as defined by IETF. The IETF is a large open international community of network designers, operators, vendors, and researchers concerned with the evolution of the Internet architecture and the smooth operation of the Internet. SNMP consists of a set of standards for network management, including an application layer protocol, a database schema, and a set of data objects ("Simple Network Management Protocol," 2014).

Figure 5.1 identifies patterns from the assembled data sets that correspond to the milestones of the life cycle of the technology (Abercrombie et al., 2012) defined below, as milestones 1- 9, and abbreviated in the figure as M1 - M9.

Milestone 1: *Initial discovery* is the genesis of a specific subject domain and is built on previous work that can be traced via initial original scientific and conference literature.

Milestone 2: *Critical discoveries* are those breakthrough discoveries that can also be traced via initial original scientific, conference literature.

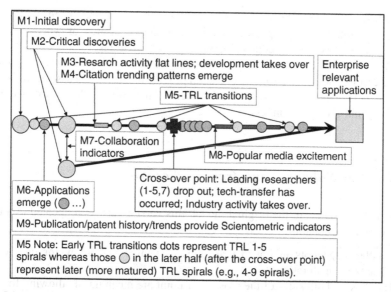

Fig. 5.1 Generalized technology maturation milestones (defined as M1 through M9) over system life cycle

Milestone 3: R&D activity is exhibited. The rate of growth in initial (i.e., original) scientific, conference, and patent literature activity can be traced as identified by trend analysis.

Milestone 4: (*Corollary to Milestone 3*): The trending patterns of citations follow the R&D activity. This phenomenon may be exhibited by a measure of the currency of citations (i.e., a vitality score, mean reference age normalized in relation to the sub-field set (Sandström & Sandström, 2009)) which may show an aging vitality score.

Milestone 5: *Technology Readiness Level (TRL) transitions* occur and initial scientific, conference and patent activity identifications are made. Literature trends initially up then down. Patent trend patterns (up and down) should be identifiable. Conference progression from papers to topics to sessions to independent conferences is a notable trend pattern. Topic moves across journal types from basic to applied research type publications may be prevalent.

Milestone 6: *Applications emerge* from proposed and viable initial scientific and conference literature, and patents. Prototyped and commercial applications, which originate from patents and business white papers, become visible via scholarly literature and popular media searches.

Milestone 7: *Collaboration indicators* become evident as coauthors from different fields (unrelated or otherwise disparate) and group-to-group collaboration patterns come to light. Such collaboration patterns aid in identifying subject matter trends as will trending international collaboration.

Milestone 8: *Sentiment and excitement* points determine when excitement waxes (originates) in the popular media, trade journals, etc., or when it wanes (exits from) in the same media and how long excitement last.

Milestone 9: *Publication/patent history/trends* of critical players (initial scientific literature and conferences, and patents) provide measures and associated rate of change help identify players and subsequent commercial industry involvement (initial point, growth, changes).

5.5.1   Key Enhancements from Earlier Works

From enhancements to the original model, this version has been updated to allow for multidimensionality:

1. Smaller sized dots in Fig. 5.1 are closer together and allow for interactive releases for maturation.
2. The dots on both sides of the cross-over point are meaningful, allowing for

   (a) Deep dive analysis both on the left and right side.
   (b) Allowing for TRL finer resolution (see Expansion/Evolution of Milestone 5 Concerning Technology Readiness Levels section).

3. The dots are more contiguous, which in reality the distances can be analyzed via:

   (c) Normalization
   (d) Vitality scoring
   (e) Application versioning emergence
   (f) Collaboration indicator
   (g) Sentiment and excitement tracking-both positive and/or negative

5.6   Expansion/Evolution of Milestone 5 Concerning Technology Readiness Levels

Technology Readiness Level (TRL) has been defined as a means to assess the maturity of a system (devices, materials, components, software, work processes, etc.) during various stages of its life cycle and in some cases during early operations ("Technology Readiness Level," 2013).

Generally speaking, new inventions or discoveries are not easily converted to immediate application. Instead, new discoveries are usually subjected to experimentation, refinement, and their derived technologies (if any) are subjected to increasingly rigorous testing prior to any implementation life cycle. Once the discovery or invention moves into a technology development/implementation life cycle, it can be suitable for use. In the earlier published model (Abercrombie et al.,

Table 5.1 Technology Readiness Levels in the department of defense (DoD)-Adapted from: ("Defense Acquisition Guidebook," 2012) and ("Technology Readiness Assessment (TRA) Guidance," 2011)

Technology Readiness Level	Description	Supporting information
1. Basic principles observed and reported	A lowest level of technology readiness. Scientific research begins to be translated into applied research and development (R&D). Examples include paper studies of a technology's basic properties.	Published research that identifies the principles that underlie this technology. References to who, where, when.
2. Technology concept and/or application formulated	Invention begins. Once basic principles are observed, practical applications can be invented. Applications are speculative, and there may be no proof or detailed analysis to support the assumptions. Examples are limited to analytic studies.	Publications or other references that outline the application being considered and that provide analysis to support the concept.
3. Analytical and experimental critical function and/or characteristic proof of concept	Active R&D is initiated. This includes analytical studies and laboratory studies to physically validate the analytical predictions of separate elements of the technology. Examples include components that are not yet integrated or representative.	Results of laboratory tests performed to measure parameters of interest and comparison to analytical predictions for critical subsystems. References to who, where, and when these tests and comparisons were performed.
4. Component and/or breadboard validation in laboratory environment	Basic technological components are integrated to establish that they will work together. This is relatively "low fidelity" compared with the eventual system. Examples include integration of "ad hoc" hardware in the laboratory.	System concepts that have been considered and results from testing laboratory-scale breadboard(s). References to who did this work and when. Provide an estimate of how breadboard hardware and test results differ from the expected system goals.
5. Component and/or breadboard validation in relevant environment	Fidelity of breadboard technology increases significantly. The basic technological components are integrated with reasonably realistic supporting elements so they can be tested in a simulated environment. Examples include "high-fidelity" laboratory integration of components.	Results from testing laboratory breadboard system are integrated with other supporting elements in a simulated operational environment. How does the "relevant environment" differ from the expected operational environment? How do the test results compare with expectations? What problems, if any, were encountered? Was the breadboard system refined to more nearly match the expected system goals?

(continued)

Table 5.1 (continued)

Technology Readiness Level	Description	Supporting information
6. System/subsystem model or prototype demonstration in a relevant environment	Representative model or prototype system, which is well beyond that of TRL 5, is tested in a relevant environment. Represents a major step up in a technology's demonstrated readiness. Examples include testing a prototype in a high-fidelity laboratory environment or in a simulated operational environment.	Results from laboratory testing of a prototype system that is near the desired configuration in terms of performance, weight, and volume. How did the test environment differ from the operational environment? Who performed the tests? How did the test compare with expectations? What problems, if any, were encountered? What are/were the plans, options, or actions to resolve problems before moving to the next level?
7. System prototype demonstration in an operational environment.	Prototype near or at planned operational system. Represents a major step up from TRL 6 by requiring demonstration of an actual system prototype in an operational environment (e.g., in an aircraft, in a vehicle, or in space).	Results from testing a prototype system in an operational environment. Who performed the tests? How did the test compare with expectations? What problems, if any, were encountered? What are/were the plans, options, or actions to resolve problems before moving to the next level?
8. Actual system completed and qualified through test and demonstration.	Technology has been proven to work in its final form and under expected conditions. In almost all cases, this TRL represents the end of true system development. Examples include developmental test and evaluation (DT&E) of the system in its intended weapon system to determine if it meets design specifications.	Results of testing the system in its final configuration under the expected range of environmental conditions in which it will be expected to operate. Assessment of whether it will meet its operational requirements. What problems, if any, were encountered? What are/were the plans, options, or actions to resolve problems before finalizing the design?

2012, 2014), this would be Milestone 5. In this current model, we have expanded that definition and allow the TRLs to span across multiple milestones. Accomplishing this expansion, the model evolves and can be adapted to the standards as illustrated in the definitions in Table 5.1.

5.7   Application of TRL Logic to the Modified Model

The correlation of TRL model logic with respect to Milestones of the modified model is as follows:

- *First Publications*-TRL 1 Basic principles observed and reported.
- *Publications become prominent and citations begin*-TRL 2 Technology concept and/or application formulated.
- *Standards-Patents-Publications prevalent and media hype*-TRL 3 Analytical and experimental critical function and/or characteristic proof-of concept.
- *First Patents*-TRL 4 Component and/or breadboard validation in laboratory environment.
- *Patents continued*-TRL 5 Component and/or breadboard validation in relevant environment.
- *Patents and/or patent trending may be used*-TRL 6 System/subsystem model or prototype demonstration in a relevant environment (ground or space).
- *Patent and First Product Offering Overlap*-TRL 7 System prototype demonstration in a space environment.
- *First Product offering*-TRL 8 Actual system completed and "flight qualified" through test and demonstration (ground or space).
- *Multiple Product offering*-TRL 9 Actual system "flight proven" through successful mission operations.

The first application of applying the TRL model logic to modified model follows. The example is adapted from the SNMP baseline data set (Abercrombie et al., 2012, 2014), reevaluated, and tabularized to reflect activity of four separate data sources to determine associated TRLs:

- *First Publications*-TRL 1 Basic principles observed and reported:

  - Initially observed and reported via Web News Sources 1987 and continuing through 2008.
  - Academic publications observed and reported in 1992 (Version 1-1992) and continuing through 2008.

- *Publications become prominent and citations begin*-TRL 2 Technology concept and/or application formulated:

  - Scholarly academic articles begin in 1992 and continue through 2008, peaking that year.
  - Academic Citations begin in 1992 and peak 2008 at end of study indicating that technology has emerged and is still active.

- *Standards-Patents-Publications prevalent and media hype*-TRL 3 Analytical and experimental critical function and/or characteristic proof of concept:

  - Patents peak 2008 at end of study (Fig. 5.2)-also indicating that technology has emerged and is still active and may evolve further.

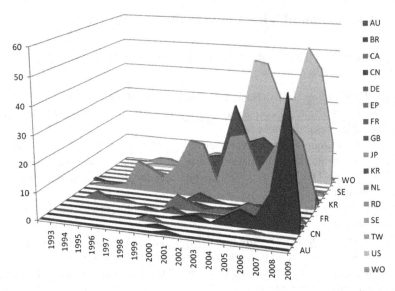

Fig. 5.2 Number of patents per year per country/multi-national organizations identified—SNMP data set. Legend key: Australia (AU), Brazil (BR), Canada (CA), China (CN), Germany (DE), European Patent Office (EP), France (FR), the UK (GB), Japan (JP), South Korea (KR), the Netherlands (NL), Research Disclosure (RD), Sweden (SE), Taiwan (TW), the USA (US), World Intellectual Property Organization (WO)

– Trending.

SNMPv1-1988.
SNMPv2-1993 and then revised in 1996.
SNMPv3-appeared 2002 and then standardized in 2004 (Table 5.2).

- *First Patents*-TRL 4 Component and/or breadboard validation in laboratory environment-1993 as illustrated in Table 5.3. This can further examined as a breakout by country per year which results in an initial TRL 4 (first patent) for each country as follows: Australia (AU)-2001, Brazil (BR)-2000, Canada (CA)-1999, China (CN)-2002, Germany (DE)-1998, European Patent Office (EP)-1994, France (FR)-2001, the UK (GB)-2001, Japan (JP)-1993, South Korea (KR)-1997, the Netherlands (NL)-1996, Research Disclosure (RD)-1998, Sweden (SE)-1996, Taiwan (TW)-2000, the USA (US)-1995, and World Intellectual Property Organization (WO)-1995.
- *Patents continued*-TRL 5 Component and/or breadboard validation in relevant environment. Patents continue throughout the time period of the data set through 2008 into 2009. What is of particular interest is that different patterns emerge for different countries as follows: Australia (AU)-2001 only, Brazil (BR)-2000 only, Canada (CA)-1999, 2001, and 2006, China (CN)-2002, 2004–2008, Germany (DE)-intermittent 1998–2006, European Patent Office (EP)-intermittent 1994–2008, France (FR)-2001, 2004, 2006, the UK (GB)-intermittent 2001–2006,

Table 5.2 SNMP data set from (Abercrombie et al., 2012) reevaluated and tabularized to reflect activity of disparate data sources to determine associated TRLs

Year	Patents	Web news sources SNMP	Total of SNMP versions	Web news sources SNMP Version 1	Web news sources SNMP Version 2	Web news sources SNMP Version 3	Academic citations	Academic articles	Application specific product emergence
1987		14							
1988		23							2
1989		9							4
1990		56							4
1991		92							1
1992		120	2	1	1		2	2	3
1993	1	99	4		4			4	2
1994	4	106	6	1	5		1		2
1995	6	165	3	1	2				4
1996	17	176	1		1		1	3	1
1997	13	129	5	1	3	2	1	5	6
1998	12	165	5	3	2	2	3	3	2
1999	25	204	10	3	1	6	8	6	8
2000	42	284	17	2	2	12	10	3	7
2001	59	311	6	2	1	3	8	2	3
2002	56	546	28	16	5	7	11	6	1
2003	116	400	19	2	2	15	18	9	2
2004	102	374	14	3	2	9	19	7	2
2005	87	307	6		3	3	28	13	1
2006	93	285	9	2		7	24	9	
2007	129	290	6	2		4	39	9	1
2008	142	247	12	6		6	49	4	
Grand total	904	4,402	153	43	34	76	222	85	56

Table 5.3 SNMP patent data set from (Abercrombie et al., 2012) reevaluated and tabularized to reflect number of patents per country per year to provide data to determine TRL 4–7

Country	1993	1994	1995	1996	1997	1998	1999	2000	2001	2002	2003	2004	2005	2006	2007	2008	Totals
AU									1								1
BR								1									1
CA							2		1					1			4
CN										2		3	6	4	13	47	75
DE						1		2		1	1	1		2			8
EP		3	1	1				5	3	1	2	2	5	4	8	4	39
FR									1			1		1			3
GB									3	1			1	1			6
JP	1	1	2	11	9	6	11	21	20	7	24	25	12	21	30	24	225
KR					3		2	1	14	15	34	21	23	19	13	15	160
NL				1													1
RD						1	1		1	2		2	2	1	1	2	13
SE				1		1											2
TW								1		2							3
US			1	1	1	2	6	8	8	20	46	45	32	32	52	44	298
WO			2	2		1	3	3	7	5	9	2	6	7	12	6	65
Totals by years 1993–2008	1	4	6	17	13	12	25	42	59	56	116	102	87	93	129	142	904

Country key: Australia (AU), Brazil (BR), Canada (CA), China (CN), Germany (DE), European Patent Office (EP), France (FR), the UK (GB), Japan (JP), South Korea (KR), Netherlands (NL), Research Disclosure (RD), Sweden (SE), Taiwan (TW), the USA (US), World Intellectual Property Organization (WO)

Japan (JP)-1993–2008, South Korea (KR)-1997, 1999–2008, the Netherlands (NL)-1996, Research Disclosure (RD)-intermittent 1998–2008, Sweden (SE)-1996, 1998, Taiwan (TW)-2000, 2002, the USA (US)-1995–2008, and World Intellectual Property Organization (WO)-1995–1996, 1998–2008.

- Trending follows versions evolution and emergence as in TRL3.
- Peak Plotted-See milestones and radar chart (Abercrombie et al., 2012).

• *Patents and/or patent trending may be used*-TRL 6 System/subsystem model or prototype demonstration in a relevant environment (ground or space).

- Peak-See milestones and radar chart (Abercrombie et al., 2012).

• *Patent and First Product Offering Overlap*-TRL 7 System prototype demonstration in a space environment.

- Sentiment-SNMPv1-1988, SNMPv2-1993 and then revised in 1996, SNMPv3-appeared 2002 and then standardized in 2004.
- Peaks in 2001.

• *First Product offering*-TRL 8 Actual system completed and "flight qualified" through test and demonstration (ground or space).

- In 1988, there are two product offerings, based on earlier version of SNMP and then continue to be fairly consistent thereafter, but strong new product showings in mid-late 1990s. These are correlated with standardization of SNMPv2 and then new product offerings with the standardization of SNMPv3.

• *Multiple Product offering*-TRL 9 Actual system "flight proven" through successful mission operations.

- Peaks in 1999.

The second case of applying the TRL model logic to the current modified model follows with the example of updated data used to create the popular "tire tracks" model (Lee et al., 2012). This data set captures mutually reinforcing developments between university research and development, industrial research and development, and industrial growth (Madhavan et al., 2012). The data set was selected because it is well documented, originally included in (Evolving the High Performance Computing and Communications Initiative to Support the Nation's Information Infrastructure, 1995), with an update in (Innovation in Information Technology, 2003), and more recent update (Lee et al., 2012).

From Table 5.4 the areas of fundamental research in IT and the industry interest analog are as follows:

• Digital Communications (Broadband and Mobile).
• Computer Architecture (Microprocessors).
• Software Technologies (Personal Computing).
• Networking (Internet and Web).

Table 5.4 Generalized data set of research tracks and events as summarized in (Lee et al., 2012) and mapping of the source data set to determine associated TRLs

Data source: Continuing innovation in information technology							Mapping of data to TRLs		
Item-areas of fundamental research in IT	Item-industry interest analog	Univer. research begun (analog to scholarly pursue)	Industry R&D (analog to patents and trade secrets)	Products	$1 Billion Market	$10 Billion Market	TRL1 and 2	TRL 3, 4, and 5	TRL 6, 7, 8, and 9
Digital Communications	Broadband and Mobile	1965	1981	1991	1996	2007	1965	1981	1991
Computer Architecture	Micro-processors	1976	1976	1981	1986	1994	1976	1976	1981
Software Technologies	Personal Computing	1965	1978	1969	1980	1997	1965	1978	1969
Networking	Internet and Web	1968	1979	1977	1983	1999	1968	1979	1977
Parallel and Distributed Systems	Cloud Computing	1968	1972	1985	1980	2008	1968	1972	1985
Databases	Enterprise Systems	1974	1972	1981	1990	1998	1974	1972	1981
Computer Graphics	Entertainment and Design	1965	1965	1970	1976	2004	1965	1965	1970
AI and Robotics	Robotics and Assistive Technologies	1969	1972	1990	1994	N/A	1969	1972	1990

- Parallel and Distributed Systems (Cloud Computing).
- Databases (Enterprise Systems).
- Computer Graphics (Entertainment and Design).
- AI and Robotics (Robotics and Assistive Technologies).

Applying the TRL logic in this data set results in the following:

- *First Publications*-TRL 1 Basic principles observed and reported and *Publications become prominent and citations begin*-TRL 2 Technology concept and/or application formulated are found to be:

  - Digital Communications (Broadband and Mobile)-1965, Computer Architecture (Microprocessors)-1976, Software Technologies (Personal Computing)-1965, Networking (Internet and Web)-1968, Parallel and Distributed Systems (Cloud Computing)-1968, Databases (Enterprise Systems)-1974, Computer Graphics (Entertainment and Design)-1965, and AI and Robotics (Robotics and Assistive Technologies)-1969.

- *Standards-Patents-Publications prevalent and media hype*-TRL 3 Analytical and experimental critical function and/or characteristic proof-of-concept, *First Patents*-TRL 4 Component and/or breadboard validation in laboratory environment, and *Patents continued*-TRL 5 Component and/or breadboard validation in relevant environment are found to be:

  - Digital Communications (Broadband and Mobile)-1981, Computer Architecture (Microprocessors)-1976, Software Technologies (Personal Computing)-1978, Networking (Internet and Web)-1979, Parallel and Distributed Systems (Cloud Computing)-1972, Databases (Enterprise Systems)-1972, Computer Graphics (Entertainment and Design)-1965, and AI and Robotics (Robotics and Assistive Technologies)-1972.

- *Patents and/or patent trending may be used*-TRL 6 System/subsystem model or prototype demonstration in a relevant environment (ground or space), *Patent and First Product Offering Overlap*-TRL 7 System prototype demonstration in a space environment, *First Product offering*-TRL 8 Actual system completed and "flight qualified" through test and demonstration (ground or space), *Multiple Product offering*-TRL 9 Actual system "flight proven" through successful mission operations are found to be:

  - Digital Communications (Broadband and Mobile)-1991, Computer Architecture (Microprocessors)-1981, Software Technologies (Personal Computing)-1969, Networking (Internet and Web)-1977, Parallel and Distributed Systems (Cloud Computing)-1985, Databases (Enterprise Systems)-1981, Computer Graphics (Entertainment and Design)-1970, and AI and Robotics (Robotics and Assistive Technologies)-1990.

One sees that several TRLs can be grouped together in this data set. This can be explained as only three of the five parameters apply to the TRL model logic. However, the last two parameters ($1B Market and $10B Market) are extremely important. What is quite impressive is the market space that these technologies are seen to occupy, upwards to $10B.

5.8 Discussion

The first data set consists of four distinct and separately searchable online sources (i.e., scholarly publications and citation, patents, news archives, and online mapping networks). The data reflects a time line of the technology transitions categorized by TRLs as shown in Fig. 5.3. Analyses of these results convinced us to adapt the model (Abercrombie et al., 2012) so that the TRLs are used to determine the stage in the sequence of the life cycle of the particular technology being investigated (Abercrombie et al., 2013).

The second case study uses the modified model to study a data set created from survey data from within the IT Sector (Lee et al., 2012) measuring the relationships among universities, industry, and governments' innovations and leadership. Figure 5.4 identifies the TRL transitions on eight subject areas of fundamental research in IT and industry-specific categories (Abercrombie et al., 2013).

Applying the definitions of the TRL model logic transitions provides a stepwise explanation for the subject system's (or technology) evolution. This then adds insight into its maturity and market impact (i.e., innovation in terms of the new Triple Helix model).

The modeling of the TRL within the context of the life cycle examines and clarifies how to systematically classify the system/technology evolution. This model starts with an initial discovery via original scientific and conference

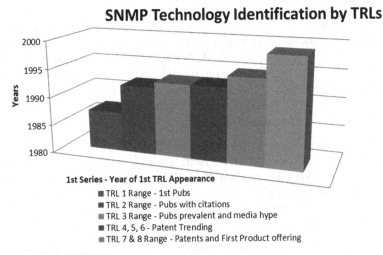

Fig. 5.3 SNMP technology identification by TRLs

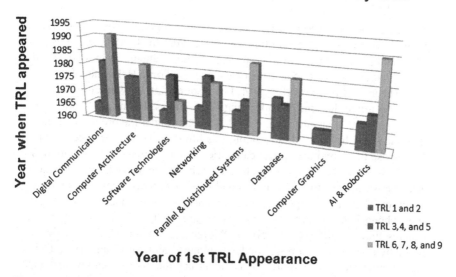

Fig. 5.4   Area of fundamental IT research technology identification by TRLs

literature (the academic strand of the Triple Helix model). The model accounts for technology development and implementation, often critical to its evolving and maturing outcome. This process is accounted for via original scientific and conference literature (the academic strand of the Triple Helix model funded by government grants and in some cases by direct industry funding). Patents are derived primarily by the influence of industry or government research (multiple strands of the Triple Helix model). Finally, the model fully accounts for transitioning through TRLs to commercial application and significant global economic impact (the industry strand of the Triple Helix model).

Conclusion and Future Research

The relationships among independent sources of information were addressed as a way to systematically explain the identifiable states of new systems or technologies, from innovation through to commercial application. In the first case study, we selected a very well-known, documented technology to test the TRL model logic transitioning hypothesis. In the second case study, the TRL transitioning hypothesis was validated by selecting a cross-section of fundamental IT research domains and their corresponding industry interest category spanning 1965–2011. In both cases, applying the TRL transitioning technique to the documented subject areas resulted in trends that clarified and refined the identification of the technology within the definition of a life cycle. The TRL transitions, in the modified model, are useful in the creation

(continued)

(continued)

of business intelligence in a knowledge-based economy. The Triple Helix model has great importance in this context (Leydesdorff, 2010, 2013). Business intelligence assists in providing a basis for strategic business decision (s) for innovation, a strategic goal of the current Triple Helix model ("Concept: Theoretical Framework," 2014; Leydesdorff, 2012). Further research is needed to refine the critical underlying data sources. In this study only two economic impact categories were used. To better under the progression (i.e., enabling breakthroughs) of a technology's life cycle, it will be necessary to decompose the economic impact categories. Another area of investigation is to address the associated supply chains among the industry interest categories. We would like to better understand how inventions from one category affect (overlap) the emergence (development) properties in another category (e.g., phone hardware versus iTunes software) within the context of both systems and technology life cycles. Moreover, we intend to investigate alternative techniques to better understand key agents of change (i.e., TRL transitioning) toward a more robust identification of technology and its systems life cycle.

This work examined an expanded model that clarifies and identifies the life cycle of a system or a collection of systems (i.e., a technology) from initial discovery (via original scientific and conference literature), through critical evolution and development (via original scientific, conference literature and patents), transitioning through TRLs and ultimately on to commercial application. We modified a previously documented model (Abercrombie et al., 2012) with the goal to utilize the TRL terms to address systems life cycle and the relationship between the TRL model concepts and the Triple Helix model. What we discovered is that this method allows for beneficial use of disparate data sources. Once the data sets have been normalized, evaluations can proceed. The stepwise process of this modified model (Abercrombie et al., 2014) is as follows:

1. Search external databases or commercial sources for data associated with a particular technology (the initial search string).
2. An ad-hoc network formed from a variety of sources, when investigated in-total, becomes an integrated network.

    (a) This allows the disparate online data set sources to be analyzed.

3. Normalize the different disparate data sets.
4. Determine the TRL transition(s) inherent in the data sets from step 3.
5. The TRL transition(s), as indicated in the modified model, result(s) in the creation of business intelligence.

(continued)

(continued)

6. Business Intelligence findings in a knowledge based economy provide/ assist in providing a basis for business decision(s).

This enhanced model has the potential to provide for tracking the life cycle of a system or its associated technology from initial discovery (via original scientific and conference literature), through critical discoveries (via original scientific, conference literature and patents), transitioning through TRLs, to, ultimately, commercial applications. The impact of scholarly works, patents and online web news sources are identified during this process. Currently, one of the most comprehensive and useful approaches is to search external databases for data associated with a particular technology or a series of key phrases. The raw data, however, can be confusing and may not provide insight into a trend. But, when normalized, what actually is happening is formation of an ad hoc network that when investigated as a whole can be recognized as integrated. As integrated, these once disparate online data set sources can be analyzed. In the future, we intend to apply a richer set of machine learning techniques to the indicators for the TRL transitioning. We anticipate such new work will result in a more robust identification of systems/technology with respect to both current state in their respective life cycles and clarification of their relationship to the Triple Helix model.

Acknowledgments The views expressed in this chapter are those of the authors and do not reflect the official policy or position of our sponsoring institutions and/or sponsors, the Department of Energy or the U.S. Government. This manuscript has been authored by a contractor of the U.S. Government (USG) under contract DE-AC05-00OR22725. Accordingly, the USG retains a nonexclusive, royalty-free license to publish or reproduce the published form of this contribution, or allow others to do so, for USG purposes.

References

Abercrombie, R. K., Schlicher, B. G., & Sheldon, F. T. (2013). Analysis of search results for the clarification and identification of technology emergence (AR-CITE). In J. Gorraiz, E. Schiebel, C. Gumpenberger, M. Hörlesberger, & H. Moed (Eds.), *Proceedings of 14th International Society of Scientometrics and Informetrics Conference-ISSI 2013* (Vol. 2, pp. 1854–1856). Vienna, Austria: Facultas Verlags und Buchhandels AG.
Abercrombie, R. K., Schlicher, B. G., & Sheldon, F. T. (2014). Scientometric methods for identifying emerging technologies. *US Patent Publication No. US 20140006332 A1, US Patent Application No. 13/927.141* (p. 18). USPTO: Google Patents.
Abercrombie, R. K., Udoeyop, A. W., & Schlicher, B. G. (2012). A study of scientometric methods to identify emerging technologies via modeling of milestones. *Scientometrics, 91* (2), 327–342.
Boehm, B. (1988). A spiral model of software development and enhancement. *IEEE Computer, 21* (5), 61–72.

CA Labs and Royal Melbourne Institute of Technology win Australian Government Grant for Research into Insider Threat. (2011). *Press releases from CA Technologies.* Retrieved from http://prwire.com.au/print/ca-labs-and-royal-melbourne-institute-of-technology-win-australian-government-grant-for-research-into-insider-threat-1

Case, J. D., Fedor, M., Schoffstall, M. L., & Davin, J. (1990). RFC1157-Simple Network Management Protocol (SNMP): RFC Editor.

Coccia, M. (2014a, January 27). Blurring boundaries to strengthen identities: challenges and opportunities in the Triple Helix model. *The Triple Helix article series-2.* Retrieved March 31, 2014, from http://www.oxbridgebiotech.com/review/business-development/blurring-boundaries-strenghten-identities-challenges-opportunties-triple-helix-model/

Coccia, M. (2014b, January 7). Collaborations between industry and academia: Managing the Triple-Helix Model. *The Triple Helix article series-1* Retrieved March 31, 2014, from http://www.oxbridgebiotech.com/review/business-development/managing-triple-helix/

Concept: Theoretical Framework. (2014). Retrieved March 6, 2014, from http://tha2014.org/concept

Defense Acquisition Guidebook. (2012). Retrieved February 21, 2014, from https://acc.dau.mil/CommunityBrowser.aspx?id=518692&lang=en-US

Drogo, D. (2012). Article comment. Retrieved February 26, 2014, from http://www.computerworld.com/s/article/71151/System_Development_Life_Cycle

Etzkowitz, H., & Leydesdorff, L. (2000). The dynamics of innovation: From National Systems and "Mode 2" to a Triple Helix of university–industry–government relations. *Research Policy, 29* (2), 109–123.

Etzkowitz, H., & Ranga, M. (2010). *A Triple Helix System for knowledge-based regional development: From "Spheres" to "Spaces".* Paper presented at the Triple Helix VIII Conference 2010, Madrid, Spain.

Evolving the high performance computing and communications initiative to support the Nation's Information Infrastructure. (1995). Washington, DC: The National Academies Press.

Frye, R., Levi, D., Routhier, S., & Lucent, W. (2003). RFC3584-coexistence between Version 1, Version 2, and Version 3 of the Internet-standard Network Management Framework: RFC Editor.

Harper, L., Van-Nostrnad, A., Pennock, M., & Algosos, W. (2010). *System of systems technology analysis and selection methodology.* Paper presented at the 13th Annual Systems Engineering Conference "Achieving Acquisition Excellence Via Effective Systems Engineering". Retrieved from http://www.dtic.mil/ndia/2010systemengr/ThursdayTrack7_10823Harper.pdf

Information Resources Management, The Department of Justice Systems Development Life Cycle Guidance Document. (2003). Retrieved from http://www.justice.gov/jmd/irm/lifecycle/ch1.htm

Innovation in Information Technology. (2003). Washington, DC: The National Academies Press.

Kay, R. (2002). QuickStudy: System development life cycle. *Computerworld.* Retrieved from http://www.computerworld.com/s/article/71151/System_Development_Life_Cycle.

Kostalova, J., & Tetrevova, L. (2013). Triple Helix Model and partnerships of technical faculties of universities. In P. Lorca, C. Popescu, M. Pervan, & C. Lu (Eds.), *Proceedings of the 5th International Conference on Applied Economics, Business and Development (AEBD '13)* (Recent researches in applied economics and management-business administration and financial management-volume 1, pp. 89–94). Chania, Crete Island, Greece: World Scientific and Engineering Academy and Society Press.

Kujaswki, E. (2010). *The trouble with the system readiness level (SRL) index for managing the acquisition of defense systems.* Paper presented at the 13th Annual Systems Engineering Conference "Achieving Acquisition Excellence Via Effective Systems Engineering". Retrieved from http://www.dtic.mil/ndia/2010systemengr/ThursdayTrack7_10745Kujawski.pdf

Lee, P., Dean, M. E., Estrin, D. L., Kajiya, J. T., Raghavan, P., & Viterbi, A. J. (2012). *Continuing innovation in information technology.* Washington, DC: The National Academies Press.

Leydesdorff, L. (2010). The knowledge-based economy and the triple helix model. *Annual Review of Information Science and Technology, 44*(1), 365–417.

Leydesdorff, L. (2012). The Triple Helix of university-industry-government relations. Retrieved from http://www.leydesdorff.net/th12/th12.pdf from http://citeseerx.ist.psu.edu/viewdoc/summary?doi=10.1.1.302.3991

Leydesdorff, L. (2013). Triple Helix of university-industry-government relations. In E. G. Carayannis (Ed.), *Encyclopedia of creativity, innovation, and entrepreneurship* (pp. 1844–1851). New York, NY: Springer.

Leydesdorff, L., & Etzkowitz, H. (1996). Emergence of a Triple Helix of university—industry—government relations. *Science and Public Policy, 23*(5), 279–286.

Madhavan, K. P. C., Vorvoreanu, M., Elmqvist, N., Johri, A., Ramakrishnan, N., Wang, G. A., et al. (2012). Portfolio mining. *Computer, 45*(10), 95–99.

McConkie, E., Mazzuchi, T. A., Sarkani, S., & Marchette, D. (2013). Mathematical properties of System Readiness Levels. *Systems Engineering, 16*(4), 391–400.

Mohapatra, P. K. J. (2010). *Software engineering (a lifecycle approach)*. Daryaganj, Delhi: New Age International.

Park, H. W. (2014). An interview with Loet Leydesdorff: The past, present, and future of the triple helix in the age of big data. *Scientometrics, 99*(1), 199–202.

Pugh, R. (2013). *The good, the bad and the ugly: Triple Helix policies and programmes in Wales*. Paper presented at the Triple Helix XI International Conference 2013. Retrieved from http://tha2013.org/index.php/tha/2013/paper/view/230

Sandström, U., & Sandström, E. (2009). Meeting the micro-level challenges: Bibliometrics at the individual level. *12th International Conference on Scientometrics and Informetrics* (pp. 845–856).

Schach, R. (1999). *Software engineering* (4th ed.). Boston, MA: McGraw-Hill.

Simple Network Management Protocol. (2014, last updated: March 20, 2014). *Wikipedia*. Retrieved March 31 2014, from http://en.wikipedia.org/wiki/Simple_Network_Management_Protocol

Smallwood, C. (2011). The triple helix of innovation. Retrieved from http://blogs.ca.com/innovation/2011/07/28/the-triple-helix-of-innovation/

Taylor, J. (2004). *Managing information technology projects: Applying project management strategies to software, hardware, and integration initiatives*. New York, NY: American Management Association.

Technology Readiness Assessment (TRA) Guidance. (2011). Retrieved from https://acc.dau.mil/CommunityBrowser.aspx?id=18545

Technology Readiness Level. (2013, last update date: February 27, 2014). *Wikipedia*. Retrieved March 5, 2014, from http://en.wikipedia.org/wiki/Technology_readiness_level

Tetřevová, L., & Kostalova, J. (2012). Problems with application of the Triple Helix in the Czech Republic. In Z. Panian (Ed.), *Proceedings of the 4th WSEAS World Multiconference on Applied Economics, Business and Development (AEBD '12)* (Recent researches in business and economics, pp. 41–46). Porto, Portugal: World Scientific and Engineering Academy and Society Press.

VIII Triple Helix International Conference on University, Industry and Government Linkages Book of Abstracts. (2010). In M. Amaral, I. Sender, M. Cendoya & R. Zaballa (Eds.), *Triple Helix VIII Conference*. Retrieved from http://www.triplehelixconference.org/th/8/doc/Abstract_book_Triple_Helix.pdf

XI Triple Helix International Conference Program—Listing of Papers. (2013). *Triple Helix XI Conference*. Retrieved from http://www.triplehelixconference.org/th/11/papers.htm

Chapter 6
Spatial Scientometrics and Scholarly Impact: A Review of Recent Studies, Tools, and Methods

Koen Frenken and Jarno Hoekman

Abstract Previously, we proposed a research program to analyze spatial aspects of the science system which we called "spatial scientometrics" (Frenken, Hardeman, & Hoekman, 2009). The aim of this chapter is to systematically review recent (post-2008) contributions to spatial scientometrics on the basis of a standardized literature search. We focus our review on contributions addressing spatial aspects of scholarly impact, particularly, the spatial distribution of publication and citation impact, and the effect of spatial biases in collaboration and mobility on citation impact. We also discuss recent dedicated tools and methods for analysis and visualization of spatial scientometric data. We end with reflections about future research avenues.

6.1 Introduction

One of the main trends in scientometrics has been the increased attention to spatial aspects. Parallel to a broader interest in the "geography of science" in fields as history of science, science and technology studies, human geography and economic geography (Barnes, 2001; Finnegan, 2008; Frenken, 2010; Livingstone, 2010; Meusburger, Livingstone, & Jöns, 2010; Shapin, 1998), the field of scientometrics has witnessed a rapid increase in studies using spatial data. In an earlier review, Frenken et al. (2009, p. 222) labelled these studies as "spatial scientometrics" and

K. Frenken (✉)
Innovation Studies, Copernicus Institute for Sustainable Development, Utrecht University, Utrecht, The Netherlands
e-mail: k.frenken@uu.nl

J. Hoekman
Department of Pharmaceutical Sciences and Innovation Studies, Copernicus Institute for Sustainable Development, Utrecht University, Utrecht, The Netherlands
e-mail: j.hoekman@uu.nl

© Springer International Publishing Switzerland 2014
Y. Ding et al. (eds.), *Measuring Scholarly Impact*,
DOI 10.1007/978-3-319-10377-8_6

defined this subfield as "quantitative science studies that explicitly address spatial aspects of scientific research activities."

The chapter provides an update of the previous review on spatial scientometrics by Frenken et al. (2009), specifically focusing on contributions from the post-2008 period that address spatial aspects of scholarly impact. We will do so by reviewing contributions that describe the spatial distribution of publication and citation impact, and the effect of spatial biases in collaboration and mobility on citation impact, as two spatial aspects of scholarly impact. We then discuss recent efforts to develop tools and methods that visualize scholarly impact using spatial scientometric data. At the end of the chapter, we look ahead at promising future research avenues.

6.2 Selection of Reviewed Papers

6.2.1 Scope of Review

We conducted a systematic review of contributions to spatial scientometrics that focused on scholarly impact by considering original articles published since 2008. Following the definition of spatial scientometrics introduced by Frenken et al. (2009), we only included empirical papers that made use of spatial information as it can be retrieved from publication data. Moreover, we paid special attention to three bodies of research within the spatial scientometrics literature. *First*, studies were eligible when they either describe or explain the distribution of publication or citation output across spatial units (e.g., cities, countries, world regions). *Second*, studies were considered when they explain scholarly impact of articles based on the spatial organization of research activities (e.g., international collaboration). We refer to this body of research as "geography of citation impact." *Third*, the review considered studies that report on tools and methods to visualize the publication and citation output of spatial units on geographic maps.

Given the focus of this book on scholarly impact, we chose not to provide a comprehensive overview of *all* spatial scientometrics studies published since 2008. Hence we did not consider contributions focused on the spatial organization of research collaboration or the localized emergence of research fields. For notable advancements in these subfields of spatial scientometrics we refer amongst others to Hoekman, Frenken, and Tijssen (2010); Waltman, Tijssen, and Eck (2011); Leydesdorff and Rafols (2011); Boschma, Heimeriks, and Balland (2014).

6.2.2 Search Procedure

The procedure to select papers for review followed three steps. *First*, we retrieved all papers that were citing the 2009 spatial scientometrics review paper (Frenken et al., 2009) either in Thomson Reuters Web of Science (from now on: WoS) or Elsevier Scopus (from now on: Scopus). *Second*, we queried WoS to get a comprehensive overview of all spatial scientometric articles published since 2008. The search was limited to WoS subject categories "information science & library science," "geography," "planning & development," and "multidisciplinary sciences." The following search query consisting of a combination of spatial and scientometric search terms was performed on March 1, 2014:

TS = (spatial* OR "space" OR spatio* OR geograph* OR region* OR "cities" OR "city" OR international* OR countr* OR "proximity" OR "distance" OR "mobility") AND TS = ("publications" OR "co-publications" OR "articles" OR "papers" OR "web of science" OR "web of knowledge" OR "science citation index" OR "scopus" OR scientometr* OR bibliometr* OR citation*)[1]

Third, a number of additional articles were included after full-text reading of key contributions and evaluation of the cited and citing articles therein.

A total of 1,841 articles met the search criteria of the first two steps. Titles and abstracts of all articles were evaluated manually to exclude articles that (1) did not make use of publication data (n = 1,082) or (2) did not report on the spatial organization of research (n = 405). All other 354 publications were manually evaluated and selected for review when they focused on one of three research topics on scholarly impact mentioned above. Subsequently, articles not identified in the WoS search, but cited in or citing key contributions were added.

6.3   Review

We organize our review in three topics. First, we focus on contributions that analyze the spatial distribution of publication and citation impact accross world regions, countries and subnational regions. We subsequently pay attention to the geography of citation impact and provide an exhaustive review of all contributions that have analyzed the effect of the spatial organization of research activities on scholarly impact. In a third section we focus on the development of tools and methods to support the analysis and visualization of spatial aspects of science.

6.3.1   Spatial Distribution of Publication and Citation Impact

The spatial distribution of research activities continues to be a topic of major interest for academic scholars and policy makers alike. In *Nature News*, Van

---

[1] Document type=Article; Indexes=SCI-EXPANDED, SSCI, A&HCI, CPCI-S, CPCI-SSH; Timespan = 2009–2014.

Noorden (2010) discussed recent developments in the field by focusing on the strategies of urban regions to be successful in the production of high-quality scientific research. Another recent initiative that received attention is the Living Science initiative (http://livingscience.inn.ac/) which provides real-time geographic maps of the publication activity of more than 100,000 scientists.

The interest in the spatial distribution of research activities is also noted in our systematic search for spatial scientometric contributions in this sub-field. We found more than 200 papers that analyzed distributions of bibliometric indicators such as publication or citation counts across countries and regions. Space does not allow us to review all these papers. What follows are a number of outstanding papers on the topic organized according to the spatial level of analysis.

6.3.1.1 World Regions and Countries

Two debates have dominated recent analyses of the distribution of research activities over world regions and countries. *First*, it is well known that for many decades scientific research activities were disproportionally concentrated in a small number of countries, with the USA and the UK consistently ranking on top in terms of absolute publication output and citation impact. Yet, in recent years this "hegemony" has been challenged by a number of emerging economies such as China, India, and Brazil.

In particular the report of the Royal Society in 2011, *"Knowledge, Networks and Nations"* emphasizes the rapid emergence of new scientific powerhouses. Using data from Scopus covering the period 1996–2008, the report concludeds that *"Meanwhile, China has increased its publications to the extent that it is now the second highest producer of research output in the world. India has replaced the Russian Federation in the top ten, climbing from 13th in 1996 to tenth between 2004 and 2008"* (Royal Society, 2011, p. 18). Based on a linear extrapolation of these observations, the report also predicts that China is expected to surpass the USA in terms of publication output before 2020. The prediction was widely covered in the media, yet also criticized on both substantial (Jacsó, 2011) and empirical grounds (Leydesdorff, 2012). For instance, Leydesdorff (2012) replicated the analysis using the WoS database for the period 2000–2010 and finds considerable uncertainty around the prediction estimates, suggesting that the USA will be the global leader in publication output for at least another decade. Moreover, Moiwo and Tao (2013) show that China's normalized publication counts for overall population, population with tertiary education and GDP, is relatively low, while smaller countries such as Switzerland, The Netherlands and Scandinavian countries are world leaders on these indicators.

Huang, Chang, and Chen (2012) also analyze changes in the spatial concentration of national publication and citation output for the period 1981–2008 using several measures such as the Gini coefficient and Herfindahl index. Using National Science Indicators data derived from WoS, they show that publication activity continues to be concentrated in a small number of countries including the USA, UK, Germany, and France. Yet, their analysis also reveals that the degree of concentration is gradually decreasing over time due to the rapid growth of publication output in China and other Asian countries such as Taiwan and Korea. What is more, when the USA is removed from the analysis, concentration indicators drop significantly and a pluralist map of publications and citations becomes visible.

A second issue concerning the spatial distribution of research activities that has received considerable interest in recent years is the debate about the European Paradox. For a long time it was assumed that European countries were global leaders in terms of impact as measured by citation counts, but lagged behind in converting this strength into innovation, economic growth and employment (Dosi, Llerena, & Labini, 2006). The idea originated from the European Commission's White Paper on Growth, Competitiveness and Employment which stated that "*the greatest weakness of Europe's research base however, is the comparatively limited capacity to convert scientific breakthroughs and technological achievements into industrial and commercial successes*" (Commission of the European Communities, 1993, p. 87). This assumption about European excellence became a major pillar of the Lisbon Agenda and creation of a European Research Area.

To scrutinize the conjecture of Europe's leading role in citation output, Albarrán, Ortuño, and Ruiz-Castillo (2011a) compared the citation distributions of 3.6 million articles published in 22 scientific fields between 1998 and 2002. The contributions of the EU, USA and Rest of the World are partitioned to obtain two novel indicators of the distribution of the most and least cited papers as further explained in a twin paper (Albarrán, Ortuño, & Ruiz-Castillo, 2011b). They observe that the USA "*performs dramatically better than the EU and the RW on both indicators in all scientific fields*" (Albarrán, Ortuño, & Ruiz-Castillo, 2011a, p. 122), especially when considering the upper part of the distribution. The results are confirmed using mean citation rates instead of citation distributions, although the gap between the USA and Europe is smaller in this case (Albarrán, Ortuño, & Ruiz-Castillo, 2011c). Herranz and Ruiz-Castillo (2013) further refine the analysis by comparing the citation performance of the USA and EU in 219 subfields instead of 22 general scientific fields. They find that on this fine-grained level the USA outperforms the EU in 189 out of 219 subfields. They do not find a particular cluster of subfields in which the EU outperforms the USA. On the basis of this finding they conclude that the idea of the European Paradox can definitely be put to rest.

In addition to studies *describing* the spatial distribution of research activities across countries and world regions, a number of studies have focused on *explaining* these distributions using multivariate models with national publication output as the dependent variable. For instance, Pasgaard and Strange (2013) and Huffman et al. (2013) explain national distributions of publication output in climate change research and cardiovascular research, while Meo, Al Masri, Usmani, Memon, and Zaidi (2013) build a similar model to explain the overall publication count of a set of Asian countries. All three studies observe significant positive effects of Research and Development (R&D) investments, Gross Domestic Product (GDP) and population on publication output. Pasgaard and Strange (2013) and Huffman et al. (2013) also find that field-related variables such as burden of disease in the case of cardiovascular research and CO_2 emissions in the case of climate change research explain national publication output.

Focusing specifically on European countries, Almeida, Pais, and Formosinho (2009) explain publication output of countries on the basis of specialization patterns. Using principal component analysis on national publication and citation distributions of countries by research fields, they find that countries located in close physical proximity to each other also show similar specialization patterns. This suggests that these countries profit from each other through knowledge spillovers. We return to this issue in the next section where we discuss a number of papers that explain publication output at the sub-national instead of national level.

6.3.1.2 Regions and Cities

The interest in describing the spatial distribution of scientific output at the level of sub-national regions and cities has been less than in analysis of the output of countries or aggregates of countries. This is likely due to the fact that larger data efforts are required to accurately classify addresses from scientific papers into urban or regional categories as well as from the fact that science policy is mainly organized at national and transnational levels. Nevertheless, the number of studies addressing the urban and regional scales has been rapidly expanding after 2008 and this trend is likely to continue given the increased availability of tools and methods to conduct analysis at fine-grained spatial levels (see Sect. 6.3.3).

Matthiessen and Schwarz (2010) study the 100 largest cities in the world in terms of publication and citation output for two periods: 1996–1998 and 2004–2006. Even during this short period they observe a rapid rise of cities in Southeast Asia as major nodes in the global science system when considering either publication or citation impact. They also note the rise of Australian, South American, and Eastern European cities. These patterns all indicate that the traditional dominance of cities in North America and Western Europe is weakening, although some of these cities remain major world-city hubs such as the San Francisco Bay Area, New York, London-Cambridge, and Amsterdam.

Cho, Hu, and Liu (2010) analyze the regional development of the Chinese science system in great detail for the period 2000–2006. They observe that the regional distribution of output and citations is highly skewed with coastal regions dominating. However, mainland regions have succeeded in quickly raising their scientific production, but still have low citation impact, exceptions aside. Tang and Shapira (2010) find very similar patterns for the specific field of nanotechnology.

An interesting body of research has analyzed whether publication and citation output within countries is concentrating or de-concentrating over time. A comprehensive study by Grossetti et al. (2013) covering WoS data for the period

1987–2007 finds that in most countries in the world, the urban science system is de-concentrating, indicating that the largest cities are undergoing a relative decline in a country's scientific output (see also an earlier study on five countries by Grossetti, Milard, and Losego (2009)). The same trend is observed by looking at citation instead of publication output. Two other studies on France, with a specific focus on small and medium-sized cities (Levy & Jégou, 2013; Levy, Sibertin-Blanc, & Jégou, 2013), and a study on Spain (Morillo & De Filippo, 2009), support this conclusion. The research results are significant in debunking the popular policy notion that the spatial concentration of recourses and people supports scientific excellence.

Remarkably, the explanation of publication output at the regional level has received a great deal of attention in recent years. That is, a number of contributions address the question why certain regions generate more publication output than others. This body of research relies on a so-called knowledge production function framework where number of publications are considered as the output variables and research investment, amongst other variables, as input in the knowledge production system. Acosta, Coronado, Ferrándiz, and León (2014) look at the effect of public spending on regional publication output using Eurostat data on R&D spending in the Higher Education sector. They find a strong effect of public investment on regional publication output. Interestingly, this effect is strongest in less developed regions ("Objective 1 regions" in the European Union) when compared to more developed regions, meaning that an increase in budget has a higher payoff in less developed regions than in more developed regions. This result is in line with Hoekman, Scherngell, Frenken, and Tijssen (2013) who find that the effect of European Framework Program subsidies is larger in regions that publish less compared to regions that publish more.

Sebestyén and Varga (2013) also apply a knowledge production framework with a specific focus on the role of inter-regional collaboration networks and knowledge spillover effects between neighboring regions. They find that scientific output is dependent on embeddedness in national and international networks, while it is not supported by regional agglomeration of industry or publication activity in neighboring regions. They conclude that regional science policy should focus on networking with other regions domestically and internationally, rather than on local factors or regions in close physical proximity. Their results also confirm the results of an earlier study on Chinese regions which highlight the importance of spillovers stemming from international collaborations (Cho et al., 2010).

Finally, some papers analyze the impact of exogenous events on the publication output of regions or countries. Magnone (2012) studies the impact of the triple disaster in Japan (earthquake, tsunami and nuclear accident) on the Materials Science publication output in the cities of Sendai, Tsukuba and Kyoto (the latter being a "non-disaster situation" control). As expected, the author observes clear and consistent negative effects of the disaster on publication output in Sendai and Tsukuba, compared to Kyoto. Studies with similar research questions include Braun (2012) who studies the effect of war on mathematics research activity in Croatia; Miguel, Moya-Anegón, and Herrero-Solana (2010) scrutinizing the effect of the socio-economic crisis in Argentina on scientific output and impact, and Orduña-Malea, Ontalba-Ruipérez, and Serrano-Cobos (2010) focusing on the impact of 9/11 on international scientific collaboration in Middle Eastern countries.

6.3.2   Geography of Citation Impact

6.3.2.1   Collaboration

A topic which has received considerable attention is the effect of geography, particularly international collaboration, on the citation impact of articles. These geographical effects can be assessed at both the author level and the article level. Typically, studies use a multivariate regression method with the number of citations of a paper or article as the dependent variable, following an early study by Frenken, Hölzl, and Vor (2005). This research set-up allows to control for many other factors that may affect citation impact, such as the number of authors and country effects (e.g., English speaking countries) when explaining citations to articles, and age and gender when explaining citations to individual scientists.

He (2009) finds for 1,860 papers written by 65 biomedical scientists in New Zealand that internationally co-authored papers indeed receive more citations than national collaborations, while controlling for many other factors. More interestingly, he also finds even higher citation impact of local collaborations within the same university when compared to international collaborations. This suggests that local collaboration, which is often not taken into account in the geography of citation impact, may have much more benefits than previously assumed.

The importance of local collaboration is confirmed by Lee, Brownstein, Mills, and Kohane (2010) who consider the effect of physical distance on citation impact by analyzing collaboration patterns between first and last authors that are both located at the Longwood campus of Harvard Medical School. They find that at this microscale, physical proximity in meters and within-building collaboration is positively related to citation impact. The authors do not, however, control for alternative factors that may explain these patterns such as specialization.

Frenken, Ponds, and Van Oort (2010) test the effects of international, national and regional collaboration for Dutch publications in life sciences and physical sciences derived from WoS. They show that research collaboration in the life sciences has a higher citation impact if organized at the regional level than at the national level, while the opposite is found for the physical sciences. In both fields the citation impact of international collaboration exceeds the citation impact of both national and regional research collaboration, in particular for collaborations with the USA. Sin (2011) compares the impact of international versus national collaboration in the field of Library and Information Sciences. In line with other studies, a positive effect for international collaboration is found, while no significant effect of national collaboration as compared to single authorships is observed.

One problem in interpreting the positive effect of international collaboration on citation impact holds that this finding may indeed signal that international collaboration results in higher research quality, yet also that the results from internationally coauthored papers diffuse from centers in multiple countries, as noted by Frenken et al. (2010). These two effects are not necessarily mutually exclusive. Lancho Barrantes, Bote, Vicente, Rodríguez, and de Moya Anegón (2012) try to disentangle between the "quality" and "audience" effect by studying whether national biases on citation impact are larger in countries that produce many papers. They find that the "audience" effect is especially large in relatively small countries, while the quality effect of internationally co-authored papers seems to be a general property irrespective of country size.

Nomaler, Frenken, and Heimeriks (2013) do not look at the effect of international versus national collaboration, but at the effect of kilometric distance between collaborating authors. On the basis of all scientific papers published in 2000 and coauthored by two or more European countries, they show that citation impact increases with the geographical distance between the collaborating countries. Interestingly, they also find a negative effect for EU countries, suggesting that collaborations with a partner outside the EU are more selective, and, hence, have higher quality.

An interesting study by Didegah and Thelwall (2013a) looks at the effects of the geographical properties of references of nanotechnology papers. In particular, they test the hypothesis that papers with more references to "international" journals—defined as journals with more geographic dispersion of authors—have more citation impact. They indeed find this effect. Moreover, after controlling for the effect they no longer observe an effect of international collaboration on citation impact. However, in a related paper that studies the effect of 11 factors on citation impact, of which international collaboration is one, Didegah and Thelwall (2013b) do observe a positive effect of international collaboration on citation impact.

Finally, a study by Eisend and Schmidt (2014) is worth mentioning in this context. They study how the internationalization strategies of business research scholars affect their research performance in terms of citation impact. Their study is original in that they specifically look at how this effect is influenced by the knowledge resources of individual researchers. They find that international collaboration supports performance more if researchers lack language skills and knowledge of foreign markets. This indicates that international collaboration provides researchers with access to complementary skills. Collaboration also improves the performance of less experienced researchers with the advantage diminishing with increasing research experience.

A methodological challenge of studies that assess the effect of international collaboration on citation impact is self-selection bias. Indeed, one can expect that better scientists are more likely to engage in international collaboration. For example, Abramo, D'Angelo, and Solazzi (2011) find that Italian natural scientists who produce higher quality research tend to collaborate more internationally. The same result was found by Kato and Ando (2013) for chemists worldwide. To control for self-selection, they investigate whether the citation rate of international papers is higher than the citation rate of domestic papers, controlling for performance, that is, by looking at papers with at least one author in common. Importantly, they still find that international collaboration positively and significantly affects citation impact. Obviously, the issue of self-selection should be high on the agenda for future research.

6.3.2.2 Mobility

An emerging research topic in spatial scientometrics of scholarly impact concerns the question of whether internationally mobile researchers outperform other researchers in terms of productivity and citation impact. Although descriptive studies have noted a positive effect of international mobility on the citation impact of researchers (Sandström, 2009; Zubieta, 2009), it remains unclear whether higher performance is caused by international mobility (e.g., through the acquisition of new skills), or by self-selection (better scientists being more mobile).

In a recent study, Jonkers and Cruz-Castro (2013) explored this effect for a sample of Argentinian researchers with foreign work experience. When returning home, these researchers show a higher propensity to co-publish with their former host country than with other countries. These researchers also have a higher propensity to publish in high-impact journals as compared to their non-mobile peers, even when the mobile scientists don't publish with foreign researchers. Importantly, the study accounted for self-selection (better scientists being more mobile) by taking into account the early publication record of researchers as an explanatory variable for high-impact publications after their return to Argentina.

Another study by Trippl (2013) investigates the impact of internationally mobile star scientists on regional knowledge transfer. Here, the question holds whether a region benefits from attracting renowned scientists from abroad. It was found that mobile star scientists do not differ in their regional knowledge transfer activities from non-mobile star scientists. However, mobile scientists have more interregional linkages with firms which points to the importance of mobile scientists for complementing intraregional ties with interregional ones.

6.3.3 Tools and Methods

Besides empirical contributions to the field of spatial scientometrics, a growing group of scholars have focused on the development of tools and methods to support the analysis and visualization of spatial aspects of science. Following a more general interest in science mapping (see for instance: http://scimaps.org/maps/browse/) and a trend within the academic community to create open source analytical tools, most of the tools reviewed below are freely available for analysis and published alongside the publication material.

Leydesdorff and Persson (2010) provide a comprehensive review and user's guide of several methods and software packages that were freely available to visualize research activities on geographic maps up to 2010. They focus specifically on the visualization of collaboration and citation networks that can be created on the basis of author-affiliate addresses on publications. In their review they cover, amongst others, the strengths and weaknesses of *CiteSpace*, *Google Maps*, *Google Earth*, *GPS Visualizer* and *Pajek* for visualization purposes. One particular strength of the paper is that it provides software to process author-affiliate addresses of publication data retrieved from Web of Science or Scopus for visualization on the city level. The software has been refined over the last years and can be found on: http://www.leydesdorff.net/software.htm.

Further to the visualization of research networks, Bornmann et al. (2011) focus on the geographic mapping of publication and citation counts of cities and regions. They extract all highly cited papers in a particular research field from Scopus, and develop a method to map "excellent cities and regions" using *Google Maps*. The percentile rank of a city as determined on the base of its contribution to the total set of highly cited papers is visualized by plotting circles with different radii (frequency of highly cited papers) and colors (city rank) on a geographic map. The exact procedure including a user guide for this visualization tool is provided at: http://www.leydesdorff.net/mapping_excellence/index.htm.

A disadvantage of the approach in Bornmann et al. (2011) is that it visualizes the absolute number of highly cited papers of a particular city without normalizing for the total number of publications in that city. Bornmann and Leydesdorff (2011) provide such a methodological approach using a statistical z test that compares observed proportions of highly cited papers of a particular city with expected proportions. "Excellence" can then be defined as cities in which "*authors are located who publish a statistically significant higher number of highly cited papers than can be expected for these cities*" (Bornmann & Leydesdorff, 2011, p.1954). The authors use similar methods as in Bornmann et al. (2011) to create geographic maps of excellence for three research fields: physics, chemistry and psychology. The maps confirm the added value of normalization as cities with high publication output do not necessarily have a disproportionate number of highly cited papers. Further methodological improvement to this method is provided by Bornmann and Leydesdorff (2012) who use the Integrated Impact Indicator (I3) as an alternative to normalized citation rates. Another improvement is that they correct observed citation rates for publication years.

Researchers using the above visualization approaches should be aware of a number of caveats that are extensively discussed in Bornmann et al. (2011). Visualization errors may occur due to amongst others imprecise allocation of geo-coordinates or incomplete author-affiliate addresses. Created geographic maps should therefore be always carefully scrutinized manually.

Building on the abovementioned contributions, Bornmann, Stefaner, de Moya Anegón, and Mutz (2014a) introduce a novel web application (www.

excellencemapping.net) that can visualize the performance of academic institutions on geographic maps. The web application visualizes field-specific excellence of academic institutions that frequently publish highly-cited papers. The underlying methodology for this is based on multilevel modeling that takes into account the data at the publication level (i.e., whether a particular paper belongs to the top 10 % most cited papers in a particular institution) as well as the academic institution level (how many papers of an institution belong to the overall top 10 % of most cited papers). Using this methodology, top performers by scientific fields who publish significantly more top-10 % papers than an "average" institution in a scientific field are visualized. Results are visualized by colored circles on the location of the respective institutions on a geographic map. The web application provides the possibility to select the circles for further information about the institutions. In Bornmann, Stefaner, de Moya Anegon, and Mutz (2014b) the web application is further enhanced by adding the possibility to control for the effect of covariates (such as the number of residents of a country in which an institution is located) on the performance of institutions. Using this method one can visualize the performance of institutions under the hypothetical situation that all institutions have the same value on the covariate in question. For instance, institutions can be visualized that have a very good performance once controlled for their relatively low national GDP. In the coming years, further development of the scientific excellence tool is anticipated.

Bornmann and Waltman (2011) use a somewhat different approach to map regions of excellence based on heat maps. The visualization they propose uses density maps that can be created using the *VOSviewer* software for bibliometric mapping (Van Eck & Waltman, 2010). A step-by-step instruction to make these maps is provided on: http://www.ludowaltman.nl/density_map/. In short, the heat maps rely on kernel density estimations of the publication activity of geographic coordination and a specification of a kernel width (in kilometers) for smoothing. Research excellence is then visualized for regions instead of individual cities, especially when clusters of cities with high impact publication activity are located in close proximity to each other. The created density maps reveal clusters of excellence running from South England, over Netherlands/Belgium and Western Germany to Northern Switzerland.

An entirely different approach to visualizing bibliometric data is explored by Persson and Ellegård (2012). Inspired by the theory of time-geography which was initially proposed by Thorsten Hägerstrand in 1955, they reconstruct time-space paths of the scientific publications citing the work of Thorsten Hägerstrand. Publications are plotted on a two dimensional graph with time (years) on the vertical axis, space (longitude) on the horizontal axis and paths between a time-space location indicating citations between articles.

Conclusions and Recommendations

Clearly, the interest in analyzing spatial aspects of scientific activity using spatial scientometric data is on the rise. In this review we specifically looked at contributions that focused on scholarly impact and found a large number of such papers. While previously most studies focused solely on national levels, many scientometrics contributions now take into account regional and urban levels. What is more, in analyzing relational data, kilometric distance is increasingly taken into account as one of the determinants of scholarly impact. The research design of spatial scientometric papers is also more elaborate than for earlier papers, with theory-driven hypotheses and increasingly a multivariate regression set-up. Progress has also been made in the automatic generation of data as well as in visualization of this data on geographic maps. Having said this, we identify below some research avenues that fill some existing research gaps in theory, topics, methodology and data sources.

Little theorizing: As noted, many studies start from hypotheses rather than from data. Yet, most often, hypotheses are derived from general theoretical notions rather than from specific theories of scientific practices. Indeed, spatial scientometrics makes little reference to theories in economics, geography or science and technology studies, arguably the fields closest to the spatial scientometric enterprise. And, conversely, more theory-minded researchers have also shown little interest in developing more specific theories about the geography of science, so far (Frenken, 2010). Clearly, more interaction between theory and empirics is welcome at this stage of research in spatial scientometrics. One can think of theories from network science, including the "proximity framework" and social network analysis, which aim to explain both the formation of scientific collaboration networks and their effect on scholarly impact (Frenken et al., 2009). A second possibility is to revive the links with Science and Technology Studies, which have exemplified more strategic and discursive aspects of science and scientific publishing (Frenken, 2010). Thirdly, modern economic geography offers advanced theories of localization, specifically, regarding the source of knowledge spillovers that may underlie the benefits of clustering in knowledge production (Breschi & Lissoni, 2009; Scherngell, 2014). Lastly, evolutionary theorizing may be useful to analyze the long-term dynamics in the geography of science, including questions of where new fields emerge and under what conditions existing centers lose their dominance (Boschma et al., 2014; Heimeriks & Boschma, 2014). Discussion of such possibilities in further detail is, however, beyond the purpose and scope of this chapter.

Self-selection as methodological challenge: As repeatedly stressed, a major problem in assessing the effect of geography (such as mobility or long-distance collaboration) on researchers' scholarly impact arises from

(continued)

(continued)

self-selection effects. One can expect that more talented researchers are more internationally oriented, if only because they search for more specialized and state-of-the-art knowledge. Hence, the positive effects of internationalization on performance may not reflect, or only partially, the alleged benefits from international collaboration as such. We have highlighted some recent attempts to deal with self-selection in the case of international mobility (Jonkers & Cruz-Castro, 2013) and international collaboration (Kato & Ando, 2013). Obviously, more research in this direction is welcome.

Mobility as an underdeveloped topic: A research topic which remains underinvestigated in the literature, despite its importance for shaping spatial aspects of the science system, is scientific mobility. Although we identified a number of papers (e.g. Jonkers & Cruz-Castro, 2013; Trippl, 2013) dealing with the topic, the total number of papers is relatively low and most papers focus on theoretical rather than empirical questions. One of the reasons for this state of affairs is the known difficulty in disambiguating author names purely based on information derived from scientific publications. The challenge in these cases is to determine whether the same or similar author names on different publications refer to the same researcher (for an overview see: Smalheiser and Torvik (2009)). Arguably, the increase in authors with a Chinese last name has made such disambiguation even more difficult, due to the large number of scholars sharing only a few family names such as Zhang, Chen or Li (Tang & Walsh, 2010). To deal with this issue scholars have started to develop tools and methods to solve the disambiguation problems. Recent examples include but are not limited to Tang and Walsh (2010); D'Angelo, Giuffrida, and Abramo (2011); Onodera et al. (2011); Wang et al. (2012); Wu and Ding (2013). Most of these studies now agree on the necessity to rely on external information (e.g. name lists) for a better disambiguation or to complement bibliometric data with information from other sources (e.g. surveys, curriculum vitae). For an overview of author name disambiguation issues and methods, please see the Chap. 7.

Data source dependency: All spatial scientometric analyses are dependent on the data sources that are being used. It is important to note in this context that there are differences between the set of journals that are covered in Web of Science and Scopus, with Scopus claiming to include more 'regional' journals. Moreover, the coverage of bibliometric databases changes over time, which may have an effect on longitudinal analyses of research activities. A telling example of this is the earlier mentioned predictions of the rise of China in terms of publication output. Leydesdorff (2012) showed in this respect that predictions of China's growth in publication output differ considerably between an analysis of the Scopus or Web of

(continued)

(continued)

Science database. Basu (2010) also observes a strong association between the number of indexed journals from a particular country and the total number of publications from that country. Other notable papers focusing on this issue include Rodrigues and Abadal (2014); Shelton, Foland, and Gorelskyy (2009) and Collazo-Reyes (2014). A challenge for further research is therefore to distinguish between changes in publication output of a particular spatial unit due to changes in academic production or changes in coverage of scientific journals with a spatial bias. Following Martin and Irvine (1983) and Leydesdorff (2012) we suggest relying on "partial indicators" where results become more reliable when they indicate the same trends and results across a number of databases.

Alternative data sources: Finally, a limitation of our review is that we only focused on spatial scientometric papers of scholarly impact and papers that made use of spatial information as it can be retrieved from individual publications. As noted in Frenken et al. (2009) there are a number of other topics that analyze spatial aspects of research activities, including the spatial analysis of research collaboration and the localized emergence of new research fields. Moreover, in addition to publication data there are other large datasets to analyze spatial aspects of science including but not limited to Framework Programme data (Autant-Bernard, Billand, Frachisse, & Massard, 2007; Scherngell & Barber, 2009) and student mobility flows (Maggioni & Uberti, 2009). Due to space limitations we were not able to review all these contributions. Yet, while performing the systematic search of the scientometrics literature, we came across a number of innovative research topics such as those focusing on spatial aspects of editorial boards (Bański & Ferenc, 2013; García-Carpintero, Granadino, & Plaza, 2010; Schubert & Sooryamoorthy, 2010); research results (Fanelli, 2012); authorships (Hoekman, Frenken, de Zeeuw, & Heerspink, 2012); journal language (Bajerski, 2011; Kirchik, Gingras, & Larivière, 2012); and the internationality of journals (Calver, Wardell-Johnson, Bradley, & Taplin, 2010; He & Liu, 2009; Kao, 2009). They provide useful additions to the growing body of spatial scientometrics articles.

References

Abramo, G., D'Angelo, C. A., & Solazzi, M. (2011). The relationship between scientists' research performance and the degree of internationalization of their research. *Scientometrics, 86*(3), 629–643.

Acosta, M., Coronado, D., Ferrándiz, E., & León, M. D. (2014). Regional scientific production and specialization in Europe: the role of HERD. *European Planning Studies, 22*(5), 1–26. doi:10.1080/09654313.2012.752439.

Albarrán, P., Ortuño, I., & Ruiz-Castillo, J. (2011a). High-and low-impact citation measures: empirical applications. *Journal of Informetrics, 5*(1), 122–145. doi:10.1016/j.joi.2010.10.001.

Albarrán, P., Ortuño, I., & Ruiz-Castillo, J. (2011b). The measurement of low-and high-impact in citation distributions: Technical results. *Journal of Informetrics, 5*(1), 48–63. doi:10.1016/j.joi. 2010.08.002.

Albarrán, P., Ortuño, I., & Ruiz-Castillo, J. (2011c). Average-based versus high-and low-impact indicators for the evaluation of scientific distributions. *Research Evaluation, 20*(4), 325–339. doi:10.3152/095820211X13164389670310.

Almeida, J. A. S., Pais, A. A. C. C., & Formosinho, S. J. (2009). Science indicators and science patterns in Europe. *Journal of Informetrics, 3*(2), 134–142. doi:10.1016/j.joi.2009.01.001.

Autant-Bernard, C., Billand, P., Frachisse, D., & Massard, N. (2007). Social distance versus spatial distance in R&D cooperation: Empirical evidence from European collaboration choices in micro and nanotechnologies. *Papers in Regional Science, 86*(3), 495–519. doi:10.1111/j.1435-5957.2007.00132.x.

Bajerski, A. (2011). The role of French, German and Spanish journals in scientific communication in international geography. *Area, 43*(3), 305–313. doi:10.1111/j.1475-4762.2010.00989.x.

Bański, J., & Ferenc, M. (2013). "International" or "Anglo-American" journals of geography? *Geoforum, 45*, 285–295. doi:10.1016/j.geoforum.2012.11.016.

Barnes, T. J. (2001). 'In the beginning was economic geography'–a science studies approach to disciplinary history. *Progress in Human Geography, 25*(4), 521–544.

Basu, A. (2010). Does a country's scientific 'productivity' depend critically on the number of country journals indexed? *Scientometrics, 82*(3), 507–516.

Bornmann, L., & Leydesdorff, L. (2011). Which cities produce more excellent papers than can be expected? A new mapping approach, using Google Maps, based on statistical significance testing. *Journal of the American Society for Information Science and Technology, 62*(10), 1954–1962.

Bornmann, L., & Leydesdorff, L. (2012). Which are the best performing regions in information science in terms of highly cited papers? Some improvements of our previous mapping approaches. *Journal of Informetrics, 6*(2), 336–345.

Bornmann, L., Leydesdorff, L., Walch-Solimena, C., & Ettl, C. (2011). Mapping excellence in the geography of science: An approach based on Scopus data. *Journal of Informetrics, 5*(4), 537–546.

Bornmann, L., Stefaner, M., de Moya Anegón, F., & Mutz, R. (2014a). Ranking and mapping of universities and research-focused institutions worldwide based on highly-cited papers: A visualisation of results from multi-level models. *Online Information Review, 38*(1), 43–58.

Bornmann, L., Stefaner, M., de Moya Anegon, F., & Mutz, R. (2014b). What is the effect of country-specific characteristics on the research performance of scientific institutions? Using multi-level statistical models to rank and map universities and research-focused institutions worldwide. *arXiv preprint arXiv:1401.2866.*

Bornmann, L., & Waltman, L. (2011). The detection of "hot regions" in the geography of science—a visualization approach by using density maps. *Journal of Informetrics, 5*(4), 547–553.

Boschma, R., Heimeriks, G., & Balland, P. A. (2014). Scientific knowledge dynamics and relatedness in biotech cities. *Research Policy, 43*(1), 107–114.

Braun, J. D. (2012). Effects of war on scientific production: mathematics in Croatia from 1968 to 2008. *Scientometrics, 93*(3), 931–936.

Breschi, S., & Lissoni, F. (2009). Mobility of skilled workers and co-invention networks: An anatomy of localized knowledge flows. *Journal of Economic Geography, 9*, 439–468. doi:10.1093/jeg/lbp008.

Calver, M., Wardell-Johnson, G., Bradley, S., & Taplin, R. (2010). What makes a journal international? A case study using conservation biology journals. *Scientometrics, 85*(2), 387–400.

Cho, C. C., Hu, M. W., & Liu, M. C. (2010). Improvements in productivity based on co-authorship: a case study of published articles in China. *Scientometrics, 85*(2), 463–470.

Collazo-Reyes, F. (2014). Growth of the number of indexed journals of Latin America and the Caribbean: the effect on the impact of each country. *Scientometrics, 98*(1), 197–209.

Commission of the European Communities. (1993). *White paper on growth, competitiveness and employment.* Brussels: COM(93) 700 final.

D'Angelo, C. A., Giuffrida, C., & Abramo, G. (2011). A heuristic approach to author name disambiguation in bibliometrics databases for large-scale research assessments. *Journal of the American Society for Information Science and Technology, 62*(2), 257–269.

Didegah, F., & Thelwall, M. (2013a). Determinants of research citation impact in nanoscience and nanotechnology. *Journal of the American Society for Information Science and Technology, 64*(5), 1055–1064.

Didegah, F., & Thelwall, M. (2013b). Which factors help authors produce the highest impact research? Collaboration, journal and document properties. *Journal of Informetrics, 7*(4), 861–873.

Dosi, G., Llerena, P., & Labini, M. S. (2006). The relationships between science, technologies and their industrial exploitation: An illustration through the myths and realities of the so-called 'European Paradox'. *Research Policy, 35*(10), 1450–1464.

Eisend, M., & Schmidt, S. (2014). The influence of knowledge-based resources and business scholars' internationalization strategies on research performance. *Research Policy, 43*(1), 48–59.

Fanelli, D. (2012). Negative results are disappearing from most disciplines and countries. *Scientometrics, 90*(3), 891–904.

Finnegan, D. A. (2008). The spatial turn: Geographical approaches in the history of science. *Journal of the History of Biology, 41*(2), 369–388.

Frenken, K. (2010). Geography of scientific knowledge: A proximity approach. *Eindhoven Center for Innovation Studies (ECIS) working paper series* 10-01. Eindhoven Center for Innovation Studies (ECIS). Retrieved from http://econpapers.repec.org/paper/dgrtuecis/wpaper_3a1001.htm

Frenken, K., Hardeman, S., & Hoekman, J. (2009). Spatial scientometrics: Towards a cumulative research program. *Journal of Informetrics, 3*(3), 222–232.

Frenken, K., Hölzl, W., & Vor, F. D. (2005). The citation impact of research collaborations: the case of European biotechnology and applied microbiology (1988–2002). *Journal of Engineering and Technology Management, 22*(1), 9–30.

Frenken, K., Ponds, R., & Van Oort, F. (2010). The citation impact of research collaboration in science-based industries: A spatial-institutional analysis. *Papers in Regional Science, 89*(2), 351–371.

García-Carpintero, E., Granadino, B., & Plaza, L. M. (2010). The representation of nationalities on the editorial boards of international journals and the promotion of the scientific output of the same countries. *Scientometrics, 84*(3), 799–811.

Grossetti, M., Eckert, D., Gingras, Y., Jégou, L., Larivière, V., & Milard, B. (2013, November). Cities and the geographical deconcentration of scientific activity: A multilevel analysis of publications (1987–2007). *Urban Studies,* 0042098013506047.

Grossetti, M., Milard, B., & Losego, P. (2009). La territorialisation comme contrepoint à l'internationalisation des activités scientifiques.*L'internationalisation des systèmes de recherche en action. Les cas français et suisse.* Retrieved from http://halshs.archives-ouvertes.fr/halshs-00471192

He, Z. L. (2009). International collaboration does not have greater epistemic authority. *Journal of the American Society for Information Science and Technology, 60*(10), 2151–2164.

He, T., & Liu, W. (2009). The internationalization of Chinese scientific journals: A quantitative comparison of three chemical journals from China, England and Japan. *Scientometrics, 80*(3), 583–593.

Heimeriks, G., & Boschma, R. (2014). The path- and place-dependent nature of scientific knowledge production in biotech 1986–2008. *Journal of Economic Geography, 14*(2), 339–364. doi:10.1093/jeg/lbs052.

Herranz, N., & Ruiz-Castillo, J. (2013). The end of the 'European Paradox'. *Scientometrics, 95*(1), 453–464. doi:10.1007/s11192-012-0865-8.

Hoekman, J., Frenken, K., de Zeeuw, D., & Heerspink, H. L. (2012). The geographical distribution of leadership in globalized clinical trials. *PLoS One, 7*(10), e45984.

Hoekman, J., Frenken, K., & Tijssen, R. J. (2010). Research collaboration at a distance: Changing spatial patterns of scientific collaboration within Europe. *Research Policy, 39*(5), 662–673.

Hoekman, J., Scherngell, T., Frenken, K., & Tijssen, R. (2013). Acquisition of European research funds and its effect on international scientific collaboration. *Journal of Economic Geography, 13*(1), 23–52.

Huang, M. H., Chang, H. W., & Chen, D. Z. (2012). The trend of concentration in scientific research and technological innovation: A reduction of the predominant role of the US in world research & technology. *Journal of Informetrics, 6*(4), 457–468.

Huffman, M. D., Baldridge, A., Bloomfield, G. S., Colantonio, L. D., Prabhakaran, P., Ajay, V. S., ... & Prabhakaran, D. (2013). Global cardiovascular research output, citations, and collaborations: A time-trend, bibliometric analysis (1999–2008). *PLoS One, 8*(12), e83440.

Jacsó, P. (2011). Interpretations and misinterpretations of scientometric data in the report of the Royal Society about the scientific landscape in 2011. *Online Information Review, 35*(4), 669–682.

Jonkers, K., & Cruz-Castro, L. (2013). Research upon return: The effect of international mobility on scientific ties, production and impact. *Research Policy, 42*(8), 1366–1377.

Kao, C. (2009). The authorship and internationality of industrial engineering journals. *Scientometrics, 81*(1), 123–136.

Kato, M., & Ando, A. (2013). The relationship between research performance and international collaboration in chemistry. *Scientometrics, 97*(3), 535–553.

Kirchik, O., Gingras, Y., & Larivière, V. (2012). Changes in publication languages and citation practices and their effect on the scientific impact of Russian science (1993–2010). *Journal of the American Society for Information Science and Technology, 63*(7), 1411–1419.

Lancho Barrantes, B. S., Bote, G., Vicente, P., Rodríguez, Z. C., & de Moya Anegón, F. (2012). Citation flows in the zones of influence of scientific collaborations. *Journal of the American Society for Information Science and Technology, 63*(3), 481–489.

Lee, K., Brownstein, J. S., Mills, R. G., & Kohane, I. S. (2010). Does collocation inform the impact of collaboration? *PLoS One, 5*(12), e14279.

Levy, R., & Jégou, L. (2013). Diversity and location of knowledge production in small cities in France. *City, Culture and Society, 4*(4), 203–216.

Levy, R., Sibertin-Blanc, M., & Jégou, L. (2013). La production scientifique universitaire dans les villes françaises petites et moyennes (1980–2009). *M@ ppemonde*, (110 (2013/2))

Leydesdorff, L. (2012). World shares of publications of the USA, EU-27, and China compared and predicted using the new Web of Science interface versus Scopus. *El profesional de la información, 21*(1), 43–49.

Leydesdorff, L., & Persson, O. (2010). Mapping the geography of science: Distribution patterns and networks of relations among cities and institutes. *Journal of the American Society for Information Science and Technology, 61*(8), 1622–1634.

Leydesdorff, L., & Rafols, I. (2011). Local emergence and global diffusion of research technologies: An exploration of patterns of network formation. *Journal of the American Society for Information Science and Technology, 62*(5), 846–860.

Livingstone, D. N. (2010). *Putting science in its place: Geographies of scientific knowledge.* Chicago: University of Chicago Press.

Maggioni, M. A., & Uberti, T. E. (2009). Knowledge networks across Europe: Which distance matters? *The Annals of Regional Science, 43*(3), 691–720.

Magnone, E. (2012). An analysis for estimating the short-term effects of Japan's triple disaster on progress in materials science. *Journal of Informetrics, 6*(2), 289–297.

Martin, B. R., & Irvine, J. (1983). Assessing basic research: Some partial indicators of scientific progress in radio astronomy. *Research Policy, 12*(2), 61–90.

Matthiessen, C. W., & Schwarz, A. W. (2010). World cities of scientific knowledge: Systems, networks and potential dynamics. An analysis based on bibliometric indicators. *Urban Studies, 47*(9), 1879–1897.

Meo, S. A., Al Masri, A. A., Usmani, A. M., Memon, A. N., & Zaidi, S. Z. (2013). Impact of GDP, spending on R&D, number of universities and scientific journals on research publications among Asian countries. *PLoS One, 8*(6), e66449.

Meusburger, P., Livingstone, D. N., & Jöns, H. (2010). *Geographies of science*. Dordrecht: Springer.

Miguel, S., Moya-Anegón, F., & Herrero-Solana, V. (2010). The impact of the socio-economic crisis of 2001 on the scientific system of Argentina from the scientometric perspective. *Scientometrics, 85*(2), 495–507.

Moiwo, J. P., & Tao, F. (2013). The changing dynamics in citation index publication position China in a race with the USA for global leadership. *Scientometrics, 95*(3), 1031–1050.

Morillo, F., & De Filippo, D. (2009). Descentralización de la actividad científica. El papel determinante de las regiones centrales: el caso de Madrid. *Revista española de documentación científica, 32*(3), 29–50.

Nomaler, Ö., Frenken, K., & Heimeriks, G. (2013). Do more distant collaborations have more citation impact? *Journal of Informetrics, 7*(4), 966–971.

Onodera, N., Iwasawa, M., Midorikawa, N., Yoshikane, F., Amano, K., Ootani, Y., ... & Yamazaki, S. (2011). A method for eliminating articles by homonymous authors from the large number of articles retrieved by author search. *Journal of the American Society for Information Science and Technology, 62*(4), 677–690.

Orduña-Malea, E., Ontalba-Ruipérez, J. A., & Serrano-Cobos, J. (2010). Análisis bibliométrico de la producción y colaboración científica en Oriente Próximo (1998–2007). *Investigación bibliotecológica, 24*(51), 69–94.

Pasgaard, M., & Strange, N. (2013). A quantitative analysis of the causes of the global climate change research distribution. *Global Environmental Change, 23*(6), 1684–1693.

Persson, O., & Ellegård, K. (2012). Torsten Hägerstrand in the citation time web. *The Professional Geographer, 64*(2), 250–261.

Rodrigues, R. S., & Abadal, E. (2014). Ibero-American journals in Scopus and Web of Science. *Learned Publishing, 27*(1), 56–62.

Royal Society. (2011). *Knowledge, networks and nations: Global scientific collaboration in the 21st century*. London: The Royal Society.

Sandström, U. (2009). Combining curriculum vitae and bibliometric analysis: Mobility, gender and research performance. *Research Evaluation, 18*(2), 135–142.

Scherngell, T. (Ed.). (2014). *The geography of networks and R&D collaborations*. Berlin: Springer.

Scherngell, T., & Barber, M. J. (2009). Spatial interaction modelling of cross-region R&D collaborations: Empirical evidence from the 5th EU framework programme. *Papers in Regional Science, 88*(3), 531–546.

Schubert, T., & Sooryamoorthy, R. (2010). Can the centre–periphery model explain patterns of international scientific collaboration among threshold and industrialised countries? The case of South Africa and Germany. *Scientometrics, 83*(1), 181–203.

Sebestyén, T., & Varga, A. (2013). Research productivity and the quality of interregional knowledge networks. *The Annals of Regional Science, 51*(1), 155–189.

Shapin, S. (1998). Placing the view from nowhere: historical and sociological problems in the location of science. *Transactions of the Institute of British Geographers, 23*(1), 5–12.

Shelton, R. D., Foland, P., & Gorelskyy, R. (2009). Do new SCI journals have a different national bias? *Scientometrics, 79*(2), 351–363.

Sin, S. C. J. (2011). International coauthorship and citation impact: A bibliometric study of six LIS journals, 1980–2008. *Journal of the American Society for Information Science and Technology, 62*(9), 1770–1783.

Smalheiser, N. R., & Torvik, V. I. (2009). Author name disambiguation. *Annual Review of Information Science and Technology, 43*(1), 1–43.

Tang, L., & Shapira, P. (2010). Regional development and interregional collaboration in the growth of nanotechnology research in China. *Scientometrics, 86*(2), 299–315.

Tang, L., & Walsh, J. P. (2010). Bibliometric fingerprints: Name disambiguation based on approximate structure equivalence of cognitive maps. *Scientometrics, 84*(3), 763–784.

Trippl, M. (2013). Scientific mobility and knowledge transfer at the interregional and intraregional level. *Regional Studies, 47*(10), 1653–1667.

Van Eck, N. J., & Waltman, L. (2010). Software survey: VOSviewer, a computer program for bibliometric mapping. *Scientometrics, 84*(2), 523–538.

Van Noorden, R. (2010). Cities: Building the best cities for science. *Nature, 467*(7318), 906–908.

Waltman, L., Tijssen, R. J., & Eck, N. J. V. (2011). Globalisation of science in kilometres. *Journal of Informetrics, 5*(4), 574–582.

Wang, J., Berzins, K., Hicks, D., Melkers, J., Xiao, F., & Pinheiro, D. (2012). A boosted-trees method for name disambiguation. *Scientometrics, 93*(2), 391–411.

Wu, J., & Ding, X. H. (2013). Author name disambiguation in scientific collaboration and mobility cases. *Scientometrics, 96*(3), 683–697.

Zubieta, A. F. (2009). Recognition and weak ties: Is there a positive effect of postdoctoral position on academic performance and career development? *Research Evaluation, 18*(2), 105–115.

Chapter 7
Researchers' Publication Patterns and Their Use for Author Disambiguation

Vincent Larivière and Benoit Macaluso

Abstract In recent years we have been witnessing an increase in the need for advanced bibliometric indicators for individual researchers and research groups, for which author disambiguation is needed. Using the complete population of university professors and researchers in the Canadian province of Québec (N = 13,479), their papers as well as the papers authored by their homonyms, this paper provides evidence of regularities in researchers' publication patterns. It shows how these patterns can be used to automatically assign papers to individuals and remove papers authored by their homonyms. Two types of patterns were found: (1) at the individual researchers' level and (2) at the level of disciplines. On the whole, these patterns allow the construction of an algorithm that provides assignment information for at least one paper for 11,105 (82.4 %) out of all 13,479 researchers—with a very low percentage of false positives (3.2 %).

7.1 Introduction

Since the creation of the Science Citation Index in the 1960s—and the subsequent online availability of Thomson's various citation indexes for the sciences, social sciences, and the humanities through the Web of Science (WoS)—most large-scale bibliometric analyses have mainly been performed using the address (institutions, countries, etc.) journal, paper, or discipline field. Analyses made using the author

---

V. Larivière (✉)
École de bibliothéconomie et des sciences de l'information, Université de Montréal,
C.P. 6128, Succ., Centre-Ville, Montréal, QC, Canada, H3C 3J7
e-mail: vincent.lariviere@umontreal.ca

B. Macaluso
Observatoire des sciences et des technologies (OST), Centre interuniversitaire
de recherche sur la science et la technologie (CIRST), Université du Québec à Montréal,
Case Postale 8888, Succursale Centre-Ville, Montréal, QC, Canada, H3C 3P8
e-mail: macaluso.benoit@uqam.ca

© Springer International Publishing Switzerland 2014
Y. Ding et al. (eds.), *Measuring Scholarly Impact*,
DOI 10.1007/978-3-319-10377-8_7

field are much rarer, and typically have used small samples of researchers.[1] There is, thus, an important part of the bibliometric puzzle that was missing: the individual researcher, to which we can attribute socio-demographic characteristics (gender, age, degree, etc.). Until the last few years, issues related to the attribution of papers to individuals had not been discussed extensively in the bibliometric community (Enserink, 2009). However, the advent of h-indexes (Hirsch, 2005) and its numerous variants (Egghe, 2006; Schreiber, 2008; Zhang, 2009) aimed at evaluating individual researchers, as well as the need for more advanced bibliometric data compilation methods for measuring the research output of research groups whose names do not appear on papers (e.g. interuniversity groups, departments) or for measuring the effect of funding on researchers' output and impact (Campbell et al., 2010), has increased the need for author disambiguation.

The main challenge for author-level analyses is the existence of homonyms (or the inexistence of a researcher unique identifier), which makes the attribution of papers to distinct individuals quite difficult. Two general types of problems can be found at the level of authors (Smalheiser & Torvik, 2009). First and foremost, two or more individuals can share the same name (homonyms). Second, one researcher can sign papers in more than one manner (with or without initial(s), maiden name, etc.). These difficulties are exacerbated by two characteristics of the Web of Science (WoS). First, prior to 2006, only the surname and initial(s) of authors' first name(s) were indexed, for a maximum of three initials. Hence, researchers sharing the same surname and initial(s)—for example, John Smith and Jane Smith—were grouped under the same distinct string (Smith-J). Although the complete given name of authors is now indexed in the WoS, it only does so for journals providing this information in the author section of their papers,[2] in addition to the fact that it obviously does not solve the problem for papers published before 2006. Similarly, prior to 2008, no link was made in the database between an author and his/her institutional address. Although this was not a problem for sole authored papers—which only represent a slight fraction of papers published—it was more problematic for coauthored papers. More specifically, for a paper authored by three researchers and on which three institutional addresses are signed, it is impossible to know the exact institutional affiliation of each author, as several combinations are possible. Hence, the search for "Smith-J" among papers on which McGill University appears will, for example, retrieve papers from John Smith and Jane Smith, but also from Joseph Smith who, albeit not from McGill University, has collaborated with an author from McGill (homonymy of collaborators). Along these lines, there is a dearth of information on the extent of the homographic problem in the scientific community. Apart from Aksnes (2008) and Lewison (1996) who, respectively,

---

[1] The recent collection of Scientometrics papers dealing with individual researchers published by Academia Kiado (Braun, 2006) illustrates this trend: the study with the highest number of researchers included has less than 200. Similarly, notable studies in the sociology of science by Cole and Cole (1973), Merton (1973) and Zuckerman (1977) analyzed small datasets.

[2] Physics journals, for instance, often having very long author lists, only provide initial(s) of author (s) given name(s).

compiled data on the extent of homonyms among Norwegian researchers and on the frequency of authors' initial(s)—but did not test directly their effect on the compilation of bibliometric data on individual researchers—there is very little information on the extent to which researchers share the same name and its effect on the compilation of bibliometric data at the level of individual researchers.

This papers aims to contribute to this literature by presenting regularities found in papers manually assigned to the entire population of university professors and researchers (N = 13,479) in the Canadian province of Québec[3] as well as all the papers that were authored by their homonyms. It first reviews some of the relevant literature on the topic, and then presents a series of regularities found in researchers' publication patterns and how these can be used to automatically assign papers to individual researchers. Two types of patterns are presented: (1) individual researchers' past publication behavior and how it determines subsequent behavior and (2) the relationship between researchers' departmental affiliation and the disciplines in which they publish. These patterns are then used, in a reverse engineering manner, to automatically assign papers to these individuals. Results in terms of both false positives and false negatives are presented and discussed in the conclusion.

7.2   Previous Studies on the Attribution of Individual Authors' Publications

Over the last few years, several studies have provided algorithms for the disambiguation of individual researchers. However, most studies—with the notable exception of Reijnhoudt, Costas, Noyons, Borner, and Scharnhorst (2013) and Levin, Krawczyk, Bethard, and Jurafsky (2012) have been performed using relatively small datasets (Gurney, Horlings, & van den Besselaar, 2012; Wang et al., 2012) and, quite often, without actually having clean data on the papers authored by homonyms and papers authored by the "real" researcher. Jensen, Rouquier, Kreimer, and Croissant (2008) attempted to compile publication and citation files for 6,900 CNRS researchers using the Web of Science. Instead of removing, in each researchers' publication file, the papers written by homonyms, they evaluated the probability that a given researcher has homonyms, and, if this probability was high, they completely removed the researcher from the sample. More precisely, they first measured, by comparing the surname and initials of each researcher (VLEMINCKX-S) with some of its variants (VLEMINCKX-SG, VLEMINCKX-SP, etc.), the probability that the researcher has homonyms. If the researcher had too many variants, it was removed. Their second criterion was related to the number of papers published: if a researcher had too many papers, it was considered as an

---

[3] See for example Gingras, Larivière, Macaluso, and Robitaille (2008) and Larivière, Macaluso, Archambault, and Gingras (2010) for the some results based on this population.

indication that more than one scientist was behind the records. Hence, researchers had to publish between 0.4 and 6 papers per year to be considered in the sample. Their third criterion was that the first paper of each researcher to have been published when the researcher was between 21 and 30 years old. Their resulting database contained 3,659 researchers (53 % of the original sample).

This method has at least two major shortcomings. First, the fact that a name (e.g. VLEMINCKX-S) is unique does not imply that it represents only one distinct researcher. In this particular case, it could be the surname and initial combination for Serge Vleminckx, Sylvain Vleminckx, Sophie Vleminckx, etc. Second, it removes from the sample highly active researchers (who published more than six papers per year), which obviously distorts their results. This method is similar to that of Boyack and Klavans (2008), who used researchers with uncommon surnames to reconstitute individual researchers' publication and patenting activities. Using the combination of the name of the author/inventor and the research institution[4] signed on the paper, they calculated the odds that the paper belonged to the given author.

Another method is that of Han, Zha, and Giles (2005) who, using K-means clustering algorithms and Naïve Bayes probability models, managed to categorize 70 % of the papers authored by the very common "strings" Anderson-J and Smith-J into distinct clusters. The variables they used were the names of coauthors, the name of journals, and the title of the papers. The assumption behind this algorithm is that researchers generally publish papers on the same topics, in the same journals, and with the same coauthors. A similar method was also used by Torvik, Weeber, Swanson, and Smalheiser (2005) using Medline. Similarly, Wooding, Wilcox-Jay, Lewison, and Grant (2006) used coauthors for removing homonyms from a sample of 29 principal investigators funded by the Arthritis Research Campaign. For each author, they first found a core of papers which, without a doubt, belonged to the right researcher. Using this "core" subset of papers in the specialty of arthritis, they created for each researcher, a list of coauthors which were used to gather papers in areas other than arthritis. A novel aspect of this study is that several rounds of coauthor inclusion were performed, increasing between each round the number of coauthors in the core. After three rounds of the algorithm, 99 % of the authors' papers were assigned—which could be considered as the recall of papers—with only 3 % of false positives (97 % precision). This method is very similar to that used by Kang et al. (2009), and has been expanded by Reijnhoudt et al. (2013), to include additional heuristics, such as email address and reprint author, among others. Cota, Ferreira, Nascimento, Gonçalves, and Laender (2010) also used similar heuristics, coauthor, title of paper, and publication venue) and manage to disambiguate authors of about 4,500 papers of the DBLP and BDBComp collections.

---

[4] The bibliometric part of their paper used the Scopus database, which, contrary to Thomson Reuters' databases, links names of authors with institutional addresses for papers published since 1996.

Aswani, Bontcheva, and Cunningham (2006) used, in addition standard bibliographic information (abstract, initials of the authors, titles, and coauthors), automatic web-mining for grouping papers written by the same author. The web-mining algorithm searches for the full names of the authors and tries, for example, to find their own publications' page. Their results show that web-mining improves the clustering of papers into distinct authors, but the small sample used in the study makes the results less convincing. On the whole, most of these studies indeed manage to (1) automatically disambiguate authors or to (2) automatically assign papers to authors, although most of them do so with very small datasets, and often without a thorough analysis of false negatives, false positives, and various error rates. Finally, Levin et al. (2012) developed a large-scale citation-based bootstrapping algorithm—with an emphasis on self-citations—to disambiguate 54 million WoS author–paper combinations. They show that, when combined with emails, author names, and language, self-citations were the best bootstrapping element. They then manually disambiguated 200 authors—which, in this context, is not a very large sample—to assess the precision and recall of the algorithm, and found values of 0.832 and 0.788, respectively.

7.3 Methods

Contrary to most existing studies on the topic, this study uses, as a starting point, a list of distinct university based researchers ($N = 13{,}479$), including their department and university (Larivière et al., 2010). The database of university researchers' papers and of those authored by homonyms was thus obtained by matching the surname and initials of these researchers contained in the list to the surname and initials of authors of Quebec's scientific articles indexed in the Web of Science. This first match resulted in a database of 125,656 distinct articles and 347,421 author–article combinations. Each article attributed to each researcher was then manually validated in order to remove the papers authored by homonyms. This manual validation is generally made by searching the title of each of the papers on Google to find their electronic versions on which, generally, the complete names of the authors are written. This often helps to decide if the papers belong to the researcher. Another method is to search the name of the researcher on Google to find his/her website to get an indication of his/her publications' list or CV. After a few papers, one generally understands the publication pattern of the researcher and correctly attributes his/her papers. This essential but time-consuming step reduced the number of distinct papers by 51 % to 62,026 distinct articles and by 70 % to 103,376 author–article combinations. Analysis of this unique dataset, including the characteristics of both assigned and rejected papers, sheds light on the extent of homonyms in Quebec's scientific community.

To assess the reliability and reproducibility of the manual validation of university researchers and professors' publication files, tests with different individual "attributors" were performed for a sample of 1,380 researchers (roughly 10 % of the researchers). It showed that for most publication files, the two coders manually assigned exactly the same papers. More specifically, for 1,269 files (92 %) researchers had exactly the same papers assigned. A difference of one paper was found in 72 cases (5.2 %), two papers in 15 cases (1.1 %), three papers in 9 cases (0.7 %) and four papers in 3 cases (0.2 %). The remaining 12 files had a maximum difference of 12 papers each. In terms of author–article combinations, the error rates are even lower. Out of the 12,248 author–article links obtained the first time, 12,124 (or 99 %) remained unchanged the second time. Manual validation is thus quite reliable and reproducible.

In order to find patterns in researchers' publication output, this study uses, for each of Quebec's university researchers and professors, a dataset of all the WoS-Indexed papers with authors that matched their authors' name (for example, Smith-J) amongst all papers with at least one Canadian address for the 2000–2007 period. These papers were manually categorized as belonging to the right researcher or as belonging to a homograph, which allows—contrary to most studies presented in the preceding section—to test how these patterns could be used to discriminate false positives from papers that correctly belong to a researcher. The difference between researchers' papers that were manually assigned and those rejected allows the testing of the algorithm.

In order to help the search for patterns, each journal indexed in the WoS was assigned a discipline and a specialty according to the classification scheme used by U.S. National Science Foundation (NSF) in its Science and Engineering Indicators series (Appendix 1).[5] The main advantage of this classification scheme over that provided by Thomson Reuters is that (1) it has a two-level classification (discipline and specialty), which allows the use of two different levels of aggregation and, (2) it categorizes each journal into only one discipline and specialty, which prevents double counts of papers when they are assigned to more than one discipline. Similarly, a discipline was assigned to each of the researchers' departments (Appendix 2). These disciplines were assigned based on the 2000 revision of the U.S. Classification of Instructional Programs (CIP) developed by the U.S. Department of Education's National Center for Education Statistics (NCES).[6] This dataset serves as the backbone for finding the relationships between the disciplinary affiliation of university researchers and the discipline of their publications.

---

[5] More details on the classification scheme can be found at: http://www.nsf.gov/statistics/seind06/c5/c5s3.htm#sb1

[6] For more details on the CIP, see: http://nces.ed.gov/pubs2002/cip2000/.

7.4   Regularities in Researchers' Publication Patterns

A first interesting piece of information found in this dataset is the percentage of papers of each researcher that was retained after manual validation. More specifically, these cleaned publication files made it possible to estimate the extent of homonym problems for all Quebec university researchers for whom at least one article was automatically matched (N = 11,223) using the name of the researcher within papers having at least one Quebec institutional address.[7] With an automatic matching of researchers' names, compared to a cleaned publication file (Fig. 7.1):

- The papers matched for 2,972 researchers (26.5 %) were all rejected which, in turn, meant that they had not actually published any papers (all papers were written by homonyms).
- Between 0.1 and 25 % of the papers matched were assigned to 1,862 researchers (16.6 %).
- Between 25.1 and 50 % of the papers matched were assigned to 975 researchers (8.7 %).

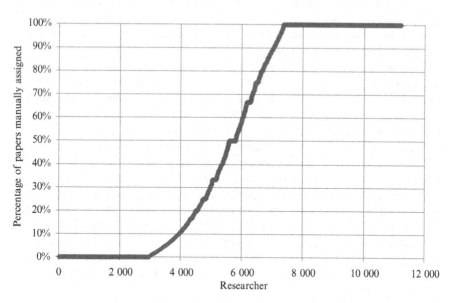

Fig. 7.1 Percentage of papers assigned after manual validation, by researcher

---

[7] Thus, 2,256 of Quebec's researchers did not publish any papers during that period nor had any of their homonyms.

– Between 50.1 and 75 % of the papers matched were assigned to 722 researchers (6.4 %).
– Between 75.1 and 99.9 % of the papers matched were assigned to 818 researchers (7.3 %).
– The papers matched for 3,874 researchers (34.5 %) were all conserved after manual validation (i.e., they had no homonyms within the subset of Quebec papers).

Crude matching—without removing papers authored by homonyms—is thus valid for slightly more than a third of the researchers. On the other hand, the scientific production of the remaining two-thirds was overestimated. Since it is impossible to know a priori which researchers will be overestimated and which ones will not, the validation of each paper from each researcher is, theoretically, needed. As mentioned previously, papers of these publication files were all manually validated (assigned or rejected) and serve, in a reverse engineering manner, as a test bed for finding patterns in the publications of researchers.

In a manner similar to that of Wooding et al. (2006) for arthritis research, papers were then analyzed in order to find characteristics which could help isolating a core of papers for each researcher—i.e. a subset of all of each researcher's papers that we are sure are not those of homonyms. This was more complex in the context of this paper, as core papers had to be found for researchers that could be active in any field of science and not only in arthritis. After several rounds of empirical analysis, the combination of three variables optimized the ratio between the number of papers found and the percentage of false positives. Figures 7.2 and 7.3 present the two sets of criteria with which a core set of papers could be found for university-based researchers. Figure 7.2 presents the first matching criteria: the complete name of researchers matched with the complete name of authors—including the complete given name (available in the Web of Science since 2006)—and the name of the researcher's university matched with the name of the university on the paper.

Figure 7.3 presents the second matching criteria. Firstly, the name of the author of the paper had to be written exactly in the same manner as the name of the researcher in the list. Secondly, the institution appearing on the paper (or its affiliation, e.g. Royal Victoria Hospital is affiliated to McGill University) had to be the same as the institution appearing on the list and, thirdly, the discipline of the journal in which the paper is published, the department or the institution of the authors had to be similar[8] to the department of the researcher as it appeared on the list of university professors and researchers or the discipline of the paper had to be among the five disciplines in which researchers from this department published most.

---

[8] The similarity threshold (MinSimilarity) was set at 0.25 in Microsoft SQL Server SQL Server Integration Services (SSIS). More details on the system can be found at: http://technet.microsoft.com/en-US/library/ms345128(v=SQL.90).aspx

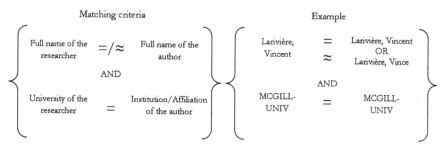

Fig. 7.2 First matching criteria for creating the core of papers

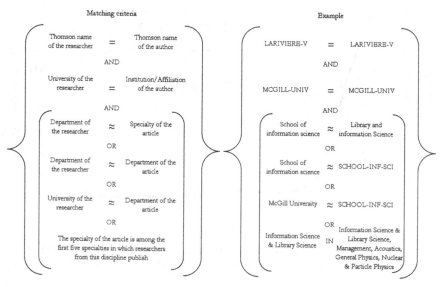

Fig. 7.3 Second matching criteria for creating the core of papers

Following Boyack and Klavans (2008), an analysis of rare surnames was also performed, which were defined as surnames only belonging to one individual in the list of university researchers. Hence, all papers authored by researchers having a rare name, and on which their institution of affiliation appears, were included in core papers. As shown in Table 7.1, these three criteria allow the creation of a core set of papers for more than 75 % of the individual researchers for which at least one paper has been manually assigned (8,081), matches 56.4 % of their distinct papers and 47.5 % of the author–paper combinations, e.g. LARIVIERE-V and paper 'X'. At each level of analysis, the number of false positives is rather low; and is especially low at the level of author–paper combinations (less than 1 %).

Another set of regularities was found in individual researchers' publication patterns. The idea behind this search for patterns for individual researchers was to be able, using subset of papers in the core, to find other papers that belonged to the researchers but that did not exhibit the characteristics found in Figs. 7.2 and 7.3.

Table 7.1 Results of the matching of core papers at the levels of university researchers, articles, and author–paper combinations

	Manual validation (N)	Automatic assignment		False positives	
Unit of analysis		N	%	N	%
Researchers	8,081	6,117	75.7 %	344	4.3 %
Articles	62,629	35,353	56.4 %	772	1.2 %
Author–paper combinations	97,850	46,472	47.5 %	809	0.8 %

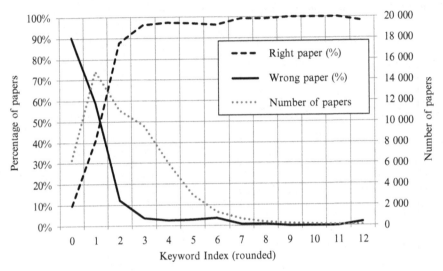

Fig. 7.4 Percentage of rightly assigned and wrongly assigned papers, as a function of the keywords previously used by a university researcher, 2000–2003 and 2004–2007

To do so, each researcher's publication record was divided into two distinct time periods: 2000–2003 and 2004–2007.

Using the characteristics of the papers published by each given researcher during the first time period, we then tried to automatically assign to the same researcher the papers published during the second time period. Two indicators were quite successful in doing so: 1) the use of the same words in the title, author keywords, and abstract fields of upcoming publications (Fig. 7.4) and 2) the citation of the same references (Fig. 7.5) of papers for which the Thomson name [e.g. LARIVIERE-V] and the institution [MCGILL-UNIV] also matched. Figure 7.4 presents the percentage of rightly and wrongly attributed papers, as a function of the keyword index. The keyword index is a simple indicator compiled for each 2004–2007 paper matched to a researcher, based on the keywords of the papers assigned to the researcher for the period 2000–2003. Its calculation is as follows:

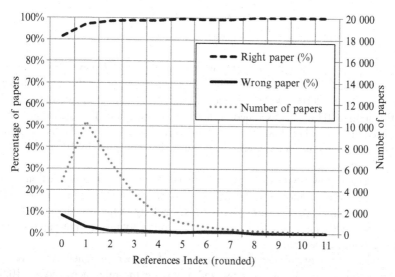

Fig. 7.5 Percentage of rightly assigned and wrongly assigned papers, as a function of the references previously made by a university researcher, 2000–2003 and 2004–2007

$$Ki_{pr} = \left(\frac{N_{kpm}}{N_{kp}} \times \frac{1}{\sqrt{N_{kt}}} \right) \times 100 \qquad (7.1)$$

where N_{kpm} is the number of keywords of a 2004–2007 paper that match the keywords used in the 2000–2003 papers of a researcher, N_{kp} is the total number of keywords of the 2004–2007 paper and N_{kt} is the total number of keywords used in all the 2000–2003 papers assigned to the researcher. The square root of N_{kt} was used instead of N_{kt} alone in order to obtain an overall number of keywords (denominator) that is not too high—especially for very productive researchers. The result is multiplied by 100 in order to be closer to an integer.

Figure 7.4 shows that when the keyword index is at 2, about 90 % of the papers rightly belong to the researcher and that slightly greater than 10 % are false positives. When the keyword index is greater than 2 (3 or more), the percentage of rightly assigned papers rises above 95 %, and stays at this level until 7, where about 100 % of the papers are assigned to the right researcher. These numbers mean that it is possible to rightly assign papers to a researcher using the regularities found in the title words, keywords, and words of the abstract.

Figure 7.5 presents the references index for 2004–2007 papers, based on papers published between 2000 and 2007. The references index is very similar to the keyword index previously presented; it is based on the pool of references made previously (2000–2003) by the researcher. Its calculation is as follows:

$$Ri_{pr} = \left(\frac{N_{rpm}}{N_{rp}} \times \frac{1}{\sqrt{N_{rt}}} \right) \times 100 \qquad (7.2)$$

Where N_{rpm} is the number of references of a 2004–2007 paper that match the references used in the 2000–2003 papers of a researcher, N_{rp} is the total number of references of the 2004–2007 paper and N_{rt} is the total number of references used in all the 2000–2003 papers assigned to the researcher. Again, The square root of N_{rt} was used instead of N_{rt} alone in order to have an overall number of cited references (denominator) that is not too high. The result is also multiplied by 100 in order to be closer to an integer.

Figure 7.5 shows that as soon as a signal is obtained, i.e. that at least one of the referenced works of the 2004–2007 paper was previously made in the 2000–2003 dataset, more than 90 % of the papers rightly belong to the researcher. When the references index increases to 1 or above, the quasi-totality of the papers rightly belong to the researcher.

Using the keywords and references found in the papers assigned in the core (set at 2 or more for the keyword index and at >0 for the references index), 10,892 additional papers were assigned, with only 236 papers being false positives (2.2 % of the added papers), for an overall error rate of 2.2 % for papers and 1.7 % for author–paper combinations (Table 7.2). Since this matching of papers can only be made for researchers for which a certain number of core papers were matched, the number of researchers stays the same, but slightly more researchers have at least one paper wrongly assigned (6.7 %).

Another round of automatic matching of papers was also performed with the same references and keywords (set at the same thresholds), but using only the Thomson name [e.g. LARIVIERE-V] and the province [QC], but not the institution [MCGILL-UNIV]. Using this method, 3,645 additional papers were retrieved, of which 674 were false positives (Table 7.3). Although this percentage seems quite high, the overall proportion of false positives at the level of articles remains quite low (3.2 %) and is even lower for author–paper combinations (2.3 %).

In order to increase the number of researchers for which a certain number of core papers could be found, the relationship between the discipline of the researchers and the specialty of the papers was analyzed. An increase in the number of researchers for which core papers could be found is important because core papers

Table 7.2 Results of the matching of core papers and papers with the same keywords or cited references, at the levels of university researchers, articles, and author–paper combinations

Unit of analysis	Manual validation (N)	Automatic assignment		False positives	
		N	%	N	%
Researchers	8,081	6,117	75.7 %	407	6.7 %
Articles	62,629	46,245	73.8 %	1,008	2.2 %
Author–paper combinations	97,850	64,765	66.2 %	1,078	1.7 %

Table 7.3 Results of the matching of core papers and papers with the same keywords or cited references, without the 'same institution' criteria, at the levels of university researchers, articles, and author–paper combinations

Unit of analysis	Manual validation (N)	Automatic assignment		False positives	
		N	%	N	%
Researchers	8,081	6,117	75.7 %	576	9.4 %
Articles	62,629	49,890	79.7 %	1,577	3.2 %
Author–paper combinations	97,850	72,918	74.5 %	1,682	2.3 %

are the starting point for the automatic assignment of several other papers. For each of the 5,615 existing combinations of disciplines of publications (Appendix 1) and of discipline of departments (Appendix 2), a matrix of the percentage of papers from each discipline of publications that rightly belonged to researchers from each department was calculated. Unsurprisingly, it was found that papers published in the main specialty in which researchers from a given specialty publish were more likely to belong to the right researchers. For example, 100 % of the 186 papers published in geography journals that matched the names of authors of geography departments belonged to the right researcher. The same is true for several other obvious department–specialty relationships, such as university researchers from chemical engineering departments publishing in chemical engineering journals (99 % of the 1,017 papers rightly assigned), but also for less obvious relationships such as researchers in civil engineering publishing in Earth and planetary science journals (95 % of the 316 papers rightly assigned).

On the other hand, all the 333 papers published in biochemistry and molecular biology journals that matched authors' names from the disciplines of anthropology, archaeology, and sociology belonged to the wrong researcher. The same is also true for the 202 papers published in organic chemistry that matched authors from business departments. Given that no university-affiliated researcher from these domains has ever published in journals of these specialties during the period studied, there are low chances that researchers from the same domain will do so in the future.

Figure 7.6 presents the matrix of the percentage of assigned papers, for each combination of discipline of departments (x-axis) and specialty of publication (y-axis). Darker zones are combinations of specialties of publications and discipline of departments where a larger proportion of papers was accepted during manual validation; lighter zones are combinations where a majority of papers were rejected during manual validation. This figure illustrates that there is a majority of discipline of department/specialty of publication combinations where the quasi totality of papers were authored by homonyms (light zones), and a few darker zones where a large proportion of papers belonged to the right researcher. Unsurprisingly, zones where most of the papers were assigned are generally cases where the discipline of the department is related with the discipline of the journal—for example, researchers from departments of information science and library science publishing in journals of library and information science. The presentation of this landscape clearly shows

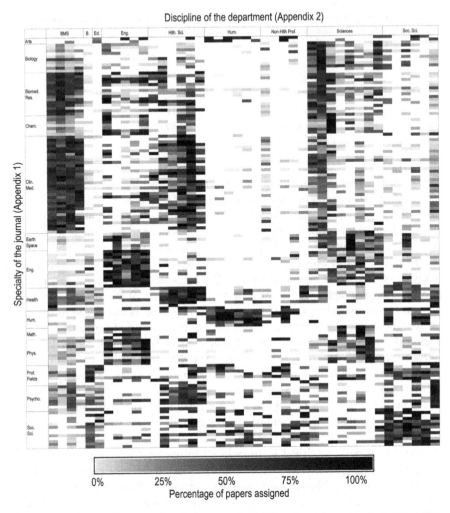

Fig. 7.6 Percentage of papers assigned after manual validation, for each combination of discipline of departments (x-axis) and specialty of publication (y-axis)

that there are some combinations where the majority of papers were assigned during manual validation and others where only a minority of papers was assigned during the process. We can, thus, focus on these light zones to automatically exclude papers from a given department published in given specialty, and on dark zones to automatically include papers from other department/specialty combinations.

Figure 7.7 aggregates, by rounded percentage of properly attributed papers, the numbers of rejected and of accepted author–paper combinations. We see that the number of wrongly assigned papers drops significantly for department/specialty combinations greater than 80 %, and even more after 95 %. These percentages were thus used to automatically assign papers in specific disciplines that matched

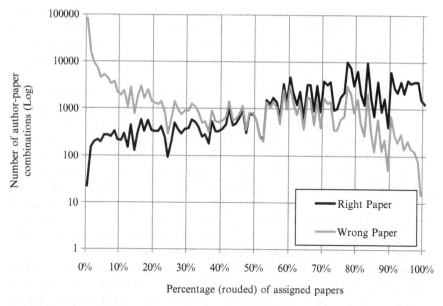

Fig. 7.7 Number of rightly and wrongly assigned author–paper combinations, as a function of assignment percentage of papers from a discipline to authors from a department

Table 7.4 Overall results of the automatic matching of papers, using core papers, keywords, and references previously made, and the matrix of discipline of departments and specialty of papers

Unit of analysis	Manual validation (N)	Automatic assignment		False positives	
		N	%	N	%
Researchers	8,081	6,427	79.5 %	610	9.5 %
Articles	62,629	50,353	80.4 %	1,633	3.2 %
Author–paper combinations	97,850	73,771	75.4 %	1,750	2.4 %

researchers from given departments. In order to reduce the number of false positives, a 95 % assignment rate was chosen for papers on which the institution of the author does not appear (but only its province). This includes a total of 17,002 papers, of which 16,518 are properly attributed and only 484 are inaccurately attributed (2.8 %). For papers on which the institution of the researcher appears, an 80 % attribution rate was used. This attributed 68,785 papers, of which 10.7 % were false positives.

One must note that all these processes were performed in parallel; a paper assigned with one of these criteria could have been already attributed during another step of the matching process. Hence, the numbers of papers presented here include several papers that were already matched using one of the criteria previously presented in this section. Table 7.4 presents the error rates for all of the steps combined. The inclusion of the algorithm based on the matrix of departments and disciplines of publication added 310 researchers in the subset of those with at least one paper in the core. On the whole, the multiple algorithms used so far for automatically attributed papers for 6,427 researchers, for a total of 50,353 papers and 73,331 author–paper combinations.

Patterns presented so far in this paper allow the creation of a dataset of papers that is likely to belong to the right researcher and assign at least one paper to almost 80 % of the researchers. They make possible the creation of a core set of papers, as well as of a few other layers of papers, based on the similarity of their characteristics to those included in the core. The following algorithm does the opposite and aims at finding indications that the paper clearly does not belong to the researcher.

As shown on Fig. 7.6, there are several combinations of discipline of departments and specialties of papers where the vast majority of papers were rejected during manual validation. Indeed, if no university researcher from department X has ever published in the specialty Y, no researcher is likely to do so. Papers falling into these combinations could thus automatically be rejected.

These patterns not only allow the rejection of papers, but also to close researchers' publication files, as all of their papers can either all have been assigned—using the methods previously presented—or rejected using the department/specialty matrix. Using a 50 % threshold was optimal, as it automatically rejected 202,928 author–paper combinations, of which 183,656 were real negatives (91 %), and only 19,272 were false negatives (9 %). These rejected author–paper combinations account for a significant share (90 %) of all rejected combinations (226,325).

After all these steps, 5,036 publications files out of the 13,479 (37.4 %) were automatically marked as closed (including the 2,256 files for which no paper, either authored by a researcher in the list or by a homograph), as all of their papers were either all assigned or all rejected. Another 6,069 researchers had at least one of their papers automatically assigned (45 %), for a total of 50,353 papers, with 1,633 being false positives (3.2 %). On the whole, this algorithm provides attribution information on at least one paper for 11,105 (82.4 %) out of all 13,479 researchers, or on 8,849 out of the 11,223 researchers (78.8 %), when one excludes the 2,256 files for which no paper matched, either authored by the researcher or a homograph. Hence, there are still 2,374 researchers for which no automatic decision on any of their matched papers can be made (attribution or rejection) and, hence, for which a complete manual validation needs to be performed. This algorithm can nonetheless be very helpful, as it automatically assigns a large proportion of papers, excludes an even larger proportion and reduces from 11,223 to 2,374 (79 %) the number of researchers for which a complete manual validation has to be performed.

Conclusion

This paper has provided evidence of regularities in researchers' publication patterns, and that these patterns could be used to automatically assign papers to individual and remove papers authored by their homonyms. Two types of patterns were found: (1) at the individual researchers' level and (2) at the collective level.

At the level of individuals, we found that researchers were quite regular in their referencing practices. This could be expected: as shown elsewhere, researchers tend to cite the same material throughout their careers (Barnett & Fink, 2008; Gingras et al., 2008). We thus tested this finding for the subset of Quebec researchers and found that papers with the same surname and initial were always those of the "right" researcher when at least one of the references of the paper had already been made in one of the papers previously assigned to the researcher. Similarly, researchers also tend to work on the same topics. Using the pool of keywords previously used by researchers and comparing them with papers subsequently published, we found that the use of the same keywords meant in most of the cases that the paper belonged to the same researcher.

At the collective level, two general patterns emerged. The first pattern we found was that the institution of affiliation of a given researcher appeared on most of the papers that rightly belonged to him/her. This simple regularity allowed the creation of a core subset of papers, which could then be used to gather researchers' other papers using the previous references and previous keywords methods. The other pattern relates to the relationship between the department discipline and the specialty of the journal in which papers are published. For some departments/specialty combinations, a majority of papers belonged to the "right" researcher, while for other combinations, a majority belonged to homonyms. Thus, the former combinations allowed the automatic attribution of papers, while the latter made automatic rejection of author–paper combinations possible.

Compared with most existing studies on author disambiguation, which were generally performed for a small subset of researchers (Aswani et al., 2006; Han et al., 2005; Wooding et al., 2006) or for specific author–article combinations (Boyack & Klavans, 2008) this is an important step forward. That being said, the recent developments in bibliographic databases used in bibliometrics—such as the researcher ID, ORCID, the link between each of the authors and their addresses as well as the indexation of the complete given names of authors—are perhaps even more important, as they are likely to make this attribution easier in the future.

Appendix 1: List of Disciplines Assigned to Journals

Arts
Fine Arts & Architecture
Performing Arts

Biology
Agricultural & Food Sciences
Botany
Dairy & Animal Science
Ecology Entomology
General Biology
General Zoology
Marine Biology & Hydrobiology
Miscellaneous Biology
Miscellaneous Zoology

Biomedical Research
Anatomy & Morphology
Biochemistry & Molecular Biology
Biomedical Engineering
Biophysics
Cellular Biology Cytology & Histology
Embryology
General Biomedical Research
Genetics & Heredity
Microbiology Microscopy
Miscellaneous Biomedical Research
Nutrition & Dietetic
Parasitology
Physiology
Virology

Chemistry
Analytical Chemistry
Applied Chemistry
General Chemistry
Inorganic & Nuclear Chemistry
Organic Chemistry
Physical Chemistry
Polymers

Clinical Medicine
Addictive Diseases
Allergy
Anesthesiology
Arthritis & Rheumatology
Cancer
Cardiovascular System
Dentistry

Dermatology & Venereal Disease
Endocrinology
Environmental & Occupational Health
Fertility
Gastroenterology
General & Internal Medicine
Geriatrics
Hematology
Immunology
Miscellaneous Clinical Medicine
Nephrology
Neurology & Neurosurgery
Obstetrics & Gynecology
Ophthalmology
Orthopedics
Otorhinolaryngology
Pathology
Pediatrics
Pharmacology
Pharmacy
Psychiatry
Radiology & Nuclear Medicine
Respiratory System
Surgery
Tropical Medicine
Urology
Veterinary Medicine

Earth and Space
Astronomy & Astrophysics
Earth & Planetary Science
Environmental Science
Geology
Meteorology & Atmospheric Science
Oceanography & Limnology

Engineering and Technology
Aerospace Technology
Chemical Engineering
Civil Engineering
Computers
Electrical Engineering & Electronics
General Engineering
Industrial Engineering
Materials Science
Mechanical Engineering
Metals & Metallurgy
Miscellaneous Engineering & Technology

(continued)

Appendix 1 (continued)

Nuclear Technology
Operations Research

Health
Geriatrics & Gerontology
Health Policy & Services
Nursing
Public Health
Rehabilitation
Social Sciences, Biomedical
Social Studies of Medicine
Speech-Language Pathology and Audiology

Humanities
History
Language & Linguistics
Literature
Miscellaneous Humanities
Philosophy
Religion

Mathematics
Applied Mathematics
General Mathematics
Miscellaneous Mathematics
Probability & Statistics

Physics
Acoustics
Applied Physics
Chemical Physics
Fluids & Plasmas
General Physics
Miscellaneous Physics
Nuclear & Particle Physics
Optics
Solid State Physics

Professional Fields
Communication
Education
Information Science & Library Science
Law
Management
Miscellaneous Professional Field
Social Work

Psychology
Behavioral Science & Complementary
    Psychology
Clinical Psychology
Developmental & Child Psychology
Experimental Psychology
General Psychology
Human Factors
Miscellaneous Psychology
Psychoanalysis
Social Psychology

Social Sciences
Anthropology and Archaeology
Area Studies
Criminology
Demography
Economics
General Social Sciences
Geography
International Relations
Miscellaneous Social Sciences
Planning & Urban Studies
Political Science and Public Administration
Science studies
Sociology

Appendix 2: List of Disciplines Assigned to Departments

Basic Medical Sciences
General Medicine
Laboratory Medicine
Medical Specialties
Surgical Specialties

Business & Management
Education
Engineering
Chemical Engineering
Civil Engineering
Electrical & Computer Engineering
Mechanical & Industrial Engineering
Other Engineering

Health Sciences
Dentistry
Kinesiology/Physical Education
Nursing
Other Health Sciences
Public Health & Health Administration
Rehabilitation Therapy

Humanities
Fine & Performing Arts
Foreign Languages, Literature, & Linguistics
Area Studies
French/English
History

Philosophy
Religious Studies & Vocations

Non-health Professional
Law & Legal Studies
Library & Information Sciences
Media & Communication Studies
Planning & Architecture
Social Work

Sciences
Agricultural & Food Sciences
Biology & Botany
Chemistry
Computer & Information Science
Earth & Ocean Sciences
Mathematics
Physics & Astronomy
Resource Management & Forestry

Social Sciences
Anthropology, Archaeology,
   & Sociology
Economics
Geography
Other Social Sciences & Humanities
Political Science
Psychology

References

Aksnes, D. W. (2008). When different persons have an identical author name. How frequent are homonyms? *Journal of the American Society for Information Science and Technology, 59*(5), 838–841.

Aswani, N., Bontcheva, K., & Cunningham, H. (2006). Mining information for instance unification. *Lecture Notes in Computer Science, 4273*, 329–342.

Barnett, G. A., & Fink, E. L. (2008). Impact of the internet and scholar age distribution on academic citation age. *Journal of the American Society for Information Science and Technology, 59*(4), 526–534.

Boyack, K. W., & Klavans, R. (2008). Measuring science–technology interaction using rare inventor–author names. *Journal of Informetrics, 2*(3), 173–182.

Braun, T. (Ed). (2006). Evaluations of Individual Scientists and Research Institutions: Scientometrics Guidebooks Series. Budapest, Hungary : Akademiai Kiado.

Campbell, D., Picard-Aitken, M., Côté, G., Caruso, J., Valentim, R., Edmonds, S., ... & Archambault, É. (2010). Bibliometrics as a performance measurement tool for research evaluation: The case of research funded by the National Cancer Institute of Canada. *American Journal of Evaluation, 31*(1), 66–83.

Cole, J. R., & Cole, S. (1973). *Social stratification in science*. Chicago, IL: University of Chicago Press.

Cota, R. G., Ferreira, A. A., Nascimento, C., Gonçalves, M. A., & Laender, A. H. F. (2010). An unsupervised heuristic-based hierarchical method for name disambiguation in bibliographic citations. *Journal of the American Society for Information Science and Technology, 61*(9), 1853–1870.

Egghe, L. (2006). Theory and practice of the g-index. *Scientometrics, 69*(1), 131–152.

Enserink, M. (2009). Are you ready to become a number? *Science, 323*, 1662–1664.

Gingras, Y., Larivière, V., Macaluso, B., & Robitaille, J. P. (2008). The effects of aging on researchers' publication and citation patterns. *PLoS One, 3*(12), e4048.

Gurney, T., Horlings, E., & van den Besselaar, P. (2012). Author disambiguation using multi-aspect similarity measures. *Scientometrics, 91*(2), 435–449.

Han, H., Zha, H., & Giles, C. L. (2005). Name disambiguation in author citations using a K-way spectral clustering method. *Proceedings of the 5th ACM/IEEE-CS Joint Conference on Digital Libraries* (pp. 334–343). Retrieved from http://clgiles.ist.psu.edu/papers/JCDL-2005-K-Way-Spectral-Clustering.pdf.

Hirsch, J. E. (2005). An index to quantify an individual's scientific research output. *Proceedings of the National Academy of Science, 102*(46), 16569–16572.

Jensen, P., Rouquier, J. B., Kreimer, P., & Croissant, Y. (2008). Scientists who engage in society perform better academically. *Science and Public Policy, 35*(7), 527–541.

Kang, I. S., Seung-Hoon, N., Seungwoo, L., Hanmin, J., Pyung, K., Won-Kyung, S., & Jong-Hyeok, L. (2009). On co-authorship for author disambiguation. *Information Processing and Management, 45*(1), 84–97.

Larivière, V., Macaluso, B., Archambault, E., & Gingras, Y. (2010). Which scientific elites? On the concentration of research funds, publications and citations. *Research Evaluation, 19*(1), 45–53.

Levin, M., Krawczyk, S., Bethard, S., & Jurafsky, D. (2012). Citation-based bootstrapping for large-scale author disambiguation. *Journal of the American Society for Information Science and Technology, 63*(5), 1030–1047.

Lewison, G. (1996). The frequencies of occurrence of scientific papers with authors of each initial letter and their variation with nationality. *Scientometrics, 37*(3), 401–416.

Merton, R. K. (1973). *The sociology of science: Theoretical and empirical investigations*. Chicago, IL: Chicago University Press.

Reijnhoudt, L., Costas, R., Noyons, E., Borner, K., & Scharnhorst, A. (2013). "Seed + Expand": A validated methodology for creating high quality publication oeuvres of individual researchers. *arXiv preprint arXiv:1301.5177*.

Schreiber, M. (2008). A modification of the h-index: The hm-index accounts for multi-authored manuscripts. *Journal of Informetrics, 2*(3), 211–216.

Smalheiser, N. R., & Torvik, V. I. (2009). Author name disambiguation. In B. Cronin (Ed.), *Annual review of information science and technology* (Vol. 43, pp. 287–313). Medford, NJ: ASIST and Information Today.

Torvik, V. I., Weeber, M., Swanson, D. R., & Smalheiser, N. R. (2005). Probabilistic similarity metric for Medline records: A model for author name disambiguation. *Journal of the American Society for Information Science and Technology, 56*(2), 140–158.

Wang, J., Berzins, K., Hicks, D., Melkers, J., Xiao, F., & Pinheiro, D. (2012). A boosted-trees method for name disambiguation. *Scientometrics, 93*(2), 391–411.

Wooding, S., Wilcox-Jay, K., Lewison, G., & Grant, J. (2006). Co-author inclusion: A novel recursive algorithmic method for dealing with homonyms in bibliometric analysis. *Scientometrics, 66*(1), 11–21.

Zhang, C. T. (2009). The e-index, complementing the h-index for excess citations. *PLoS One, 5*(5), e5429.

Zuckerman, H. A. (1977). *Scientific elite: Nobel laureates in the United States*. New York, NY: Free Press.

Chapter 8
Knowledge Integration and Diffusion: Measures and Mapping of Diversity and Coherence

Ismael Rafols

Abstract In this chapter, I present a framework based on the concepts of diversity and coherence for the analysis of knowledge integration and diffusion. Visualisations that help to understand insights gained are also introduced. The key novelty offered by this framework compared to previous approaches is the inclusion of cognitive distance (or proximity) between the categories that characterise the body of knowledge under study. I briefly discuss different methods to map the cognitive dimension.

8.1 Introduction

Most knowledge builds on previous knowledge—given the cumulative nature of science and technology. But the fact that knowledge mainly draws on previous knowledge also means that it does not build on "other" types of knowledge. This is what in an evolutionary understanding of science is called a cognitive trajectory—which is often associated with lock-in.[1] Under such conditions, the combination of different types of knowledge (perspectives, but also data, tools, etc.) has long been seen as a way to leap out of stagnation and create new knowledge. This perspective has been emphasised in the case of research aiming to solve social and economic problems—seen as requiring interdisciplinary efforts, both in terms of sources (i.e. requiring the integration of different types of knowledge) and it terms of

---

[1] It should be noted that the evolutionary view of science and technology is prevalent both among constructivist sociologists such as Bijker (Pinch & Bijker, 1984) and positivist economists such as Dosi (1982).

I. Rafols (✉)
Ingenio (CSIC-UPV), Universitat Politècnica de València, València, (Spain)

SPRU (Science Policy Research Unit), University of Sussex, Brighton, (UK)
e-mail: i.rafols@ingenio.upv.es

© Springer International Publishing Switzerland 2014
Y. Ding et al. (eds.), *Measuring Scholarly Impact*,
DOI 10.1007/978-3-319-10377-8_8

impacts (i.e. diffusion over different areas of research and practice) (Lowe & Phillipson, 2006; Nightingale & Scott, 2007).

Changes in science in the last two decades have been characterised as a progressive blurring of the well-defined categories of post war science. Science has shifted towards a so-called Mode 2 of knowledge production that is presented as more interdisciplinary, more heterogeneous, closer to social actors and contexts, and more susceptible to social critique (Gibbons et al., 1994; Hessels & van Lente, 2008).

In Mode 2 research, knowledge integration and diffusion play a crucial role as the processes that bridge the gaps between disciplines, organisations, institutions and stakeholders. Building on Boschma's notion of multiple dimensions of proximity (Boschma, 2005), Frenken, Boschma, and Hardeman (2010) proposed to:

> Reformulate the concept of Mode 2 knowledge production analytically as a mode of distributed knowledge production, where we operationalize the notion of distribution in five proximity dimensions [i.e. cognitive, organisational, social, institutional, geographical] (...) Mode 1 stands for scientific knowledge production in which actors are distributed, yet proximate, while Mode 2 knowledge production stands for distributed knowledge production processes, in which actors are distant.

While cognitive proximity is the primary dimension to analyse knowledge integration and diffusion in science, it is worth realising that other dimensions of proximity mediate the possibility of knowledge integration and diffusion.[2] These other dimensions are important to understand how changes in cognitive proximity happen. Policy and management instruments such as personnel recruitment, organisational reforms or incentives directly address these others dimensions (social, organisational or institutional) and it is through them that decision makers aim to influence the cognitive dimension. The Triple Helix framework, for example, investigates the institutional-cognitive-organisational relations (Etzkowitz & Leydesdorff, 2000). One can study "translational research institutes", which increase geographical and organisational proximity between, for example a cell biologist and an oncologist, as efforts to favour integration and diffusion of knowledge between basic research and practice related to cancer (Molas-Gallart, Rafols, D'Este, & Llopis, 2013).

In this chapter, I review quantitative methods and some visualisation techniques developed in recent years in order to assess where, how and to which extent knowledge integration and diffusion took place regarding specific organisation, problem-solving efforts or technologies. While this chapter focuses on mapping of the cognitive dimension, I invite the reader to think, following Boschma and Frenken's proposal, that the understanding of the dynamics of science consists in being able to relate the different analytical dimensions.

---

[2] The use of these five dimensions is an expedient simplification. One may easily conceive more dimensions within each of the dimensions listed, making a more fine-grained description of cognitive or social dimensions, for example.

8.2   Conceptual Framework: Knowledge Integration and Diffusion as Shifts in Cognitive Diversity and Coherence[3]

Let me start by defining knowledge integration as the process bringing into relation bodies of knowledge or research practice hitherto unrelated or distant. Similarly, I define knowledge diffusion as the movement or translation of a piece of knowledge to bodies of knowledge where it had not been used before. These "specialised bodies" of knowledge can refer to perspectives, concepts, theories but also to tools, techniques or information and data sources (National Academies, 2004). For example, some of the key contributions of the very successful US National Center for Ecological Analysis and Synthesis (NCEAS) in UCSB were based precisely on cross-fertilisation of methods and data sources used in different fields within ecology (Hackett, Parker, Conz, Rhoten, & Parker, 2008).

The difference between integration and diffusion is mainly one of perspective. For example, from the perspective of a Valencian laboratory working on breast cancer, RNA interference (RNAi) is *integrated* to their portfolio of methods for genetic manipulation, i.e. a piece of knowledge is integrated into the knowledge base of an organisation. However, from the perspective of an emergent technology such as RNAi, it is the RNAi technique which has *diffused* into a laboratory—a laboratory which is a point in a space that may be characterise by geography (València), discipline (oncology) or research problem (breast cancer). In this chapter, the emphasis is given to integration, but the frameworks proposed and many of the tools used can be used as well to analyse knowledge diffusion (Carley & Porter, 2012).

Both integration and diffusion are dynamic processes and, therefore, they should be analysed over time (Leydesdorff & Rafols, 2011a; Liu & Rousseau, 2010). It is, nevertheless also possible to make a static comparison of the degree of integration represented in different entities such as publications (Porter & Rafols, 2009), researchers (Porter, Cohen, Roessner, & Perreault, 2007) or university departments (Rafols, Leydesdorff, O'Hare, Nightingale, & Stirling, 2012).

The framework proposed here analyses separately the two key concepts necessary for the definition of knowledge integration. On the one hand, *diversity* describes the differences in the bodies of knowledge that are integrated, and on the other hand, *coherence* describes the intensities of the relations between these bodies of knowledge. Notice that the concept of diversity is interpreted in the same way in the case of integration and of diffusion. However, for coherence the interpretation differs for integration and diffusion. More coherence can be interpreted as an increase in integration (because knowledge has become more

---

[3] This framework was first introduced in Rafols and Meyer (2010), then represented in more general form in Liu, Rafols, and Rousseau (2012) and again in an empirical case in Rafols, Leydesdorff, O'Hare, Nightingale, and Stirling (2012) with some substantial changes. Here I try to make a further generalisation of the concept of coherence with some incremental improvements.

Diversity: property of apportioning elements into categories   Coherence: property of relating categories via elements

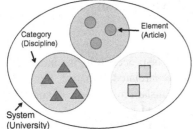 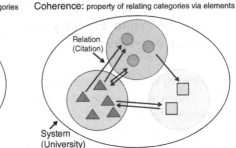

Fig. 8.1 Illustration of definitions of diversity (*left*) and coherence (*right*). In *parenthesis*, an example of the concept: the disciplinary diversity of a university by assigning articles to disciplines, and the disciplinary coherence by means of cross-disciplinary citations. *Large circles* represent categories. Small figures (*triangles*, *squares* and *small circles*) represent elements

related). In the case of diffusion, more coherence does not mean necessarily more diffusion, but a specific type of diffusion: spread over topics in which these topics have become related.

Another way of studying knowledge integration (and interdisciplinarity) is to focus on the bridging role, or *intermediation* role, of some specific scientific contributions, typically using notions from social network analysis such as betweenness centrality (Chen et al., 2009; Leydesdorff, 2007). In Rafols et al. (2012), we developed intermediation as a framework, complementary to diversity and coherence, which is useful to explore fine-grained, bottom-up perspectives of dynamics. However, given space constraints, I will leave intermediation outside of the scope of this chapter.

Given that integration can be analysed at different levels, let us first make a rather abstract description of diversity and coherence. We will consider the *system* or unit of analysis (e.g., university department), the *elements* (e.g. articles), the *categories* (e.g. Web of Science (WoS) categories) and the *relations* between categories (e.g. citations from one WoS category to another).

Diversity is a "property of the apportioning of elements or options in any system" (Stirling, 1998, 2007, p. 709). For example, the disciplinary diversity of a university (system) can be proxied by the distribution of the articles (elements) published in WoS categories (categories) (as shown in Fig. 8.1). Diversity can have three distinct attributes as illustrated in Fig. 8.2:

- *Variety*: number of categories into which the elements are apportioned (N).
- *Balance*: evenness of the distribution of elements across categories.
- *Disparity*: degree to which the categories of the elements are different.

The novelty and key contribution in Stirling's heuristic for diversity (1998, 2007) is the introduction of a distance metrics d_{ij} between categories. The idea, as illustrated in Fig. 8.2, is that diversity of a system increases not only with more categories (higher *variety*), or with a more balanced distribution (higher *balance*), but also if the elements are allocated to more different categories (higher *disparity*).

All other things being equal, there is more diversity in a project including cell biology and sociology than in one including cell biology and biochemistry. While measuring the proportion p_i of elements in a category is straight-forward, providing an estimate of cognitive distance d_{ij} is more challenging. For this purpose, the metrics behind the global maps of science developed in the 2000s have been very useful (Boyack, Klavans, & Börner, 2005; Klavans & Boyack, 2009; Moya-Anegón et al., 2007; Rafols, Porter, & Leydesdorff, 2010).

Coherence, on the other hand, aims to capture the extent to which the various parts in the system are directly connected via some relation. For example, the disciplinary coherence of a university (system) can be proxied by the citations (relations) from articles in one WoS category to references in another WoS category (categories) (Rafols et al., 2012). Or it may be explored using network properties at the element level, such as network density or intensity (Rafols & Meyer, 2010).

Further research is needed to establish how and whether coherence can be measured. In this chapter, I tentatively propose that coherence can be thought as having the attributes of density (analogue of variety), intensity (analogue of balance) and disparity, as shown in Fig. 8.3. For this purpose, let me define M as the number of existing relations in the systems (out of $N(N-1)$ relations possible with N categories), intensity of a relation i_{ij} as the scalar representing the relative strength of a relation between categories i and j. Now we can define:

- *Density*: number of relations between categories
- *Intensity*: overall intensity of the relations in the system.
- *Disparity*: degree to which the categories of the relations are different.

Since both diversity and coherence have various aspects, one can generate different, equally legitimate measures of each depending on how these aspects are weighted, as illustrated in Table 8.1. Stirling (2007) proposed a generalised formulation for diversity which can be turned into specific measures of diversity such variety or the Simpson diversity, by assigning values to the parameters α and β. Ricotta and Szeidl (2006) achieved the same result with a different mathematical formulation (possibly more rigorous but also more cumbersome). In this chapter, I tentatively introduce the same type of generalisation for the concept of coherence.

From these considerations, it follows that none of the measures in Table 8.1 should be taken then as a "definitive" and "objective" manner of capturing diversity and coherence. Instead, all measures of diversity and coherence are subjective in the sense that they are derived from judgements about: (1) the choice of categories, (2) the assignment of elements to categories, (3) what constitutes an adequate metric of intensity i_{ij}, (4) of a cognitive distance d_{ij} and, finally (5) a judgment regarding what are the useful or meaningful values of α and β for a specific purpose of the study. For example, assuming a distance $0 < d_{ij} < 1$, the analyst would use small values of β to emphasise the importance of distance in the problem under study (this is relevant in issues such as climate change where understandings from

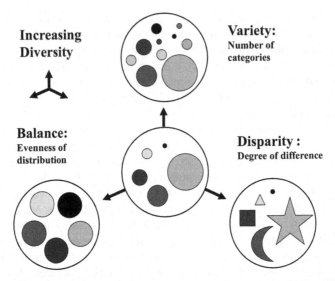

Fig. 8.2 Schematic representation of the attributes of diversity, based on Stirling (1998, p. 41). Each *full circle* represents a system under study. The *coloured figures* inside the *circle* are the categories into which the elements are apportioned. *Different shape*s indicate more difference between categories. Source: Rafols and Meyer (2010)

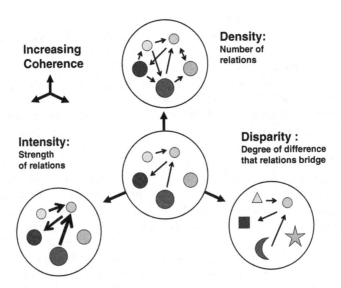

Fig. 8.3 Schematic representation of the attributes of coherence. Each circle represents the *system* under study. The figures inside the circle are the *categories* into which the *elements* are apportioned. The lines represent the relations between categories. Thicker lines indicate higher intensity in relations. Different shapes indicate more difference between categories

Table 8.1 Selected measures of diversity and coherence

Notation	
Proportion of elements in category i:	p_i
Intensity of relations between categories i *and* j:	i_{ij}
Distance between categories i and j:	d_{ij}
Diversity Indices	
Generalised Stirling diversity	$\displaystyle\sum_{i,j(i\neq j)} \left(p_i p_j\right)^{\alpha} d_{ij}^{\beta}$
Variety ($\alpha=0$, $\beta=0$)	N
Simpson diversity ($\alpha=1$, $\beta=0$)	$\displaystyle\sum_{i,j(i\neq j)} p_i p_j = 1 - \sum_i p_i^2$
Rao-Stirling diversity ($\alpha=1$, $\beta=1$)	$\displaystyle\sum_{i,j(i\neq j)} p_i p_j d_{ij}$
Coherence Indices	
Generalised coherence	$\displaystyle\sum_{i,j(i\neq j)} i_{ij}^{\gamma} d_{ij}^{\delta}$
Density ($\gamma=0$, $\delta=0$)	M
Intensity ($\gamma=1$, $\delta=0$)	$\displaystyle\sum_{i,j(i\neq j)} i_{ij} = 1 - \sum_i i_{ii}$
Coherence ($\gamma=1$, $\delta=1$)	$\displaystyle\sum_{i,j(i\neq j)} i_{ij} d_{ij}$

The two comprehensive measures which have been used and tested in the literature are highlighted

social natural sciences need to be integrated). Small values of α, on the contrary, highlight the importance of contributions by tiny proportions. Another possibility is to use various measures of diversity, each of them highlighting one single aspect, as proposed by Yegros-Yegros, Amat, D'Este, Porter, and Rafols (2013) (see also in Rafols et al. (2012) and Chavarro, Tang, and Rafols (2014)). Visualisation in overlay science maps is a way of providing a description of diversity and coherence without need to collapse the data into a single figure (Rafols et al., 2010).

For the sake of parsimony, in practice most applications have used the simplest formulations with $\alpha=1$ and $\beta=1$. This leads to the Rao-Stirling variant of diversity, $\sum_{i,j(i\neq j)} p_i p_j d_{ij}$. This measure had been first proposed by Rao (1982) and has become known in population ecology as quadratic entropy (Ricotta & Szeidl, 2006). It can be interpreted as a distance-weighted Simpson diversity (also known as Herfindahl-Hirschman index in economics-related disciplines). Rao-Stirling can be interpreted as the average cognitive distance between elements, as seen from the categorisation, since it weights the cognitive distance d_{ij} over the distribution of elements across categories p_i. Similarly, if the intensity of relations is defined as the distribution of relation (i.e. if $i_{ij}=p_{ij}$), the simplest form of

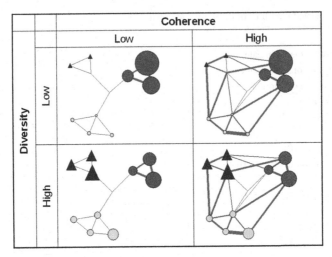

Fig. 8.4 Conceptualisation of knowledge integration as increase in cognitive diversity and coherence

coherence, for $\gamma = 1$ and $\delta = 1$, $\sum\limits_{i,j(i \neq j)} p_{ij}d_{ij}$, can be interpreted as the average distance over the distribution of relations p_{ij}, rather than the distribution of elements p_i. [4]

Each node in the network represents a cognitive category. Light grey lines show strong similarity between categories. Same shapes illustrate clusters of similar categories. The size of nodes portrays the proportion of elements in a given category. Dark (or green) lines represent *relations* between categories. Knowledge integration is achieved when an organisation becomes more diverse and establishes more relations between disparate categories. Source: Rafols et al. (2012).

The analytical framework proposed understands knowledge integration as an increase in diversity, an increase in coherence, or both. This would mean moving from top to bottom, from left to right, or in diagonal from top-left to bottom-right in Fig. 8.4. Similarly, a diffusion process would be seen as an increase in diversity. Higher coherence in diffusion means that as a research topic reaches new areas and it brings them together, whereas lower coherence means that the "topic" is used instrumentally without necessarily linking the new areas.

---

[4] To my knowledge, coherence had only been introduced in this single form, with intensity defined as the proportion of citations between WoS categories $i_{ij} = p_{ij}$. The form of coherence I adopt in this chapter follows from Soós and Kampis (2012) rather than Rafols, Leydesdorff et al. (2012). In the latter, coherence, i.e. $\sum\limits_{i,j(i \neq j)} p_{ij}d_{ij}$ was normalised (divided) by Rao-Stirling diversity, i.e. $\sum\limits_{i,j(i \neq j)} p_i p_j d_{ij}$. Such normalisation was useful to remove the correlation between the two variables, but it seems unnecessarily complicated for a general framework.

8.3   Choices on Data and Methods for Operationalisation

The framework presented so far is very general and does not presuppose a commitment to specific data or methods. Now I will operationalise the approach as it was originally developed in order to capture knowledge integration in scientific processes using bibliometric data. Let us first discuss the variety of possible choices in scientometrics regarding the system (unit of analysis), elements, categories and relations to investigate (see Liu, Rafols, & Rousseau (2012) for a previous discussion on these choices).

8.3.1   Unit of Analysis

The unit of analysis for measuring diversity can be an article, a researcher, a department or institute, a university or a research topic such as an emergent technology. One thing to notice is that for small units such as articles, diversity can sometimes be interpreted as knowledge integration, without need of further investigating coherence. For example, Alan Porter's work calls *Integration Score* the specific measure of diversity of WoS Categories in the references of an article (Porter, Cohen, Roessner, & Perreault, 2007; Porter, Roessner, & Heberger, 2008).

One needs to be cautious with choices of units of analysis that involve small numbers, such as article and researcher, because they may not have enough elements for a robust statistical analysis and the resulting measures could be very noisy, particularly when the low numbers are compounded by inaccurate assignment of elements to categories (as it happens when references are assigned to WoS categories). Thus, article or researcher level measures should be treated with caution—most of the time they will not be reliable individual descriptions, but they can be used averaged over classes—e.g., comparing interdisciplinarity between disciplines using the average disciplinary diversity of references in articles, using samples of some hundred articles (Porter & Rafols, 2009), or to carry out econometric regression models using thousands of articles to investigate the influence of diversity of references on some variables such as number of citations (Yegros-Yegros et al., 2013) or local orientation of the research (Chavarro, Tang, & Rafols, 2014).

An important consideration in choosing the unit of analysis is the recent finding by Cassi, Mescheba, and Turckheim (2014) that the Rao-Stirling diversity can be added over scales (under some plausible assumptions, in particular the use of cosine similarity). This means, that the diversity of a research institute is the sum of the diversities *within* each article it published, plus the diversity *between* the articles. This property opens up the possibility of measuring the diversity of large organisations in a modular manner.

8.3.2 Classifying Elements into Categories

Next, we need to choose the elements "contained" within the unit of analysis and the categories into which they will be classified. The choice of elements is straightforward. They will typically be articles, references, authors, organisations (as shown in the address or affiliation), or keywords that are listed in the bibliographic record. The challenge is how to classify the elements into categories. Table 8.2 provides a partial review of different choices of unit of analysis, elements and categories. Since cognitive distance is a key component in the measures of diversity and coherence, the availability of a cognitive metrics among the categories of the classification used is a relevant factor to take into account. For the choice of metrics, see reviews (Börner, Chen, & Boyack, 2003; Boyack et al., 2011).

In science, disciplines are the most conventional cognitive categories. Most database providers assign articles (usually via journals) to some type of disciplinary categories. Therefore, the most straightforward way of assigning bibliographic elements such as articles or references to categories is to rely on categories provided by databases. The most widely used classification is Thomson-Reuters' Web of Science categories, which is journal-based and very problematic (Rafols & Leydesdorff, 2009), given that articles within a journal do not necessarily share a similar topic or disciplinary perspective.

The next step is to compute the cognitive distance between categories. As in the case of the classification, the choice of a specific cognitive distance has to be based on judgement. A plausible choice is to take $d_{ij} = (1 - s_{ij})$, where s_{ij} is the cosine similarity of the WoS categories. This data is available in Excel files at Loet Leydesdorff's website (http://www.leydesdorff.net/overlaytoolkit) from 2007 onwards (Leydesdorff, Carley, & Rafols, 2012).

There is, though, the possibility of defining distance in different ways even if you start from the cosine similarity between WoS categories. For example, Soós and Kampis (2012) proposed to define d_{ij} as the sum of the $(1 - s_{ij})$ weights of edges in the shortest path from WoS categories i to j. Jensen and Lutkouskaya (2014) use $d_{ij} = 1/s_{ij}$ instead. In these two choices, more weight is given to the co-occurrence of very disparate categories (where the shortest path is long and $1/s_{ij}$ is high) than in the standard similarity. The downside of these alternatives is that the diversity measure is not any more bounded between 0 and 1.

In order to avoid using the journal classifications from data providers (lacking transparency), another possibility is to use journals as categories per se to compute diversity measures (Leydesdorff & Rafols, 2011b). The problem here is that most journals are only similar to a small set of related journals. As a result the cosine distances between most journals are practically zero. Since, in principle, the measure of cognitive distance aims to describe differences between distant areas, this result is not useful to capture cognitive distances across disciplines. To overcome this difficulty, Leydesdorff, Rafols, and Chen (2013) have recently proposed to use the distance observed in the two dimensional projection of a map of the +10,000-dimensions of the actual distance matrix. This is a very coarse

Table 8.2 Examples of different choices of systems, elements, categories and metrics used in measures of diversity

System (unit of analysis)	Elements	Category	Metrics	Examples
Article	References in article	WoS categories	Cosine similarity of WoS categories	Porter & Rafols, 2009
Article	Citations to article	WoS categories	Cosine similarity of WoS categories	Carley and Porter (2012)
Author	Articles	WoS categories	Cosine similarity of WoS categories	Porter et al. (2007)
University department or Institutes	Articles	WoS categories	Cosine similarity of WoS categories	Rafols et al. (2012); Soós and Kampis (2011)
Institutes	Articles	250 Clusters from 300,000 French publications (2007–2010)	Cosine similarity of clusters	Jensen and Lutkouskaya (2014)
Topic (emergent technology)	Articles	WoS categories	Cosine similarity of WoS categories	Leydesdorff and Rafols (2011a)
Journals	References in articles of journals	Journals	Cosine similarity of journals	Leydesdorff and Rafols (2011b)
Topic (emergent technology)	Articles	Medical subject headings (MeSH)	Co-occurrence of MeSH terms in articles	Leydesdorff, Kushnir, and Rafols (2012)
Topic (emergent technology)	Articles	Medical subject headings (MeSH)	Self-organising maps based on MeSH, titles, abstracts, references	Skupin, Biberstine, and Börner (2013)
Topic (research)	Patents	Keywords	Self-organising maps	Polanco, François, and Lamirel (2001)
Open-ended: Topic, Country, Organisation	Patents	International Patent Classification (IPC) Classes	Cosine similarity of IPC classes	Kay, Newman, Youtie, Porter, and Rafols (2012); Leydesdorff, Kushnir, and Rafols (2012)
Open-ended: Topic, Country, Organisation	Patents	Technological aggregations of IPC classes	Co-occurrence of IPC classes	Schoen et al. (2012)

approximation, but has the advantage of distributing quite evenly the distances among journals between zero and a maximum value (which we redefine as one).

Another alternative to the inaccuracies of WoS or Scopus categories to is to carry out clustering using bibliometric data (Rafols & Leydesdorff, 2009). These bottom-up categories may be more consistent with research practices, at least as seen from citation patterns. They can be based on journal clustering (e.g. more than 500 categories in the UCSD by Börner et al. (2012),[5] see also Rosvall and Bergstrom (2008)) or on paper-level clustering (e.g. about 700 categories by Waltman & van Eck, 2012).[6]

In all this analysis so far, we have relied on static classifications with stable categories of all science.[7] In the case of knowledge integration, this is useful to characterise the knowledge background from a traditional perspective of science (e.g. in terms of subdisciplines). In the case of emergent technologies, though, new research topics do not conform to these traditional categories and it is often illuminating to complement the traditional view with a more fine-grained, local, bottom-up and dynamic classification. The difficulty of this approach is that constructing very fine-grained and/or dynamic clusters that are meaningful is very demanding (Havemann, Gläser, Heinz, & Struck, 2012). Since noise increases as the sample becomes smaller, many clusters become unstable (are born, die, divide, etc.) below a threshold around 100–1,000 papers (Boyack, Klavans, Small, & Ungar, 2014), and their local structure may differ from the one obtained with a global map (Klavans & Boyack, 2011). This clustering has been the approach of Kajikawa and colleagues, using direct-citation-link clustering, for example in studies on energy technologies (Kajikawa, Yoshikawa, Takeda, & Matsushima, 2008) or bionanotechnology (Takeda, Mae, Kajikawa, & Matsushima, 2009). Boyack, Klavans, Small, and Ungar (2014) are also following this approach with very small clusters which are derived from a global data set. In principle, the framework proposed here might also work with small and dynamic categories— in practice, the challenge is constructing these categories.

Rather than relying on aggregate categories, one may try to use directly the elements as categories calculating their cognitive distance without further categorisation, as a contrast to the coarse-grained, static classification (Rafols & Meyer, 2010; Soós & Kampis, 2011). Jensen and Lutkouskaya (2014) use various measures of diversity with different categorisations in order to have a more plural view of the degree of interdisciplinarity of French national laboratories. These efforts align with the conceptualisation of scientometric advice as helping the opening-up of perspectives in science policy debate, rather than narrowing the scope of decisions (i.e. closing-down) (Barré, 2010; Rafols, Ciarli, Van Zwanenberg, & Stirling, 2012).

Finally, instead of using classifications that relate bibliometric elements with a cognitive category based on scientific point of view such as a subdiscipline, an emergent field or a research topic, as discussed above, one may instead relate the

---

[5] This classification and underlying map can be downloaded and publicly used. It is available at http://sci.cns.iu.edu/ucsdmap/.

[6] This classification is available at http://www.ludowaltman.nl/classification_system/.

[7] According to Boyack, Klavans, Small and Ungar (2014), more than 99 % of clusters are stable at a level of aggregation of about 500 clusters for all science.

elements with categories from outside science such as diseases or technologies. The Medical Subject Headings (MeSH) of PubMed offer a way of making the linkages between elements of a publication and the specific practitioner-oriented perspectives of its hierarchical classification, such as descriptors for disease, techniques and equipment, chemicals/drugs, and healthcare. Using one or more of these practitioner-oriented categories might be particularly helpful when analysing the social impact of research. Leydesdorff, Rotolo and Rafols (2012) and Skupin, Biberstine, and Börner (2013) have recently created global MeSH maps. However, unlike the global maps of science, which show consensus (Klavans & Boyack, 2009), these maps could not be matched. Hence, the underlying cognitive structure and metrics of MeSH deserves further investigation.

8.3.3 Capturing Relations

In order to measure coherence one needs to associate relations observed in the system with links among categories. Since these relations are derived from information within or between elements, the discussion in the previous subsection on the assignment of elements into categories is directly applicable to relations as well. For example, a citation allows us to relate the category of an article to the category of one its references. The challenge, as discussed, is how the article and the reference are classified into WoS categories, journals, bottom-up clusters, or MeSH terms, etc. Another straightforward way to create relations is from co-occurrences of some article attributes. For example, if MeSH terms are taken as categories, the strength of the relation between two MeSH can be estimated as their normalised number of shared publications.

An interesting point to notice regarding relations is that they do not need to be symmetrical, i.e. $i_{ij} \neq i_{ji}$. This is obvious for directed flows: it is well known, for example, that an applied research field like oncology cites cell biology proportionally more than the reverse (4.5 % vs. 7.5 % citations in 2009). In the case in which relations are non-directed (i.e. edges), such as co-occurrences, it is also possible to do an asymmetrical normalisation, i.e. to normalise i_{ij} according to counts in i category only. This raises the interesting question of whether cognitive distances, which in most studies are symmetrical ($d_{ij} = d_{ji}$), should also be taken as asymmetrical—an issue which deserves a full separate discussion.

8.3.4 Visualisation

Given that diversity and coherence are multidimensional concepts, visualisation can be helpful to intuitively present the various aspects without collapsing all the information into a single value. The method proposed here relies on the ideas of *overlaying* (projecting) the elements of the unit of analysis over the cognitive space

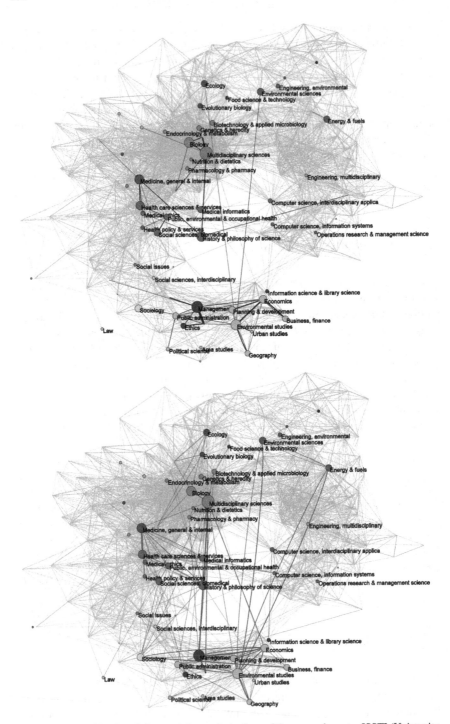

Fig. 8.5 Expected (*top*) and observed (*bottom*) citations of the research centre ISSTI (University of Edinburgh) across different Web of Science categories. The *grey lines* in the background show the global map of science (Rafols et al., 2010). The size of the nodes reflects the aggregate number

selected, an idea that I borrowed from Kevin Boyack and colleagues (Boyack, Börner, & Klavans, 2009). The visualisation has three steps. First, one builds a "basemap" representing the cognitive space selected. A widely used "basemap" is the global map of science, representing the disciplinary structure of science (freely available at Rafols et al., 2010). The map intuitively portrays the cognitive distance between its nodes, the WoS categories (or others).

Second, one projects the distribution of elements into categories over the basemap by making the size of each category (node) proportional to the frequency of elements in that specific category. This means for example that the size of a node in the global map of science is made proportional to the number of articles published in that WoS category in the sample studied. This projection or overlay allows the viewer to capture intuitively the three attributes of diversity: First, the map captures *variety* by portraying the number of categories in which a unit of analysis (e.g. university) is engaged. Second, it captures *balance* by presenting the nodes with different sizes. Third, unlike bar charts, the map conveys *disparity* among categories by illustrating the cognitive distance by means of the physical distance in the map (Rafols et al., 2010, p. 1883).

The third step is to project the relations over the map as illustrated in Fig. 8.5 (Rafols et al., 2012). This projection is perhaps the most unconventional step, since it consists of overlaying the links in the unit of analysis, over the structure of the global map, without re-positioning the nodes. The intensity of the relations is shown by the thickness of the links. It is precisely the contrast between the local relations (in thick darker lines) in comparison to the global relations (in finer lighter lines) what allows us to understand the nature and extent of knowledge integration that is being carried out. The visualisation of relations between hitherto unrelated bodies of knowledge conveys intuitively the concept of coherence.

The maps show intuitively the three aspects of coherence: whether coherence is achieved across many categories (density), the thickness of links (intensity) and whether they are linked across distant categories (disparity). Since the probability of links does not only depend on cognitive proximity, it is useful to make an overlay of the expected relations (in the case of citation, this depends both on citation sources and probability flows) and one overlay of the observed relations, as shown in Fig. 8.5.

Fig. 8.5 (continued) of citations given to a field from all ISSTI's publications. *Blue lines* show the expected citations between fields, given where ISSTI is publishing. The computation of expected citations is based on the number of publications in a field, and the average proportion of citations to other fields in all the WoS. It can be observed that the expected citations tend to be within disciplines: within biological sciences, within health services, and within social sciences. *Orange lines* show the citations between fields observed in ISSTI's publications. The citations between fields criss-cross the map of science both within disciplines and across disciplines. (Only citations larger than 0.2 % of ISSTI's total are shown). Source: Rafols, Leydesdorff et al. (2012)

8.4 How to Compute and Visualise Knowledge Integration

This section describes the protocol of the method to compute and visualise diversity and coherence. To do so, I will follow the most well-established application of this framework, based on the so-called global maps of science based on WoS categories. Since these categories are not very accurate, it is best to think this analysis as merely exploratory or illustrative. Detailed information on this method is presented in the Annex of Rafols et al. (2010). The data and basemaps used here are publicly available at Loet Leydesdorff's website http://www.leydesdorff.net/overlaytoolkit.

8.4.1 Illustrative Introduction to Measures of Diversity

This protocol illustrates how to compute diversity and coherence using excel files and Pajek maps. Supplementary files are available here: http://www.sussex.ac.uk/Users/ir28/book/excelmaps.

8.4.1.1 Data Collection

1. Delineate and download the data set from the Web of Science.

8.4.1.2 Measure of Diversity

For the sake of helping non-expert readers, the measures are presented in the spreadsheet calculations.

2. **Create a list** with the distribution of WoS categories. These are listed in the field "WC" in the file downloaded.
3. **Open** the spreadsheet file "DiversityComputation2009.xlsx."
4. **Paste** the list in the tab "INPUT." Notice that only the WoS Categories in the Journal Citation Index in 2009 are present. Other categories will not be counted.
5. **Go to tab** "OUTPUT." **Select a threshold** for the minimum proportion to be taken into account in counting variety and disparity (default = 0.01, i.e. 1 %)
6. The file provides values for Rao-Stirling diversity and other measures of diversity as described in Table 8.1.

8.4.1.3 Measure of Coherence

7. **Create a matrix** with the ordered distribution of citations from the WoS categories to WoS categories in the data set. (Unfortunately, to my knowledge

this cannot be done with publicly available software. VantagePoint[8] provides an easy template to create it).

8. **Open** the spreadsheet file "CoherenceComputation2009.xlsx."
9. **Paste** the matrix in the tab "INPUT matrix."
10. **Go to tab "OUTPUT"** to retrieve the data on coherence.

8.4.1.4 Visualisation of Diversity with Pajek

11. **Open** the Pajek file "ScienceMap2009.paj" (press F1).
12. **Upload** the vector file (.vec) with the distribution list of WoS categories "ListWoSCats.vec."

Press Ctrl-Q to visualise the overlay map (details provided in the appendix of Rafols et al. (2010)).

8.4.2 R Script for Computing Diversity of a set of Articles

This protocol provides a script for computing diversity over large data sets. Supplementary files are available here: http://www.sussex.ac.uk/Users/ir28/book/diversity.zip.

The file "diversity_measures_1.R" contains the script with the programming language R to compute the Rao-Stirling diversity for each individual article of a list of articles, based on the assignment of references to WoS categories. It requires the file with the proximity matrix ("cosine_similarity_matrix_sc.csv") and an input file with the list distribution of WoS in the reference list, as shown in "articles_sample. csv." The directory with the file needs to be written up into the script before running it.

> **Conclusions**
>
> In this chapter I have presented a framework for the analysis of knowledge integration and diffusion based on the concepts of cognitive diversity and coherence. Knowledge diffusion is seen as an increase in the cognitive diversity of the areas to which a given discovery or technology has spread. Knowledge integration is seen as an increase in cognitive diversity and/or coherence. The chapter introduced the general mathematical formulation of these concepts. It has proposed that diversity has the attributes of variety, balance and disparity, whereas coherence has the attributes of density, intensity and disparity. Diversity and coherence can be formulated in various

(continued)

---

[8] http://www.thevantagepoint.com/.

(continued)

manners depending on the relative weight of the attributes—hence their values will depend on the choice of weight given to them.

Given the importance of the choices of elements, and the relations and classifications to characterise diversity, I have discussed different approaches to classify science into categories, from the top-down and coarse classifications such as the WoS categories to more fine-grained categories. I have briefly mentioned the possibilities of characterising science with more practitioner-oriented perspectives such as those provided by MeSH terms. I have illustrated with a spreadsheet how to compute diversity and coherence using WoS categories. Since WoS categories are very inaccurate, this method should be interpreted as exploratory.

The fact that diversity and coherence can be measured using various mathematical formulations and that, for each of them, various operationalisations are possible in terms of the elements and categories chosen. This should send a serious message of caution: knowledge integration and diffusion are strongly dependent on the perspective taken. It could be that with a disciplinary perspective, a research topic has become stagnant (staying within the same discipline), but with a medical perspective, the topic is diffusing to new areas such as new diseases. Hence, the measures and maps should be read as inevitably partial perspectives—covering only a few of the possibilities for capturing knowledge dynamics. Other dynamics of knowledge integration, not covered by diversity and coherence, are also possible. For example, "intermediation" would be another way to capture knowledge integration focussing in the bridging processes (Chen et al., 2009; Rafols et al., 2012).

The framework proposed has been developed for mapping in the conventional cognitive dimension of science (disciplines and topics), but it can easily be extended to other cognitive perspectives such as those arising from medicine (via MeSH). Similarly, the approach can be easily extended to patents, using global maps of technology (Kay, Newman, Youtie, Porter, & Rafols, 2012; Leydesdorff, Kushnir & Rafols, 2012; Schoen et al., 2012), and closely related measures of diversity (Nesta & Saviotti, 2005, 2006).

Finally, I would like to highlight that while the framework has been applied to cognitive distance, it can in principle be applied as well to other analytical dimensions. For example, one might look at the geographical diversity of a collaborative project not counting the number of countries, but investigating collaborations or citations in terms of geographical distance (Ahlgren, Persson, & Tijssen, 2013). Or, one might investigate the diversity in organisations in a new topic by not just counting organisations, but taking into account the cognitive proximity of the organisations. As proposed by

(continued)

(continued)

Frenken (2010), by extending this framework to other analytical dimensions, it would be possible to investigate how knowledge integration is mediated by geographical, organisational, institutional and social networks.

Acknowledgements This chapter summarises work carried out with many collaborators, in particular with L. Leydesdorff, A.L. Porter and A. Stirling. I am grateful to D. Chavarro for writing the code in R language to compute diversity. I thank Y.X. Liu, R. Rousseau and A. Stirling for fruitful comments. I acknowledge support from the UK ESRC grant RES-360-25-0076 ("Mapping the dynamics of emergent technologies") and the US National Science Foundation (Award #1064146—"Revealing Innovation Pathways: Hybrid Science Maps for Technology Assessment and Foresight"). The findings and observations contained in this paper are those of the author and do not necessarily reflect the views of the funders.

References

Ahlgren, P., Persson, O., & Tijssen, R. (2013). Geographical distance in bibliometric relations within epistemic communities. *Scientometrics, 95*(2), 771–784.

Barré, R. (2010). Towards socially robust ST indicators: indicators as debatable devices, enabling collective learning. *Research Evaluation, 19*(3), 227–231.

Börner, K., Chen, C., & Boyack, K. W. (2003). Visualizing knowledge domains. *Annual Review of Information Science & Technology, 37*(1), 179–255.

Börner, K., Klavans, R., Patek, M., Zoss, A. M., Biberstine, J. R., Light, R. P., Larivière, V., & Boyack, K. W. (2012). Design and update of a classification system: the UCSD map of science. *PLoS One, 7*(7), e39464.

Boschma, R. A. (2005). Proximity and innovation: a critical assessment. *Regional Studies, 39*, 61–74.

Boyack, K. W., Börner, K., & Klavans, R. (2009). Mapping the structure and evolution of chemistry research. *Scientometrics, 79*(1), 45–60.

Boyack, K.W., Klavans, R., Small, H., Ungar, L. (2014). Characterizing the emergence of two nanotechnology topics using a contemporaneous global micro-model of science. *Journal of Engineering and Technology Management, 32*, 147–159.

Boyack, K. W., Klavans, R., & Börner, K. (2005). Mapping the backbone of science. *Scientometrics, 64*(3), 351–374.

Boyack, K. W., Newman, D., Duhon, R. J., Klavans, R., Patek, M., Biberstine, J. R., et al. (2011). Clustering more than two million biomedical publications: comparing the accuracies of nine text-based similarity approaches. *PLoS One, 6*, e18029.

Carley, S., & Porter, A. (2012). A forward diversity index. *Scientometrics, 90*, 407–427.

Cassi, L., Mescheba, W., Turckheim, É. (2014). How to evaluate the degree of interdisciplinarity of an institution? *Scientometrics*. In press. doi: 10.1007/s11192-014-1280-0.

Chavarro, D., Tang, P., Rafols, I. (2014). Interdisciplinarity and local issue research: evidence from a developing country. *Research Evaluation, 23*(3), 195–209. doi: 10.1093/reseval/rvu012.

Chen, C., Chen, Y., Horowitz, M., Hou, H., Liu, Z., & Pellegrino, D. (2009). Towards an explanatory and computational theory of scientific discovery. *Journal of Informetrics, 3*, 191–209.

Dosi, G. (1982). Technological paradigms and technological trajectories: a suggested interpretation of the determinants and directions of technical change. *Research Policy, 11*(3), 147–162.

Etzkowitz, H., & Leydesdorff, L. (2000). The dynamics of innovation: from national systems and "Mode 2" to a triple helix of university-industry-government relations. *Research Policy, 29*, 109–123.

Frenken, K. (2010). Geography of scientific knowledge: a proximity approach. Eindhoven Centre for Innovation Studies (ECIS). Retrieved from http://alexandria.tue.nl/repository/books/720753.pdf.

Frenken, K., Boschma, R. A., Hardeman, S. (2010). Proximity and Mode 2 knowledge production. Preprint. Retrieved from http://econ.geo.uu.nl/boschma/frenkenEcon&society.pdf.

Gibbons, M., Limoges, C., Nowotny, H., Schwartzman, S., Scott, P., & Trow, M. (1994). *The new production of knowledge: the dynamics of science and research in contemporary societies.* London: Sage.

Hackett, E. J., Parker, J., Conz, D., Rhoten, D., & Parker, A. (2008). Ecology transformed: the national center for ecological analysis and synthesis and the changing patterns of ecological research. In *The handbook of science and technology studies* (pp. 277–296). Cambridge MA: MIT.

Havemann, F., Gläser, J., Heinz, M., & Struck, A. (2012). Identifying overlapping and hierarchical thematic structures in networks of scholarly papers: a comparison of three approaches. *PLoS One, 7*, e33255.

Hessels, L. K., & van Lente, H. (2008). Re-thinking new knowledge production: a literature review and a research agenda. *Research Policy, 37*, 740–760.

Jensen, P., & Lutkouskaya, K. (2014). The many dimensions of laboratories' interdisciplinarity. *Scientometrics, 98*(1), 619–631.

Kajikawa, Y., Yoshikawa, J., Takeda, Y., & Matsushima, K. (2008). Tracking emerging technologies in energy research: toward a roadmap for sustainable energy. *Technological Forecasting and Social Change, 75*, 771–782.

Kay, L., Newman, N., Youtie, J., Porter, A. L., Rafols, I. (2014). Patent overlay mapping: visualizing technological distance. *Journal of the Association for Information Science and Technology.* In press. doi: 10.1002/asi.23146

Klavans, R., & Boyack, K. W. (2009). Toward a consensus map of science. *Journal of the American Society for Information Science and Technology, 60*, 455–476.

Klavans, R., & Boyack, K. W. (2011). Using global mapping to create more accurate document-level maps of research fields. *Journal of the American Society for Information Science and Technology, 62*, 1–18.

Leydesdorff, L. (2007). Betweenness centrality as an indicator of the interdisciplinarity of scientific journals. *Journal of the American Society for Information Science and Technology, 58*, 1303–1319.

Leydesdorff, L., Carley, S., & Rafols, I. (2012). Global maps of science based on the new web-of-science categories. *Scientometrics, 94*, 589–593.

Leydesdorff, L., Kushnir, D., Rafols, I. (2014). Interactive overlay maps for US Patent (USPTO) data based on International Patent Classifications (IPC). *Scientometrics, 98*(3), 1583–1599. doi: 10.1007/s11192-012-0923-2.

Leydesdorff, L., & Rafols, I. (2011a). Local emergence and global diffusion of research technologies: an exploration of patterns of network formation. *Journal of the American Society for Information Science and Technology, 62*, 846–860.

Leydesdorff, L., & Rafols, I. (2011b). Indicators of the interdisciplinarity of journals: Diversity, centrality, and citations. *Journal of Informetrics, 5*, 87–100.

Leydesdorff, L., Rafols, I., & Chen, C. (2013). Interactive overlays of journals and the measurement of interdisciplinarity on the basis of aggregated journal-journal citations. *Journal of the American Society for Information Science and Technology, 64*(12), 2573–2586.

Leydesdorff, L., Rotolo, D., & Rafols, I. (2012). Bibliometric perspectives on medical innovation using the medical subject headings (MeSH) of PubMed. *Journal of the American Society for Information Science and Technology, 63*, 2239–2253.

Liu, Y. X., Rafols, I., & Rousseau, R. (2012). A framework for knowledge integration and diffusion. *Journal of Documentation, 68*, 31–44.

Liu, Y. X., & Rousseau, R. (2010). Knowledge diffusion through publications and citations: a case study using ESI-fields as unit of diffusion. *Journal of the American Society for Information Science and Technology, 61*, 340–351.

Lowe, P., & Phillipson, J. (2006). Reflexive interdisciplinary research: the making of a research programme on the rural economy and land use. *Journal of Agricultural Economics, 57*, 165–184.

Molas-Gallart, J., Rafols, I., D'Este, P., Llopis, O. (2013). A framework for the evaluation of translational research based on the characterization of social networks and knowledge exchange processes. Presented at the Annual Meeting of the American Evaluation Association, Washington, DC, USA. Available at http://www.ingenio.upv.es/en/working-papers/towards-alternative-framework-evaluation-translational-research-initiatives

Moya-Anegón, F., Vargas-Quesada, B., Chinchilla-Rodríguez, Z., Corera-Álvarez, E., Munoz-Fernández, F. J., & Herrero-Solana, V. (2007). Visualizing the marrow of science. *Journal of the American Society for Information Science and Technology, 58*, 2167–2179.

National Academies. (2004). *Facilitating Interdisciplinary research*. Washington, DC: National Academies.

Nesta, L., & Saviotti, P. P. (2005). Coherence of the knowledge base and the firm's innovative performance: evidence from the U.S. pharmaceutical industry. *Journal of Industrial Economics, 8*, 123–142.

Nesta, L., & Saviotti, P. P. (2006). Firm knowledge and market value in biotechnology. *Industrial and Corporate Change, 15*, 625–652.

Nightingale, P., & Scott, A. (2007). Peer review and the relevance gap: ten suggestions for policy makers. *Science and Public Policy, 34*, 543–553.

Pinch, T. J., & Bijker, W. E. (1984). The social construction of facts and artefacts: or how the sociology of science and the sociology of technology might benefit each other. *Social Studies of Science, 14*, 399–441.

Polanco, X., François, C., & Lamirel, J. C. (2001). Using artificial neural networks for mapping of science and technology: a multi self-organizing-maps approach. *Scientometrics, 51*, 267–292.

Porter, A. L., Cohen, A. S., Roessner, J. D., & Perreault, M. (2007). Measuring researcher interdisciplinarity. *Scientometrics, 72*, 117–147.

Porter, A. L., & Rafols, I. (2009). Is science becoming more interdisciplinary? Measuring and mapping six research fields over time. *Scientometrics, 81*, 719–745.

Porter, A. L., Roessner, J. D., & Heberger, A. E. (2008). How interdisciplinary is a given body of research? *Research Evaluation, 17*, 273–282.

Rafols, I., Ciarli, T., Van Zwanenberg, P., Stirling, A. (2012). Towards indicators for opening up S&T policy. *STI Indicators Conference* (pp.675–682). Retrieved from http://2012. sticonference.org/Proceedings/vol2/Rafols_Towards_675.pdf.

Rafols, I., & Leydesdorff, L. (2009). Content-based and algorithmic classifications of journals: perspectives on the dynamics of scientific communication and indexer effects. *Journal of the American Society for Information Science and Technology, 60*, 1823–1835.

Rafols, I., Leydesdorff, L., O'Hare, A., Nightingale, P., & Stirling, A. (2012). How journal rankings can suppress interdisciplinarity. The case of innovation studies and business and management. *Research Policy, 41*, 1262–1282.

Rafols, I., & Meyer, M. (2010). Diversity and network coherence as indicators of interdisciplinarity: case studies in bionanoscience. *Scientometrics, 82*, 263–287.

Rafols, I., Porter, A. L., & Leydesdorff, L. (2010). Science overlay maps: a new tool for research policy and library management. *Journal of the American Society for information Science and Technology, 61*, 1871–1887.

Rao, C. R. (1982). Diversity and dissimilarity coefficients: a unified approach. *Theoretical Population Biology, 21*, 24–43.

Ricotta, C., & Szeidl, L. (2006). Towards a unifying approach to diversity measures: bridging the gap between the Shannon entropy and Rao's quadratic index. *Theoretical Population Biology, 70*, 237–243.

Rosvall, M., & Bergstrom, C. T. (2008). Maps of random walks on complex networks reveal community structure. *Proceedings of the National Academy of Sciences, 105*, 1118–1123.

Schoen, A., Villard, L., Laurens, P., Cointet, J. P., Heimeriks, G., Alkemade, F. (2012). *The network structure of technological developments; technological distance as a walk on the technology map*. Presented at the STI Indicators Conference, Montréal

Skupin, A., Biberstine, J. R., & Börner, K. (2013). Visualizing the topical structure of the medical sciences: a self-organizing map approach. *PLoS One, 8*, e58779.

Soós, S., & Kampis, G. (2011). Towards a typology of research performance diversity: the case of top Hungarian players. *Scientometrics, 87*, 357–371.

Soós, S., & Kampis, G. (2012). Beyond the basemap of science: mapping multiple structures in research portfolios—evidence from Hungary. *Scientometrics, 93*, 869–891.

Stirling, A. (1998). On the economics and analysis of diversity. *Science Policy Research Unit (SPRU), Electronic Working Papers Series, 28*. Retrieved from http://www.uis.unesco.org/culture/documents/stirling.pdf.

Stirling, A. (2007). A general framework for analysing diversity in science, technology and society. *Journal of The Royal Society Interface, 4*, 707–719.

Takeda, Y., Mae, S., Kajikawa, Y., & Matsushima, K. (2009). Nanobiotechnology as an emerging research domain from nanotechnology: a bibliometric approach. *Scientometrics, 80*, 23–29.

Waltman, L., & van Eck, N. J. (2012). A new methodology for constructing a publication-level classification system of science. *Journal of the American Society for Information Science and Technology, 63*, 2378–2392.

Yegros-Yegros, A., Amat, C.B., D'Este, P., Porter, A.L., & Rafols, I. (2013). *Does interdisciplinary research lead to higher citation impact? The different effect of proximal and distal interdisciplinarity*. Presented at the DRUID Conference, Barcelona. Retrieved from http://druid8.sit.aau.dk/acc_papers/54dcxbblnj9vbrlt2v4gku686mcx.pdf.

Part III
Statistical and Text-Based Methods

Chapter 9
Limited Dependent Variable Models and Probabilistic Prediction in Informetrics

Nick Deschacht and Tim C.E. Engels

Abstract This chapter explores the potential for informetric applications of limited dependent variable models, i.e., binary, ordinal, and count data regression models. In bibliometrics and scientometrics such models can be used in the analysis of all kinds of categorical and count data, such as assessments scores, career transitions, citation counts, editorial decisions, or funding decisions. The chapter reviews the use of these models in the informetrics literature and introduces the models, their underlying assumptions and their potential for predictive purposes. The main advantage of limited dependent variable models is that they allow us to identify the main explanatory variables in a multivariate framework and to estimate the size of their (marginal) effects. The models are illustrated using an example data set to analyze the determinants of citations. The chapter also shows how these models can be estimated using the statistical software Stata.

9.1 Introduction

A topic search in the Social Science Citation Index on November 13, 2013 identified over 700 journal articles in Library and Information Science (LIS) that use regression analysis. In the top 25 of source titles, we find *Scientometrics* (64 articles), *Journal of the American Society for Information Science and Technology* (46), *Information Processing & Management* (24), *Journal of Informetrics* (12), and *Journal of Documentation* (9). Until 2004 the annual number of LIS papers that implemented a regression model did not exceed 20; then, in the period 2005–2010,

N. Deschacht (✉)
Faculty of Economics and Business, KU Leuven, Campus Brussel, Warmoesberg 26, 1000 Brussel, Belgium
e-mail: Nick.Deschacht@kuleuven.be

T.C.E. Engels
Department of Research Affairs and Centre for Research & Development Monitoring (ECOOM), University of Antwerp, Middelheimlaan 1, 2020 Antwerp, Belgium

Antwerp Maritime Academy, Noordkasteel-Oost 6, 2030 Antwerp, Belgium
e-mail: Tim.Engels@uantwerpen.be

© Springer International Publishing Switzerland 2014
Y. Ding et al. (eds.), *Measuring Scholarly Impact*,
DOI 10.1007/978-3-319-10377-8_9

a gradual increase to about 50 papers per year is apparent. Since 2011 the annual number jumped to about 100, illustrating the rise of regression models in LIS. In the aforementioned journals, binary, ordinal, Poisson and negative binomial regression models are common because classification issues (e.g., authorship attribution, classification of journals, user profiles) and count data (e.g., number of papers, patents or their citations) abound in LIS. In this chapter we specify these limited dependent variable models with a view of facilitating the implementation of such models in LIS research.

Limited dependent variable models are a group of regression models in which the range of possible values of the variable of interest is limited. In some cases the outcome variable is binary, such as when the interest is in whether a journal article was cited over a certain period (yes or no). The outcome variable can also take multiple discrete values as is often the case in peer review and assessments. When frequencies are counted for a certain event the outcome variable consists of count data, such as the number of patents in a given year or the number of books published by publishing houses. In these cases the choice of the regression model may follow directly from the research question. Often, however, the choice of the regression model will be subject to careful deliberation and more than one model may be appropriate. Running multiple models on the same dataset may be instructive and can sometimes serve as a robustness check of the results. We illustrate this throughout this chapter. In the conclusions we provide the reader with some advice regarding model choice.

The strength of regression models is that they allow us to estimate the size of the "effect" of an explanatory variable on the dependent variable (the word "effect" may be misleading because it suggests causation while a regression analysis in itself does not exclude the possibility of inverse causation or spurious causation resulting from omitted variables). As opposed to association analysis, a regression analysis allows the researcher to quantify the effect of changes in the independent variables on the dependent variable. Another advantage is that regression analysis easily allows one to distinguish and isolate the effects of different explanatory variables. An interesting example in this regard is the multilevel logistic analysis of the Leiden ranking by Bornmann, Mutz, and Daniel (2013), which shows that only 5 % of the variation between universities in terms of the percentage of their publications that belong to the 10 % most cited in a certain field is explained by between university differences, whereas about 80 % is explained by differences among countries. Regression models can also be used for prediction, although the quality of such predictions is obviously conditional on the quality of the model. For most models, methods or rules of thumb to evaluate the quality of the resulting predictions are available.

The chapter introduces the main limited dependent variable models and illustrates their use to analyze the determinants of citations using data on the 2,271 journal articles published between 2008 and 2011 in the journals *Journal of Informetrics* (JOI), *Journal of the American Society for Information Science and Technology* (JASIST), *Research Evaluation* (RE), *Research Policy* (RP), and *Scientometrics* (SM). The data used in this illustration are available through the publisher's website for interested readers to experiment with on their own. The next section introduces the data set and the variables used in the illustration. Section 9.3 discusses the logit model for binary choice. The models for multiple responses and count data are discussed in Sects. 9.4 and 9.5. The final sections present some concluding remarks and practical guidance on how to estimate these models using the statistical software Stata (Long & Freese, 2006). We opted to illustrate the models in Stata because this program appears to be most commonly used in informetrics. However, all the models mentioned here may be run in R, and many in SPSS and other packages.

The aim of this chapter is primarily to demonstrate the possibilities of limited dependent variable models in LIS and to compare their strengths and weaknesses in an applied setting. The theoretical description of the various models was kept brief for reasons of space. Readers looking for more elaborate treatments are referred to econometric textbooks (Greene, 2011; Wooldridge, 2012) or specialized texts (e.g., Agresti, 2002, 2010; Hilbe, 2011).

9.2 The Data: Which Articles Get Cited in Informetrics?

Several studies have investigated intrinsic and extrinsic factors that influence the citation impact of papers. In the models in this chapter we include 12 explanatory variables—the first five of which are inspired by the literature review in Didegah and Thelwall (2013b)—to explain the number of citations (including self-citations) in the calendar year following publication. Our aim is to illustrate the applicability and the use of limited dependent variable models. The 12 variables included in the analysis are the following:

- The journal in which an article is published (8 % of the articles in our sample were published in the JOI, 33 % in SM, 21 % in RP, 6 % in RE, and 32 % in JASIST). The popularity of a journal tends to correlate positively with the impact of the articles that appear in it.
- The number of authors of the article (NumAut: min = 1, max = 11, avg = 2.40; SD = 1.34). We included this variable because collaborative articles tend to receive more citations.
- The number of countries mentioned in the address field of the article (NumCoun: min = 1, max = 9, avg = 1.31, SD = 0.60). International collaboration too tends to increase the number of citations.
- The number of cited references included in the article (NumRef: min = 0, max = 282, avg = 40.20, SD = 25.34). Papers with more references often attract more citations.
- The length in terms of number of pages of the article (NumPag: min = 1, max = 37, avg = 13.33, SD = 4.88). Longer papers can have more content, including more tables and/or figures, which in turn may translate into the receipt of more citations.

- The length of the article title in terms of number of characters (NumTitle: min = 10, max = 284, avg = 87.48, SD = 31.10). On the one hand shorter titles might be more to the point, on the other hand longer titles might occur more in article searches.
- Whether the article is the first in an issue or not (First: 8 % are first articles). An article that is the first in an issue, is likely to attract more attention and may therefore receive more citations.
- Whether funding information is included in the acknowledgments of the article or not (Fund: 20 % of the articles have funding information). Rigby (2013) reports a weak positive link between more funding information and impact of papers.
- The publication year of the article (PubYear: min = 2008, max = 2011, avg = 2009.60, SD = 1.10). Over time the number of citations tends to increase (e.g., because more source titles are added to the WoS), so we need to correct for publication year.
- The month in which the article appeared (PubMon: min = 1, max = 12, avg = 6.63, SD = 3.42). This measure is based on information in WoS on the date of the print publication and does not account for the fact that some journals may be late or that articles could be available for "early view." We included this variable because the number of citations received in the year following publication (dependent variable) is likely to be influenced by the timing of the publication of the articles.
- Whether the article deals with the h-index or not (H: 7 % of the articles are about the h-index). Articles were classified as dealing with the h-index if "h-ind*", "h ind*" and/or "Hirsch" occurred in their abstract. Among other things the h-bubble article by Rousseau, Garcia-Zorita, and Sanz-Casado (2013), which shows that h-index-related articles inflated short-term citations to a large extent, inspired us to include this variable.
- Whether the article deals with issues related to innovation and patenting or not (InnoPat: 18 % of the articles are related to these topics). Articles were classified as related to innovation and patenting if "innovation" and/or "patent*" occurred in their title and/or abstract. As innovation is high on governments' agendas, we wondered whether researching innovation would also pay off in terms of number of citations.

An issue that should be kept in mind when estimating regression models is the degree of correlation between these explanatory variables. Too much correlation (multicollinearity) inflates the standard errors on the estimated coefficients so that the estimated effects become unstable and sensitive to small variations in the data. As an indicator of the degree of collinearity, one can calculate Variance Inflation Factors (VIF) for every explanatory variable.[1] In our dataset the maximal VIF was

---

[1] If R^2_k is the coefficient of determination of a linear regression model that predicts the explanatory variable X_k as a function of the other explanatory variables, then the Variance Inflation Factor $VIF_k = \frac{1}{1-R2_k}$.

2.0, which we consider to be tolerable since thresholds of 5 or more are common in the literature (Menard, 1995; O'Brien, 2007).

9.3   Binary Regression

In bibliometrics and informetrics binary logistic models are often used for analyzing and/or predicting whether articles will be cited or not (Van Dalen & Henkens, 2005), whether patents are commercialized (Lee, 2008), used in military applications (Acosta, Coronado, Marín, & Prats, 2013) or will be infringed (Su, Chen, & Lee, 2012). These models are also used in studies of funding and editorial decisions (Fedderke, 2013), winning scientific prizes or awards (Heinze & Bauer, 2007; Rokach, Kalech, Blank, & Stern, 2011), career transitions and promotions (Jensen, Rouquier, & Croissant, 2009) and the use of public libraries by internet users (Vakkari, 2012). Many other outcomes that can be analyzed through binary regression can be thought of, e.g., whether a researcher belongs to the editorial board of a certain journal, is likely to collaborate or publish a book, will file a patent, will move to another institution, or will pass a certain threshold in terms of citations or h-index.

(a) The binary logit model

If y_i is a binary variable that can take only the values 0 and 1, then the logit model writes the probability $P(y_i = 1)$ as a function of the explanatory variables:

$$P(y_i = 1 | x_{1i} x_{2i}, \ldots, x_{ki}) = P_i = G(\beta_0 + \beta_1 x_{1i} + \ldots + \beta_k x_{ki}) \qquad (9.1)$$

where $G(z) = \frac{e^z}{1+e^z}$ is the logistic function. The range of the logistic function is between 0 and 1 which ensures that the predicted probabilities are limited to this same range. This is one of the reasons why logit models are more appropriate than OLS in the case of a binary dependent variable. OLS should also be avoided because it assumes that the error terms are normally distributed with constant variances, while neither of these conditions apply when the dependent variable is binary (for similar reasons OLS should be avoided in models of ordinal or count dependent variables). The interpretation of the coefficients is not straightforward in the logit model. This can be seen when we rewrite the model as

$$\ln\left(\frac{P_i}{1 - P_i}\right) = \text{logit}(P_i) = \beta_0 + \beta_1 x_{1i} + \ldots + \beta_k x_{ki} \qquad (9.2)$$

where $\frac{P_i}{1-P_i}$ are the odds of $y_i = 1$ (e.g., if $P_i = 0.8$ then the odds are 4 to 1). From this equation it is clear that β_i is the change in the log-odds when x_i increases by one unit and the other variables are held constant. The exponentiated coefficients e^{β_i} can then be interpreted as the factor by which the odds increase when x_i increases by one unit

(e^{β_i} is the odds ratio). However, effects in terms of odds cannot be interpreted unambiguously in terms of probabilities, because the change in probability when x_i increases by one unit depends on the level of x_i and on the values of the other explanatory variables. One way around this problem is to estimate the "marginal effect at the means" of an explanatory variable x_i, which measures the change in the prediction function if x_i increases by one unit.[2] Although such marginal effects cannot substitute for the estimated coefficients or odds ratios (which remain correct even when the explanatory variables deviate from their means), the marginal effects are usually informative.

The coefficients of the logit model are estimated by maximizing the likelihood of the data with respect to the coefficients. Most statistical software packages carry out the necessary iterative numerical optimization and calculate corresponding standard errors, which allow for significance tests on the coefficients (which test whether the estimated effects could be attributable to sampling variability). A global test on all parameters in the model tests whether the likelihood of the observed data using the estimated coefficients is significantly greater than the likelihood of a model that has no independent variables. This test is referred to as the likelihood ratio test and uses a test statistic that has an approximate chi-square distribution under the null hypothesis that all parameters are zero.

(b) Illustration

We now use the logit model to study the citation of journal articles in the field of informetrics. The dependent variable measures whether or not the article was cited in another published article during the calendar year following its publication. 66 % of all articles in our sample were cited, whereas the remaining 34 % were not. Table 9.1 presents the estimated coefficients in the model with standard errors and significance tests, and the corresponding odds ratios and marginal effects.

The results indicate that—holding all the other explanatory variables in the model constant—articles published in the JOI, SM, and RP have a significantly greater probability (than the reference category JASIST) of being cited, while that probability is lower for articles in RE. The publication month control variable has a negative effect, which was expected because the probability of citation depends on the duration since publication. Other significant effects are found for international collaboration, the number of references listed in the article and for articles about the h-index.

The coefficients, odds ratios and marginal effects give an indication of the size of these effects. The estimated coefficient for JOI is 0.85, which implies that—ceteris paribus—the log-odds of JOI articles being cited are 0.85 greater than those of JASIST articles. The corresponding odds ratio is $e^{.85} = 2.34$, which implies that the odds of JOI articles being cited are 2.34 times greater than those of JASIST articles.

---

[2] Next to the marginal effect at the means, other approaches are possible to calculate marginal effects. For a discussion and an example using bibliometric data, see Bornmann and Williams (2013).

Table 9.1 The binary logit model

	(1) Coefficients estimate/(SE)	(2) Odds ratio estimate/(SE)	(3) Marginal effects estimate/(SE)
JOI	0.851***	2.342***	0.175***
	(0.209)	(0.489)	(0.038)
RE	−0.706***	0.493***	−0.174***
	(0.212)	(0.104)	(0.052)
RP	0.510**	1.665**	0.112**
	(0.164)	(0.273)	(0.035)
SM	0.356**	1.428**	0.081**
	(0.128)	(0.183)	(0.029)
NumAut	−0.005	0.995	−0.001
	(0.036)	(0.036)	(0.008)
NumCoun	0.181*	1.198*	0.040*
	(0.086)	(0.103)	(0.019)
NumRef	0.007**	1.007**	0.002**
	(0.003)	(0.003)	(0.001)
NumPag	0.006	1.006	0.001
	(0.011)	(0.011)	(0.003)
NumTitle	−0.000	1.000	−0.000
	(0.002)	(0.002)	(0.000)
First	0.266	1.304	0.056
	(0.182)	(0.238)	(0.037)
Fund	−0.131	0.877	−0.029
	(0.124)	(0.109)	(0.028)
PubYear	0.048	1.049	0.011
	(0.043)	(0.045)	(0.010)
PubMon	−0.085***	0.918***	−0.019***
	(0.014)	(0.013)	(0.003)
H	0.953***	2.594***	0.176***
	(0.228)	(0.592)	(0.033)
InnoPat	−0.242	0.785	−0.055
	(0.139)	(0.109)	(0.032)
Constant	−96.290	0.000	
	(86.854)	(0.000)	
Chi^2	154.2	154.2	154.2
p	0.000	0.000	0.000
$Pseudo\text{-}R^2$	0.05	0.05	0.05
N	2,271	2,271	2,271

*p<0.05; **p<0.01; ***p<0.001

The marginal effect provides an indication of the effect in terms of probabilities evaluated at the means of the explanatory variables: articles in the JOI are 18 percentage points more likely of being cited than articles in JASIST (remember that the overall unconditional probability of being cited is around 66 %, so 18 percentage

Table 9.2 Prediction table for the binary logit model

		Predicted category			
		Baseline		Logit	
Observed category		Not cited	Cited	Not cited	Cited
	Not cited	0 %	100 %	17.1 %	82.9 %
	Cited	0 %	100 %	6.0 %	94.0 %

points is a substantial effect). Another sizeable marginal effect is that articles in RE are—holding all the other variables constant—17 percentage points less likely of being cited than articles in JASIST. Articles about the h-index also increase their citation probability by 18 percentage points (compared to articles that do not write about the h-index). It appears that getting published in the JOI with an article on h-indices was a strategy worth considering for scholars in the field looking to improve their own h-index!

The model can now be used to make predictions by calculating predicted citation probabilities for articles with given values on the explanatory variables. Moreover, such predicted probabilities can also be calculated for the articles in our sample. This is a way to evaluate the predictive power of our model since we know whether these articles eventually were cited or not. If we use the common decision rule to predict citation when the predicted probability of an article is greater than 0.5, then the model makes correct predictions for 17 % of the non-cited articles and 94 % of the cited articles (Table 9.2).

In order to evaluate the quality of our model, these numbers should be compared to a baseline of correct predictions that would be made in absence of the explanatory variables. Since the overall proportion of cited articles is 66 %, the best guess would then be to predict citation for any given article. In the non-cited category the proportion of correct predictions improves from 0 % (baseline) to 17 % in the logit model, while the proportion decreases from 100 to 94 % among the cited articles. The sum of the proportions of correct predictions should be greater than 100 % for a good model (Verbeek, 2008), which is the case in our example (the sum of the diagonal elements 17 % + 94 % = 111 %).[3] The most common goodness-of-fit statistic for logit models is McFadden's R^2 (reported in Table 9.1 as "Pseudo-R^2"), which is defined as the percent increase in the log-likelihood when moving from the baseline model with no explanatory variables to the full model.

Note that this evaluation of the predictive power of our model relates to the internal validity of the model ("to what extent is the model capable of reproducing the sample data?"). Good internal validity does not imply that the model would

---

[3] A related measure for model quality which is also based on the prediction table and which has the interesting property of ranging between 0 and 100 %, is the Adjusted Count R^2 (see Long & Freese, 2006).

perform equally well on new data. One way to assess external validity is to use only a subset (e.g., 90 %) of the available observations to estimate the model (the training data) and to subsequently use the excluded observations (the test data) to evaluate the model's predictive capacity.

9.4   Ordinal Regression

In the binary regression analysis of citations we lumped all articles that were cited (66 % of the sample) together in one group. However, there may be important differences between articles that have just a few citations and those that have many citations. By not using this information the tests in the binary choice model have less power, which increases the risk of failing to demonstrate a true effect (a type II error). Ordered response or ordinal regression models are appropriate when the dependent variable is an ordinal scale.

Recent examples of applications of categorical or ordered logistic models in bibliometrics and informetrics include analyses and prediction of the factors that explain information seeking behavior of academic scientists (Niu & Hemminger, 2012), of the impact of international coauthorship on citation impact (Sin, 2011), of peer assessments of research groups (Engels, Goos, Dexters, & Spruyt, 2013) and of the popularity of new Twitter hashtags (Ma, Sun, & Cong, 2013). Other examples of outcomes that can be analyzed through ordered models include the outcomes of peer review of manuscripts submitted for publication (acceptance, minor review, major review, rejection), and the rank of professors (assistant professor, associate professor, full professor). In some cases, e.g., the published outcomes of a research project (academic papers only; patent only; academic papers plus patent; academic plus popularizing papers) the response categories may not be strictly ordered. In such cases a multinomial model can be used to analyze the data.

(a) The ordered logit model

If y_i is an ordinal variable that can take only the values $j = 1, 2, \ldots, J$, then the cumulative probability is the probability that an observation i is in the jth category or lower:

$$\gamma_{ij} = P(y_i \leq j) \tag{9.3}$$

The ordered logit model is then defined as

$$\mathrm{logit}(\gamma_{ij}) = \alpha_j - \beta_1 x_{1i} - \beta_2 x_{2i} - \cdots - \beta_k x_{ki} \tag{9.4}$$

The model has a different intercept α_j for each category j (the cutpoints), whereas the slope coefficients are assumed constant over the categories.[4] β_i is then the increase in the log-odds of being in a higher category when x_i increases by one unit while the other variables are held constant. As in the binary model the odds ratio e^{β_i} is the factor by which the odds of being in a higher category increase when x_i increases by one unit.

In the ordered logit model the slope coefficients are assumed equal at every categorical level. This "proportional odds assumption" can be evaluated using the Brant test, which tests whether the slope coefficients are equal across separate binary models. In case the test does not support the assumption, an alternative model could be considered in which the coefficients are allowed to vary with the categorical levels (i.e., a multinomial logit model).

(b) Illustration

We now use the ordered logit model to study the determinants of journal article citations. The dependent variable in the analysis is an ordinal variable with three categories: (1) no citations during the year following publication, (2) few citations (i.e., one or two citations), and (3) many citations (i.e., three or more). In our sample of 2,271 articles from the field of informetrics, 34 % were not cited, 39 % received one or two citations and the remaining 27 % received three or more. Table 9.3 presents the estimated ordered logit model.

The odds ratio for JOI is 2.3, which implies that—ceteris paribus—the odds of JOI articles being in a higher category are 2.3 times greater than those of JASIST articles. It is informative to compare these results with the ones from the binary model in Table 9.3. Most of the coefficients have smaller p-values, which reflects the increased power by differentiating between articles with few and many citations. For example, the coefficient for international collaboration is now highly significant ($p < 0.001$ as opposed to $p = 0.035$ in the binary model). The dummy variable indicating whether the article is the first article published in the journal issue is now significant ($p = 0.005$) while it was not in the binary model ($p = 0.135$). A further analysis shows the reasons for this finding: while first articles have a similar probability (than other articles) of not being cited, they have a much larger probability of having many citations. This indicates that the proportional odds assumption underlying the ordered logit model may be violated here (the effect on the odds of not being cited versus being cited is not the same as the odds of receiving a few citations versus many citations). There is one variable where an inverse scenario takes place: the indicator for articles published in RP is no longer significant in the ordered logit model. The reason is that in comparison with the JASIST reference category a large proportion of its articles are not cited (producing the effect in the binary analysis), while at the same time a slightly larger proportion

---

[4] The negative signs before the coefficients are needed because cumulative probabilities were defined using a less-than or equal to symbol, while the coefficients should estimate the effect of explanatory variables on increasing levels of the dependent variable.

Table 9.3 The ordered logit model

	(1)	(2)
	Coefficients estimate/(SE)	Odds ratio estimate/(SE)
JOI	0.841^{***}	2.319^{***}
	(0.166)	(0.385)
RE	-0.796^{***}	0.451^{***}
	(0.202)	(0.091)
RP	0.234	1.264
	(0.137)	(0.173)
SM	0.310^{**}	1.363^{**}
	(0.112)	(0.152)
NumAut	-0.012	0.989
	(0.032)	(0.031)
NumCoun	0.296^{***}	1.344^{***}
	(0.072)	(0.097)
NumRef	0.009^{***}	1.009^{***}
	(0.002)	(0.002)
NumPag	-0.006	0.994
	(0.010)	(0.010)
NumTitle	-0.001	0.999
	(0.001)	(0.001)
First	0.427^{**}	1.533^{**}
	(0.153)	(0.234)
Fund	-0.127	0.881
	(0.110)	(0.097)
PubYear	0.036	1.037
	(0.037)	(0.039)
PubMon	-0.088^{***}	0.916^{***}
	(0.012)	(0.011)
H	1.026^{***}	2.789^{***}
	(0.173)	(0.482)
InnoPat	-0.177	0.838
	(0.117)	(0.098)
cut1		
Constant	72.331	2.589e+31
	(74.884)	(1.939e+33)
cut2		
Constant	74.098	1.515e+32
	(74.885)	(1.135e+34)
Chi^2	222.2	222.2
p	0.000	0.000
Pseudo-R^2	0.04	0.04
N	2,271	2,271

$*p<0.05; **p<0.01; ***p<0.001$

Table 9.4 Prediction table for the ordered logit model

		Predicted category					
		Baseline			Ordered logit		
Observed category		No citations	Few citations	Many citations	No citations	Few citations	Many citations
	No citations	0 %	100 %	0 %	43.4 %	48.7 %	7.9 %
	Few citations	0 %	100 %	0 %	29.2 %	60.2 %	10.6 %
	Many citations	0 %	100 %	0 %	16.5 %	59.9 %	23.6 %

of the RP articles have many citations (cancelling out the effect in the ordinal analysis).

The estimated ordered logit model can now be used to calculate predicted cumulative probabilities for every category. Because the cutpoints increase as the categorical level increases, the cumulative probabilities increase as well. Differences between adjacent cumulative probabilities yield predicted probabilities for each category. If we use as a decision rule to predict the category with the largest predicted probability, then the model makes correct predictions for 43 % of the non-cited articles, 60 % of the articles with few citations, and 24 % of the articles with many citations (Table 9.4).

A baseline model with no explanatory variables would predict a few citations for every article, because that is the category with the largest overall proportion (38 %). The sum of the diagonal elements for the ordered logit model in Table 8.4 (127.2 %) is greater than that in the baseline model (always 100 %), which is a minimum quality requirement for any model.

The Brant test to evaluate the proportional odds assumption (equality of the slope coefficients over the categories) results in a test statistic value of $\chi^2 = 23.5$ ($p = 0.07$). If $p > 0.05$ then the evidence against the proportional odds assumption is not significant. It may be worth to keep in mind that significance tests are all about sample sizes, which in this case implies that even small differences in slope coefficients could result in a rejection of the null hypothesis if the sample is large (while large differences may be insignificant in small samples). Because our p-value is not much greater than the significance level we also estimated a multinomial model, which consists of multiple binary logit models so that the slope coefficients are allowed to vary (the specification and estimates are not reported). For this model the sum of the diagonal elements in a prediction table (not shown) increases to 131.8 %. However, this small increase in predictive power requires the estimation of much more parameters in the model, which increases the risk of overfitting. Although both the ordered and the multinomial models have their merits, the authors favor the ordinal model in this case because of its parsimony and the fact that the proportional odds assumption is not implausible. A multinomial model would be appropriate if the proportional odds assumption is clearly violated as well as in the case of a non-ordinal dependent variable.

9.5   Count Data Models

If the variable of interest measures the frequency of an event, then count data models may be appropriate to take advantage of the cardinal (rather than ordinal) nature of the data. The standard regression framework for analyzing count data is the Poisson model, but in most practical applications extensions of this model (the quasi-Poisson and negative binomial models) are needed to overcome violations of underlying assumptions (discussed below).

Abbasi, Altmann, and Hossain (2011) implement a Poisson model to identify the effects of coauthorship networks on performance of scholars; Niu and Hemminger (2012) complemented their logistic analysis of information seeking behavior with a Poisson regression. Negative binomial regression models have been applied to model the number of papers (Barjak & Robinson, 2007; Gantman, 2012) and in the study of citation counts, for example when comparing sets of papers (Bornmann & Daniel, 2006, 2008) or the relative importance of authors and journals (Walters, 2006). Lee, Lee, Song, and Lee (2007) pioneered the use of a zero-inflated negative binomial in informetrics in their analysis of citations of patents of the Korean Institute of Science and Technology (KIST). Zero-inflated models have two parts: A binary model to predict group membership and a count model for the data in the latter group (Hoekman, Frenken, & van Oort, 2009; Long & Freese, 2006). Recently, Chen (2012), Didegah and Thelwall (2013a) and Yoshikane (2013) implemented zero-inflated negative binomial models in their studies of, respectively, predictive effects of structural variation on citation counts, of citation impact in nanoscience, and of citations of Japanese patents. Zero-inflated models assume two sources and hence different underlying causes of zeros: Perfect zeros for which structural factors explain the observation of zeros (e.g., the number of academic papers per toddler) and zeros that occur in the count distribution (e.g., some academics may have no papers during a number of years). As illustrated by Didegah and Thelwall (2013b) hurdle models may provide a good alternative, at least in the case of citations, as receiving its first citation can be considered a real hurdle for a paper after which it becomes more likely to be cited again. In the section below we limit the explanation to the standard negative binomial regression; readers interested in truncated and other variations may consult Hilbe (2011).

(a)   The Poisson, the quasi-Poisson and the negative binomial regression models

If y_i is a count variable taking only non-negative integer values ($y_i = 0, 1, 2, \ldots$) and we assume that y_i conditional on the values of the explanatory variables has a Poisson distribution:

$$P(y_i = y | x_{1i}, x_{2i}, \dots, x_{ki}) = \frac{e^{-\mu_i} \cdot \mu_i^{y}}{y!} \quad y = 0, 1, 2, \dots \quad (9.5)$$

where μ_i is the expected value of the distribution. Note that the assumption refers to the *conditional* distribution of y_i and not to the unconditional distribution of y_i. Because the latter distribution also depends on the distribution of the explanatory variables, the distribution of the observed y_i is not a valid argument for preferring this model over another. The following example makes this clear: In a model with only one binary explanatory variable in which the conditional distribution is Poisson, the unconditional observed y_i would in many cases have a bimodal distribution (and so clearly not be Poisson).

The expected value μ_i is usually modelled by

$$\mu_i = E[y_i | x_{1i}, \dots, x_{ki}] = e^{\beta_0 + \beta_1 x_{1i} + \dots + \beta_k x_{ki}} \quad (9.6)$$

For technical reasons, the log of the conditional mean of Poisson (and negative binomial) models is estimated, rather than the mean itself. The Poisson regression model can thus be defined as

$$P(y_i = y | x_{1i}, x_{2i}, \dots, x_{ki}) = \frac{e^{-\left(e^{\beta_0 + \beta_1 x_{1i} + \dots + \beta_k x_{ki}}\right)} \cdot \left(e^{\beta_0 + \beta_1 x_{1i} + \dots + \beta_k x_{ki}}\right)^{y}}{y!} \quad y = 0, 1, 2, \dots \quad (9.7)$$

of which the coefficients are usually estimated using maximum likelihood. How to interpret these coefficients becomes clear if we write the expected value as

$$\ln(E[y_i | x_{1i}, \dots, x_{ki}]) = \beta_0 + \beta_1 x_{1i} + \dots + \beta_k x_{ki} \quad (9.8)$$

which is a semi-log model familiar from linear regression. β_i is then the relative (percent) increase in μ_i when x_i increases by one unit while the other variables are held constant.[5]

A limitation of the Poisson regression model is that any Poisson distribution is completely determined by its mean and that the variance is assumed to equal that mean (the equidispersion assumption). This restriction is violated in many applications because the variance is often greater than the mean. In such cases there is overdispersion, by which we mean that the variance is greater than the variance implied by assuming a Poisson distribution. However, the maximum likelihood estimator is considered to produce consistent estimates for the coefficients regardless of the actual conditional distribution (Wooldridge, 1997). The procedure of using Poisson maximum likelihood estimation without assuming that the Poisson distribution is correct, is referred to as the quasi-Poisson model or the Poisson

---

[5] This interpretation is only approximately correct as it follows from differentiating $\ln(E[y_i | x_{1i}, \dots, x_{ki}])$ with respect to x_i. An exact interpretation is that the exponentiated coefficient e^{β_1} is the factor change in μ_i.

QMLE (quasi-maximum likelihood estimator). In the case of overdispersion the standard errors of the coefficients will be underestimated in the Poisson regression, thereby increasing the risk of making a type I error (incorrectly concluding that an effect is significant). The quasi-Poisson model adjusts the standard errors by estimating an additional parameter in the model (the quasi-Poisson assumes the variance to be a fixed multiple of the mean).[6] The Poisson and quasi-Poisson will always return the same estimates of the coefficients.

Another possibility in the case of overdispersion is to estimate a negative binomial regression model. This model also allows the conditional mean of $y_i(\mu_i)$ to differ from its variance $(\mu_i + \alpha.\mu_i^2)$ by estimating an additional parameter (the dispersion parameter α). Since the negative binomial model assumes the variance to be a quadratic function of the mean, this model allows for far greater variances for large estimates of the mean than the quasi-Poisson.[7] The Poisson model can be regarded as a special case of this more general negative binomial model when α is zero. A significance test on α can thus be regarded as a test for the presence of overdispersion in the Poisson model. The probability mass function of the negative binomial distribution differs from the Poisson distribution so that the estimated coefficients—unlike those from the quasi-Poisson model—are not the same as in the Poisson model, although they tend to be similar.

A common goodness-of-fit statistic for count data regression models is a pseudo-R^2 that is calculated as the square of the correlation between the observed y_i and the values predicted by the model $\hat{y}_i$. This R^2 is indicative of the (internal) validity of the model when it is used for making predictions. An additional measure is the Akaike information criterion (AIC), which trades off goodness-of-fit with model complexity, by adding a penalty for the number of parameters estimated in the model.[8]

(b) Illustration

We now apply the Poisson regression model to analyze journal article citations during the year following their publication. Figure 9.1 summarizes the distribution of the number of citations in our sample of 2,271 articles.

The mean number of citations is 1.94 while the variance in the distribution is 7.38. This indicates that there may be overdispersion in a Poisson regression model, so alternatives should be considered. Table 9.5 presents the estimated coefficients of a Poisson, a quasi-Poisson and a negative binomial regression model.

---

[6] The quasi-Poisson model assumes that $Var[y_i] = \varphi^2.E[y_i]$ where φ is an overdispersion parameter. An estimator for φ^2 is $\hat{\varphi}^2 = \frac{1}{n-k-1}\sum \frac{(y_i - \hat{y}_i)^2}{\hat{y}_i}$. Standard errors for the quasi-Poisson coefficients can then be obtained by multiplying those of the Poisson MLE by $\hat{\varphi}$.

[7] An extension that could further improve the fit is the Generalized Negative Binomial Regression that models the overdispersion parameter (see the gnbreg command in Stata).

[8] $AIC = -2\ln(likelihood) + 2k$, where k is the number of parameters estimated in the model. So models with lower values for the AIC are to be preferred. Unlike R^2, the AIC is a relative measure and is only useful for comparing models on the same data.

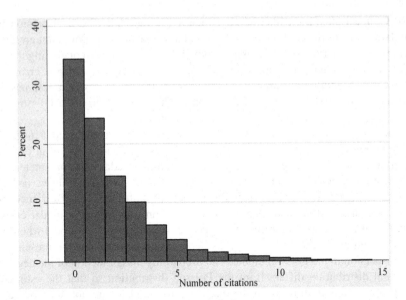

Fig. 9.1 Frequency distribution of the number of citations

In the Poisson model the estimated coefficient for the JOI is 0.47, which implies that—holding the other variables constant—the predicted number of citations of JOI articles is 47 % higher than that of JASIST articles. Articles about the h-index have 66 % more citations than other articles. Note that the coefficient estimates in the quasi-Poisson model are identical to those of the Poisson model. As could be expected the standard errors in the quasi-Poisson are substantially greater than those in the Poisson model. In fact they all are 77 % greater because the overdispersion statistic was $\hat{\varphi} = 1.77$ (not in the table). This indicates that the Poisson distribution assumption was violated and that the Poisson model should not be used for inference. For example, it would be wrong to conclude that the effect of the variable "First" is significant, since that result in the Poisson model is based on underestimated standard errors. Significance tests in the quasi-Poisson model show a positive effect of articles published in the JOI and a negative effect for articles published in RE (compared to articles in JASIST). We also find significant positive effects from international collaboration, the number of cited references, articles about the h-index and the publication month control variable. The results of the negative binomial model are very similar to those of the quasi-Poisson: the same effects are significant at the same significance levels and with very similar estimated effect sizes. For example, the predicted number of citations for an article on the h-index is 70 % higher than for other articles, whereas the effect of one additional reference is 0.7 %. Hence 100 additional references had the same effect than switching to an h-index related topic, which illustrates the effect of the h-bubble (Rousseau et al., 2013). The estimated dispersion parameter α is 0.88

Table 9.5 Count data models

	(1) Poisson estimate/(SE)	(2) Quasi-Poisson estimate/(SE)	(3) Negative binomial estimate/(SE)
JOI	0.470***	0.470***	0.549***
	(0.052)	(0.092)	(0.097)
RE	−0.716***	−0.716***	−0.690***
	(0.104)	(0.183)	(0.141)
RP	−0.079	−0.079	−0.046
	(0.053)	(0.094)	(0.090)
SM	0.094*	0.094	0.133
	(0.043)	(0.077)	(0.073)
NumAut	0.006	0.006	−0.000
	(0.012)	(0.022)	(0.020)
NumCoun	0.190***	0.190***	0.229***
	(0.022)	(0.039)	(0.044)
NumRef	0.007***	0.007***	0.007***
	(0.001)	(0.001)	(0.001)
NumPag	−0.009*	−0.009	−0.011
	(0.004)	(0.007)	(0.006)
NumTitle	−0.001	−0.001	−0.001
	(0.000)	(0.001)	(0.001)
First	0.148**	0.148	0.151
	(0.054)	(0.096)	(0.095)
Fund	−0.127**	−0.127	−0.122
	(0.045)	(0.079)	(0.072)
PubYear	0.009	0.009	0.023
	(0.014)	(0.025)	(0.024)
PubMon	−0.056***	−0.056***	−0.062***
	(0.004)	(0.008)	(0.008)
H	0.655***	0.655***	0.691***
	(0.049)	(0.087)	(0.097)
InnoPat	−0.072	−0.072	−0.099
	(0.047)	(0.083)	(0.080)
Constant	−16.849	−16.849	−46.448
	(28.634)	(50.618)	(48.594)
α			0.88
R-squared	0.09	0.09	0.08
AIC	9,854	9,854	8,312
N	2,271	2,271	2,271

*p<0.05; **p<0.01; ***p<0.001

indicating overdispersion. A likelihood-ratio test that compares the negative bino-mial model with a model where α is zero (the Poisson model) confirms that the overdispersion parameter is significant ($\chi^2 = 1544$, $p < .001$) so that the Poisson model is not reliable and the quasi-Poisson or the negative binomial should be preferred.

In order to evaluate the goodness-of-fit we calculated Pearson correlation coefficients between the observed number of citations and the predicted counts in the (quasi-) Poisson model ($r = 0.300$) and those in the negative binomial model ($r = 0.290$), resulting in values for pseudo-R^2 of 0.09 and 0.08 respectively. On the other hand, the Akaike information criterion (AIC) indicates a better fit in the negative binomial model, which has a smaller value for the AIC.

The effects that are found in the count data models are mostly the same effects that we found earlier in the categorical (binary and ordinal) models. This indicates that the main results of the analysis are robust to alterations in the model specification. Yet while the results were fairly robust, each approach did yield additional insights that might have been overlooked had only one approach been used. For example, the explanatory variable First, indicating whether an article is the first in a journal issue or not, did have a significant effect in the ordinal model, but not in the binary model nor in the count data models. With regard to the effect of the journals, one would draw similar conclusions from each of the models for JOI and for RE (the first yielding higher citation impact during the year following publication than papers in JASIST; the latter resulting in lower such citation impact). For RP and SM, however, a comparison of the results of the different models leads to a more nuanced idea as regards the citation impact of their papers in comparison with papers in JASIST.

9.6   Limited Dependent Variable Models in Stata

The data used for the analyses presented above are available via the publisher's webpage. We now show how our results can be obtained using the statistical software Stata.

To estimate the binary logit model where the dummy variable "D_cited" indicates the outcome of an article being cited and in which JASIST (the second journal) is the reference category:

```
logit D_cited ib2.Journal NumAut NumCoun NumRef NumPag NumTitle i.
    First i.Fund PubYear PubMon i.H i.InnoPat
```

A prediction table, the adjusted count R^2, odds ratios and marginal effects at the means in the binary logit model were obtained by:

```
estat classification
fitstat
logit D_cited ib2.Journal NumAut NumCoun NumRef NumPag NumTitle i.
    First i.Fund PubYear PubMon i.H i.InnoPat, or
margins, dydx(*) atmeans
```

To reduce the code needed to estimate the other models, we first define a list of independent variables which we call "indeps":

```
local indeps JOI SM RP RE NumAut NumCoun NumRef NumPag NumTitle First
    Fund PubYear PubMon H InnoPat
```

For the estimation of coefficients and odds ratios in the ordered logit model where the categorical variable "citation_categories" contains the three outcome categories:

```
ologit citation_categories 'indeps'
ologit citation_categories 'indeps', or
brant, detail
```

The Poisson, quasi-Poisson, and negative binomial models for the count variable "citations" are obtained by[9]:

```
poisson citations 'indeps'
glm citations 'indeps', family(poisson) link(log) scale(x2)
nbreg citations 'indeps'
```

Conclusion

Outcome variables that are categorical or frequency counts are common in informetrics. This chapter introduced and compared common limited dependent variable regression models that can be used to analyze such data. The use of linear models may often not be justified in informetrics, as the assumptions underlying them often do not apply in informetric datasets (Leydesdorff & Bensman, 2006). A practical issue for researchers is to decide which of the limited dependent variable models and their variations is most appropriate. In many cases the nature of the data will determine that choice (e.g., if the outcome variable is binary then there are no other options than to estimate a binary model). But sometimes the data will offer different options for modelling, as in the example of citations counts used throughout this chapter. In this case the researcher might strive to maximally exploit the information and variation in the data by avoiding to group observations into broader categories. However, there may be valid reasons for estimating categorical models

(continued)

---

[9] These models are part of a broader class of Generalized Linear Models (GLM). The quasi-Poisson model is estimated in Stata as a GLM in which the standard errors are adjusted ("scaled") using the Pearson chi-square ("x2") of the observed and predicted values in the model (i.e., the estimated overdispersion parameter $\hat{\varphi}$ that we discussed earlier).

(continued)

in those cases too (e.g., if aggregated categories are considered more appropriate for a certain research question). In such cases it may be instructive to estimate and compare different models. Yoshikane (2013), for example, used linear, logistic as well as zero-inflated negative binomial models to analyze patent citation frequencies; Niu and Hemminger (2012) ran a Poisson and two logistic models in their analysis of information seeking behavior. Altering the specification is a way to check the robustness of the main results of a study and to detect interesting anomalies in the data.

Acknowledgements The authors thank Fereshteh Didegah, Raf Guns, Edward Omey, and Ronald Rousseau for their suggestions during the writing of this chapter. We also thank Richard Williams and Paul J Wilson for their feedback and excellent suggestions.

References

Abbasi, A., Altmann, J., & Hossain, L. (2011). Identifying the effects of co-authorship networks on the performance of scholars: A correlation and regression analysis of performance measures and social network analysis measures. *Journal of Informetrics, 5*, 594–607.

Acosta, M., Coronado, D., Marín, R., & Prats, P. (2013). Factors affecting the diffusion of patented military technology in the field of weapons and ammunition. *Scientometrics, 94*, 1–22.

Agresti, A. (2002). *Categorical data analysis* (2nd ed.). New York, NY: Wiley.

Agresti, A. (2010). *Analysis of ordinal categorical data* (2nd ed.). New York, NY: Wiley.

Barjak, F., & Robinson, S. (2007). International collaboration, mobility, and team diversity in the life sciences: Impact on research performance. In D. Torres-Salinas & H. F. Moed (Eds.), *Proceedings of ISSI 2007* (pp. 63–73). Madrid: ISSI.

Bornmann, L., & Daniel, H. D. (2006). Selecting scientific excellence through committee peer review—A citation analysis of publications previously published to approval or rejection of post-doctoral research fellowship applicants. *Scientometrics, 68*, 427–440.

Bornmann, L., & Daniel, H.-D. (2008). Selecting manuscripts for a high-impact journal through peer review: A citation analysis of communications that were accepted by *Angewandte Chemie International Edition*, or rejected but published elsewhere. *Journal of the American Society for Information Science and Technology, 59*, 1841–1852.

Bornmann, L., Mutz, R., & Daniel, H.-D. (2013). Multilevel-statistical reformulation of citation-based university rankings: The Leiden ranking 2011/2012. *Journal of the American Society for Information Science and Technology, 64*, 1649–1658.

Bornmann, L., & Williams, R. (2013). How to calculate the practical significance of citation impact differences? An empirical example from evaluative institutional bibliometrics using adjusted predictions and marginal effects. *Journal of Informetrics, 7*, 562–574.

Chen, C. (2012). Predictive effects of structural variation on citation counts. *Journal of the American Society for Information Science and Technology, 63*, 431–449.

Didegah, F., & Thelwall, M. (2013a). Determinants of research citation impact in nanoscience and nanotechnology. *Journal of the American Society for Information Science and Technology, 64*, 1055–1064.

Didegah, F., & Thelwall, M. (2013b). Which factors help authors produce the highest impact research? Collaboration, journal and document properties. *Journal of Informetrics, 7*, 861–873.

Engels, T. C. E., Goos, P., Dexters, N., & Spruyt, E. H. J. (2013). Group size, h-index and efficiency in publishing in top journals explain expert panel assessments of research group quality and productivity. *Research Evaluation, 22*, 224–236.

Fedderke, J. W. (2013). The objectivity of national research foundation peer review in South Africa assessed against bibliometric indexes. *Scientometrics, 97*, 177–206.

Gantman, E. R. (2012). Economic, linguistic, and political factors in the scientific productivity of countries. *Scientometrics, 93*, 967–985.

Greene, W. H. (2011). *Econometric analysis* (7th ed.). Upper Saddle River, NJ: Prentice Hall.

Heinze, T., & Bauer, G. (2007). Characterizing creative scientists in nano-S&T: Productivity, multidisciplinarity, and network brokerage in a longitudinal perspective. *Scientometrics, 70*, 811–830.

Hilbe, J. M. (2011). *Negative binomial regression* (2nd ed.). Cambridge, UK: Cambridge University Press.

Hoekman, J., Frenken, K., & van Oort, F. (2009). The geography of collaborative knowledge production in Europe. *Annals of Regional Science, 43*, 721–738.

Jensen, P., Rouquier, J.-B., & Croissant, Y. (2009). Testing bibliometric indicators by their prediction of scientists promotions. *Scientometrics, 78*, 467–479.

Lee, Y. G. (2008). Patent licensability and life: A study of US patents registered by South Korean public research institutes. *Scientometrics, 75*, 463–471.

Lee, Y.-G., Lee, J.-D., Song, Y.-I., & Lee, S.-J. (2007). An in-depth empirical analysis of patent citation counts using zero-inflated count data model: The case of KIST. *Scientometrics, 70*, 27–39.

Leydesdorff, L., & Bensman, S. (2006). Classification and powerlaws: The logarithmic transformation. *Journal of the American Society for Information Science and Technology, 57*, 1470–1486.

Long, J. S., & Freese, J. (2006). *Regression models for categorical dependent variables using Stata*. College Station, TX: Stata Press.

Ma, Z., Sun, A., & Cong, G. (2013). On predicting the popularity of newly emerging hashtags in Twitter. *Journal of the American Society for Information Science and Technology, 64*, 1399–1410.

Menard, S. (1995). *Applied logistic regression analysis*. Thousand Oaks, CA: Sage.

Niu, X., & Hemminger, B. M. (2012). A study of factors that affect the information-seeking behavior of academic scientists. *Journal of the American Society for Information Science and Technology, 63*, 336–353.

O'Brien, R. M. (2007). A caution regarding rules of thumb for variance unflation factors. *Quality & Quantity, 41*, 673–690.

Rigby, J. (2013). Looking for the impact of peer review: Does count of funding acknowledgements really predict research impact? *Scientometrics, 94*, 57–73.

Rokach, L., Kalech, M., Blank, I., & Stern, R. (2011). Who is going to win the next Association for the Advancement of Artificial Intelligence fellowship award? Evaluating researchers by mining bibliographic data. *Journal of the American Society for Information Science and Technology, 62*, 2456–2470.

Rousseau, R., Garcia-Zorita, C., & Sanz-Casado, E. (2013). The h-bubble. *Journal of Informetrics, 7*, 294–300.

Sin, S.-C. J. (2011). International coauthorship and citation impact: A bibliometric study of six LIS journals, 1980–2008. *Journal of the American Society for Information Science and Technology, 62*, 1770–1783.

Su, H. N., Chen, C. M. L., & Lee, P. C. (2012). Patent litigation precaution method: Analyzing characteristics of US litigated and non-litigated patents from 1976 to 2010. *Scientometrics, 92*, 181–195.

Vakkari, P. (2012). Internet use increases the odds of using the public library. *Journal of Documentation, 68*, 618–638.

Van Dalen, H. P., & Henkens, K. (2005). Signals in science—On the importance of signaling in gaining attention in science. *Scientometrics, 64*, 209–233.

Verbeek, M. (2008). *A guide to modern econometrics*. New York, NY: Wiley.

Walters, G. D. (2006). Predicting subsequent citations to articles published in twelve crime-psychology journals: Author impact versus journal impact. *Scientometrics, 69*, 499–510.

Wooldridge, J. M. (2012). *Introductory econometrics: A modern approach* (5th ed.). Andover, MA: Cengage Learning.

Wooldridge, J. (1997). Quasi-likelihood methods for count data. In M.H. Pesaran and P. Schmidt (Eds.), *Handbook of applied econometrics* (Vol 2 pp. 352–406). Oxford: Blackwell.

Yoshikane, F. (2013). Multiple regression analysis of a patent's citation frequency and quantitative characteristics: The case of Japanese patents. *Scientometrics, 96*, 365–379.

Chapter 10
Text Mining with the Stanford CoreNLP

Min Song and Tamy Chambers

Abstract Text mining techniques have been widely employed to analyze various texts from massive social media to scientific publications and patents. As a bibliographic analysis tool the technique presents the opportunity for large-scale topical analysis of papers covering an entire domain, country, institution, or specific journal. For this project, we have chosen to use the Stanford CoreNLP parser due to its extensibility and enriched functionalities which can be applied to bibliometric research. The current version includes a suite of processing tools designed to take raw English language text input and output a complete textual analysis and linguistic annotation appropriate for higher-level textual analysis. The data for this project includes the title and abstract of all articles published in the Journal of the American Society for Information Science and Technology (JASIST) in 2012 ($n = 177$). Our process will provide an overview of the concepts depicted in the journal that year and will highlight the most frequent concepts to establish an overall trend for the year.

10.1 Introduction

Since Feldman and Dagan (1995) introduced *Text Data Mining*, the technique has been widely employed to analyze texts from massive social media to scientific publications and patents. Text mining is the process by which naturally occurring text is analyzed for the purpose of discovering and capturing useful information contained within. As such, text mining utilizes different algorithms to identify interesting patterns from a vast array of available data. While it employs similar exploratory data analysis functions as data mining, text mining also relies on the

M. Song (✉)
Department of Library and Information Science, Yonsei University, Seoul, South Korea
e-mail: min.song@yonsei.ac.kr

T. Chambers
Department of Information and Library Science, School of Informatics and Computing, Indiana University, Bloomington, IN, USA
e-mail: TISCHT@INDIANA.EDU

© Springer International Publishing Switzerland 2014
Y. Ding et al. (eds.), *Measuring Scholarly Impact*,
DOI 10.1007/978-3-319-10377-8_10

application of techniques and methodologies from the areas of information retrieval, information extraction, computational linguistics, and natural language processing to process the text-based corpus (Feldman & Sanger, 2007).

One, therefore, employs similar procedures as the traditional data mining process, while changing the focus of the analysis from general data to text documents. This change of focus results in new questions and challenges. The first, and main, challenge relates to problems that arise—from the data modeling perspective— because of the use of unstructured data sets. To address this, research in the field has employed various traditional techniques, such as text representation, classification, clustering, and information extraction to search for hidden patterns and important concepts or themes within the text. In this context, both the selection of characteristics and the influence of domain knowledge and domain-specific procedures play a crucial role in the performance of mining techniques.

The major methodologies of text mining include automatic classification, automatic extraction, and link analysis (Feldman & Sanger, 2007). For this reason researchers have often asserted the synergy between text mining and bibliometric study (Kostoff, del Río, Humenik, García, & Ramírez, 2001). As a bibliographic analysis technique, therefore, text mining presents the opportunity to analyze the paper production of an entire domain, country, institution, or in the case presented here, a journal. Widely used in business and the life sciences, it has also become a standard technique in the broader study of informetrics (Bar-Ilan, 2008).

In this chapter, we first present a history of the use of text mining in bibliometrics, followed by an overview of the architecture of a text mining system and its capabilities. We then introduce the Stanford CoreNLP parser and present an example of its use for bibliographic analysis by analyzing the title and abstracts of 177 papers published in the *Journal of the American Society for Information Science and Technology* in 2012. Finally, we conclude with a discussion of possible future uses of text mining for bibliographic analysis.

10.2   Text Mining in Bibliometric Research

The method of scientific domain analysis, through co-occurring keywords or title terms, was developed by Callon, Courtial, Turner, and Bauin (1983) in the late 1980s and extended by themselves (Callon, Courtial, & Laville, 1991) and other researchers throughout the early 1990s (Van Raan & Tijssen, 1993; Zitt, 1991; Zitt & Bassecoulard, 1994). However, it was Feldman and Dagan (1995) who first introduced *Text Data Mining* as a knowledge discovery tool. Their framework was based on a concept hierarchy, categorization of texts by concept, and comparison of the concept distribution to identify unexpected patterns within the Reuters-22173 text categorization test collection (Feldman & Dagan, 1995; Feldman, Klösgen, & Ziberstein, 1997).

Around the same time, Ronald Kostoff, a Navel researcher, demonstrated though multiple studies that the "frequency with which phrases appeared in full-text narrative technical documents was related to the main themes of the text" (Kostoff, Toothman, Eberhart, & Humenik, 2001, p. 225). He and his colleagues patented a technique called *database tomography* (Kostoff, Miles, & Eberhart, 1995) which mined words from article abstracts and then employed more traditional bibliometric methods to analyze the results and identify the research domain and agenda. Over the following decade Kostoff and others would use this method to analyze various domains based on their literature, including near-earth space (Kostoff, Eberhart, & Toothman, 1998), chemistry (Kostoff, Eberhart, Toothman, & Pallenbarg, 2006), and aircraft science (Kostoff, Green, Toothman, & Humenik, 2000).

The most widely used technique in bibliometric text mining focuses on co-occurring keywords extracted from titles, abstracts, or even full-text analysis, whether it is extended further to co-word, coheading or coauthor clustering (Janssens, Leta, Glänzel, & De Moor, 2006). As such, many researchers have used this method to investigate various domains from the natural sciences (De Looze & Lemarie, 1997), to information retrieval (Ding, Chowdhury, & Foo, 1999), and to medicine (Onyancha & Ocholla, 2005). Other researchers have used text mining to extract keywords from titles and combined this with co-word analysis to identify potential relationships between and the meanings of concepts within different contexts, across different domains, and through different mediums (Leydesdorff & Hellsten, 2005; Onyancha & Ocholla, 2005).

Text mining has also been combined with citation analysis to mine citations from the text in a process called citation mining. Kostoff et al. (2001) used this method to identify researcher profiles. Porter, Kongthon, and Lu (2002) proposed a similar research profiling strategy to enhance traditional literature reviews by identifying topical relationships and research trends. More recently, researchers have used both descriptor profiling and journal profiling to investigate archiving research trends (Kim & Lee, 2009) and digital library research (Lee, Kim, & Kim, 2010) in the library and information science domain. Liu, Zhang, and Guo (2012) also used citation mining identify the most significant publications by topic.

Text mining has long been a standard technique in patent analysis. Lent, Agrawal, and Srikant (1997) used a large patent collection to create the PatentMiner system which identified sequential patterns and shape queries through text mining techniques to analyze and visualize trends among patents. Bhattacharya, Kretschmer, and Meyer (2003) used text mining to identify co-words and citations in their study of the connections between scientific literature and patent documents. Others have used text mining to identify templates (Lawson, Kemp, Lynch, & Chowdhury, 1996), features (Tseng, Wang, Lin, Lin, & Juang, 2007), and citation patterns (Li, Chambers, Ding, Zhang, & Meng, 2014) within patents.

Glenisson and his colleagues (Glenisson, Glänzel, Janssens, & De Moor, 2005; Glenisson, Glänzel, & Persson, 2005) combined full-text analysis and traditional bibliometric methods to create a hybrid approach to text analysis. His work confirmed that such a methodology was effective in research evaluation. Song and Kim (2013) more recently used full-text mining to build a citation database of PubMed articles to study the knowledge structure of the bioinformatics field.

Liu et al. (2012) recently used full-text mining of citations in their study identifying the most signification publications by topic.

Bibliometric research using text mining has also moved from co-word analysis to word and document clustering. Kostoff et al. (2007) used document clustering to identify the technical structure of the Mexican science and technology literature and Janssens et al. (2006) used a variety of clustering methods to map the library and information science domain. Similarly, Kim and Lee (2008) used document clustering to explore the emerging intellectual structure of archival studies. Other researchers have focused on refining the algorithms used to cluster documents, such as Liu et al. (2010) who proposed a hybrid clustering framework to analyze journal sets and Janssens, Glänzel, and De Moor (2008) who proposed an approach based on Fisher's inverse chi-square.

10.3   Text Mining System Architecture

At the most basic level text mining systems input raw natural language documents and output patterns, connections, and tends related to those to documents. During the initial stages, unstructured natural language data is converted into analyzable structured data. Natural language processing is used to tokenize (remove punctuation), filter (removal of words with little significance), lemmatize (converting verbs to their infinitive tense and nouns to their singular form), or stem (stripping word endings *ing*, *ed*, *er*, etc.) the text prior to analysis (Hotho, Nürnberger, & Paaß, 2005).

Natural language processing (NLP) is a computational technique for the automatic analysis and representation of human language. The goal of NLP research is to implement intelligent techniques to understand normal human language (Cambria & White, 2014). As such, this research has evolved from the era of batch and manual processing to the era of Google and big data-driven companies. Although semantics advocates argue a transition from syntax-oriented NLP to semantics-oriented NLP is crucial and inevitable, the vast majority of NLP studies continue to utilize a syntax approach. Either shallow or full-text parsing may be used to assign syntactic structure. Shallow parsing includes probabilistic approaches such as memory-based parsing or the use of statistical decision trees (Rajman & Vesely, 2004), while full parsing requires constituency or dependency grammars, the latter of which uses dependency graphs where words are represented as nodes and the relationships between them exist as edges (Feldman & Sanger, 2007).

Keyword extraction is the most basic and popular approach within NLP due to its accessibility. The Penn Treebank (Marcus, Santorini, & Marcinkiewizc, 1993) is a corpus consisting of over 4.5 million words for American English annotated for part-of-speech (POS) information and is based on keyword extraction. Similarly, PageRank (Page, Brin, Motwani, & Winograd, 1999), the famous Google ranking algorithm, LexRank (Gunes & Radev, 2004), a stochastic graph-based method for computing the relative importance of textual units, and TextRank (Mihalcea & Tarau, 2004), a graph-based ranking model for text processing, are all based on NLP keyword and sentence extraction.

Statistical NLP, a version of syntax-oriented NLP, has been the mainstream NLP technique used in research since the late 1990s. The method relies on language models and is based on popular machine-learning algorithms, such as Maximum-Likelihood (Berger, Pietra, & Pietra, 1996), Expectation Maximization (Nigam, McCallum, Thrun, & Mitchell, 2000), Conditional Random Fields (Lafferty, McCallum, & Pereira, 2001), and Support Vector Machines (Joachims, 2002). By using a machine-learning algorithm over a large training corpus of annotated texts, the NLP system cannot only learn the valence of keywords (as in the keyword spotting approach), but also account for the valence of other arbitrary keywords, punctuation, and word co-occurrence frequencies. Note that advanced keyword clustering techniques such as topic modeling are discussed in Chap. 11.

Statistical methods, however, lack the semantics necessary to give the method effective predictive value individually. While statistical NLP works well for large text input, it does not work as well on smaller text units such as sentences or clauses. Semantic-based NLP is needed to focus on the intrinsic meaning associated with natural language text (Sebastiani, 2002). Rather than simply processing documents at a syntax-level, semantics-based approaches rely on implicit denotative features associated with natural language text, and thus avoid the blind use of keywords and word co-occurrence count. Linguistic processing techniques, such as part-of-speech tagging which annotates each term based on its role in the sentence (noun, verb, adjective, proper noun, etc), text chunking which groups adjacent words such as "government shutdown," and Word Sense Disambiguation (WSD) which uses the term meaning instead of the single term ("financial institution" instead of "bank") to provide deeper semantic representation for each term (Feldman & Sanger, 2007; Hotho et al., 2005). Concept-based approaches are also able to detect the semantics expressed in a more subtle manner, such as through the analysis of concepts that do not explicitly convey relevant information, but which are implicitly linked to other concepts that do so. Semantic-based NLP approaches usually either leverage techniques based on external knowledge sets process (e.g., ontologies) (Suchanek, Kasneci, & Weikum, 2007) or semantic knowledge bases (Cambria, Rajagopal, Olsher, & Das, 2013).

After this process, the system moves from a "machine readable representation of the documents to a machine understandable form of the documents" (Feldman & Sanger, 2007) by structuring the document collection using clustering, classification, relation extraction, and entity extraction techniques.

Document clustering uses an unsupervised learning process to group unlabeled documents into meaningful document clusters where documents similar to one another are grouped within the same cluster, without any prior information about the document set. Although a number of document clustering approaches have been developed over the years (Rajman & Vesely, 2004), most approaches are based on the vector space representation, hierarchical, or partitional approaches (Aggarwal

& Zhai, 2012; Cutting, Karger, & Pederson, 1993; Lin & Demner-Fushman, 2007; Wang, McKay, Abbass, & Barlow, 2002).

Initially document clustering was investigated as a means of improving information retrieval (IR) performance (Wang et al., 2002). However, it has more recently been used to facilitate nearest-neighbor search (Aggarwal, Zhao, & Yu, 2012), to support an interactive document browsing paradigm (Cutting et al., 1993), and to construct hierarchical topic structures (Ming, Wang, & Chua, 2010). In the biomedical domain, Lin and Demner-Fushman (2007) introduced an interesting semantic document clustering approach that automatically clusters biomedical literature (MEDLINE) search results into document groups for greater understanding of literature search results. Unlike traditional document clustering methods, semantic clustering techniques provide a coherent summary of the collection in the form of word-clusters (Bekkerman, El-Yaniv, Tishby, & Winter, 2001), which can be used to provide summary insight into the overall content of the underlying corpus. Variants of such methods, especially sentence clustering, can also be used for document summarization.

Classification techniques, on the other hand, assign classes to text documents using index term selection, probabilistic classifiers, nearest neighbor classifiers, decision tree classifiers, or supervised classification algorithms (Hotho et al., 2005). Document classification, which assumes categorical values for the labels, has been widely studied in the database, data mining, and information retrieval communities. Document classification has also found application in a wide variety of domains such as opinion mining, email classification, and news filtering. Opinion mining mines customer reviews or opinions, often short text documents, to determine useful information from the review (Brody & Elhadad, 2010; Ding, Liu, & Zhang, 2009). Email classification and spam filtering classifies email (Carvalho & Cohen, 2005) in order to determine either the subject or to determine if it is a junk email (Cui, Mondal, Shen, Cong, & Tan, 2005) in an automated way. News filtering, or text filtering (Du, Safavi-Naini, & Susilon, 2003), is used by most news services today to organize the large volume of news articles created by news organizations on a daily basis. Volume prohibits this being done manually, and automated methods have been very useful for categorization in a variety of Web portals (Hepple et al., 2004; Lang, 1995).

Relation extraction reduces information loss by extracting pairs of entities (employee–employer, organization–location) using morphological analysis and shallow parsing, while entity extraction assigns predefined labels to textual entities with interesting semantic properties such as company names, dates, phone numbers, etc. (Rajman & Vesely, 2004). Recent research has used text mining to extract the predefined named entities (genes, drugs, diseases) and build entity relationships based on citations to discover drug relations (Ding et al., 2013; Song, Han, Kim, Ding, & Chambers, 2013).

As a final step, many text-mining systems display results as visualizations for further exploratory data analysis. Visualizations provide analyzable view of the analytical results, as well as, a summary overview of the whole document collection. Similar to data analysis, results are most often presented in geometric representations (scatterplots, matrices, etc.), but can also be presented as pixel-oriented (recursive patterns, circle segments, etc.), icon-oriented (stick figures, shape coding, etc.), or hierarchical (treemap, Venn-diagram, etc.) representations (Rajman & Vesely, 2004). Self-organizing maps are often used to visualize document collections as they allow for low-dimensional clusters to be arranged based on a topology that preserves high-dimensional neighborhood relations. That is, documents are clustered by similarity, but are also mapped close to other similar clusters (Hotho et al., 2005).

10.4   The Stanford CoreNLP Parser

For this project, we have chosen to use the Stanford CoreNLP parser. This parser is a statistical, unlexicalized, natural language parser trained on the Wall Street Journal (De Marneff, MacCartney, & Manning, 2006; Klein & Manning, 2003a, 2003b). The current version (Stanford CoreNLP version 3.3.0) includes a suite of processing tools designed by the Stanford Natural Language Processing group with the goal of taking raw English language text input and outputting a complete textual analysis and linguistic annotation appropriate for higher-level textual analysis and understanding. As such, the suite includes tools which give "the base forms of words, their parts of speech, whether they are names of companies, people, etc., normalize dates, times, and numeric quantities, and mark up the structure of sentences in terms of phrases and word dependencies, and indicate which noun phrases refer to the same entities" (The Stanford Natural Language Processing Group, 2013). To accomplish this, the Stanford CoreNLP uses annotations, which structure and map the data, and annotators, which serve as functions over the annotations. The annotations supported by the Stanford CoreNLP are summarized in Table 10.1.

The Stanford CoreNLP is actually a suite of tools developed by the Stanford Natural Language Processing group, all of which can be downloaded as individual tools and which are summarized in Table 10.1. In addition to the parser, the suite includes the following: a part-of-speech tagger (Toutanova, Klein, Manning, & Singer, 2003) which reads in text and assigns a part-of-speech to each word (noun, verb, adjective, etc.) as denoted by an abbreviation from the Penn Treebank tag set (Marcus, Marcinkiewicz, & Santorini 1993a, 1993b); a named entity recognizer (NER), also known as CRFClassifier (Finkel, Grenager, & Manning, 2005), which uses a liner chain sequence model to identify and label sequences of words in the text which are identifiable things (people, company names, locations, etc.); a co-reference resolution system (Lee et al., 2011); and sentiment analysis tools (Socher et al., 2013), which employ a deep learning model to build a representation of the whole sentence based on structure and then computes the sentiment based on word composition meaning in longer phrases.

Table 10.1 Stanford CoreNLP-supported property annotators, the annotations they generate, and description

Property	Annotator class	Generated annotation
tokenize	PTBTokenizerAnnotator	TokensAnnotation (list of tokens), and Character-OffsetBeginAnnotation, CharacterOffsetEndAnnotation, TextAnnotation (for each token)
	Tokenizes the text.	
cleanxml	CleanXmlAnnotator	XmlContextAnnotation
	Remove xml tokens from the document.	
ssplit	WordToSentenceAnnotator	SentencesAnnotation
	Splits a sequence of tokens into sentences.	
pos	POSTaggerAnnotator	PartOfSpeechAnnotation
	Labels tokens with their POS tag.	
lemma	MorphaAnnotator	LemmaAnnotation
	Generates the word lemmas for all tokens in the corpus.	
ner	NERClassifierCombiner	NamedEntityTagAnnotation and NormalizedNamedEntityTagAnnotation
	Recognizes named (person, location, organization, misc) and numerical entities (date, time, money, number).	
regexner	RegexNERAnnotator	NamedEntityTagAnnotation
	Implements a simple, rule-based NER over token sequences using Java regular expressions to incorporate NE labels that are not annotated in traditional NL corpora.	
sentiment	SentimentAnnotator	SentimentCoreAnnotations.AnnotatedTree
	Implements Socher et al.'s sentiment model to attach a binarized tree of the sentence to the sentence level CoreMap. The nodes of the tree indicate the predicted class and scores for that subtree.	
truecase	TrueCaseAnnotator	TrueCaseAnnotation and TrueCaseTextAnnotation
	Recognizes the true case of tokens in text where this information was lost, e.g., all upper case text.	
parse	ParserAnnotator	TreeAnnotation, BasicDependenciesAnnotation, CollapsedDependenciesAnnotation, CollapsedCCProcessedDependenciesAnnotation
	Provides full syntactic analysis, using both the constituent and the dependency representations.	
dcoref	DeterministicCorefAnnotator	CorefChainAnnotation
	Implements both pronominal and nominal coreference resolution.	

Adapted from the Stanford CoreNLP website, http://nlp.stanford.edu/software/corenlp.shtml

The entire Stanford CoreNLP suite is written in Java and licensed under the GNU General Public License. It requires Java version 1.6 or higher and is recommended to run on a 64-bit machine, as it requires at least 3 GB of memory depending on the file size to parse. The suite can be downloaded from http://nlp.stanford.edu/downloads/corenlp.shtml as a 215 MB zip file.

10.5   An Example of Text Mining for Bibliometric Analysis

As an example of the use of text mining in a bibliometric study, we have chosen to use a dataset of all articles published in the *Journal of the American Society of Information Science and Technology* (JASIST) in 2012 ($n = 177$). For each of these articles we have extracted the title and abstract for further analysis. Our process will provide an overview of the concepts depicted in the journal that year and will highlight the most frequent concepts to establish an overall trend for the year.

To analyze this we will use, in addition to the Stanford CoreNLP, the MetricsConversionHandler program (http://informatics.yonsei.ac.kr/stanford_met rics/stanford_metrics.zip) and Gephi (https://gephi.org/) for visualizing the results. Both of these programs are open source and can be downloaded from the provided links.

Our workflow, as visualized in Fig. 10.1, begins with the Stanford CoreNLP. This is used to provide the text analysis of the text file **jasist_2012.txt**, which contains, as described above, the titles and abstract of all articles published by the

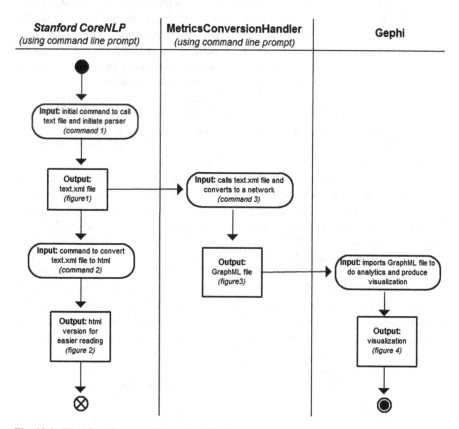

Fig. 10.1 Workflow from text file to visualization

JASIST in 2012. An output file from this process is then converted to a network file using the MetricsConversionHandler. Finally, that file is read in Gephi to identify the most connected and important nodes and ultimately to create a visualization of the topics depicted in JASIST articles in 2012.

The Stanford CoreNLP Process: After downloading and uncompressing the Stanford CoreNLP parser file, one will need to open a command line, and change the path to the folder with the uncompressed parser files. The following command (1) will be used to initiate the program.

(1) java -cp stanford-corenlp-3.3.0.jar;stanford-corenlp-3.3.0-models.jar;xom.jar;
    joda-time.jar;jollyday.jar -Xmx3g edu.stanford.nlp.pipeline.StanfordCoreNLP
    -annotators tokenize,ssplit,pos,lemma,ner,parse,dcoref -file jasist_2012.txt

The first part of command (1), "**java -cp stanford-corenlp-3.3.0.jar;stanford-corenlp-3.3.0-models.jar;xom.jar;joda-time.jar;jollyday.jar -Xmx3g edu.stanford.nlp.pipeline.StanfordCoreNLP**" calls java and the specific jars used to execute the program. The option "**-Xmx3g**" specifies the amount of computer memory that Java reserves. On a 64-bit machine, the Stanford CoreNLP typically requires 3 GB to run depending on the size of the document to parse. For Windows, the semi-colon (;) is used for separating jars whereas the colon (:) is used in Mac OS X or Linux. The next part of command (1), "**-annotators tokenize,ssplit,pos, lemma,ner,parse,dcoref**" specifies the annotators to be used on the text. This command is not mandatory if using the standard annotators as the code uses a built-in properties file, which enables the following default annotators: tokenization and sentence splitting, POS tagging, lemmatization, NER, parsing, and coreference resolution. Additional annotators are listed in Table 10.1 and instruction for their use can be found on the Stanford CoreNLP website (http://nlp.stanford.edu/down loads/corenlp.shtml). The final part of the command (1), "**-file jasist_2012.txt**" calls the input text file, which in this case is **jasist_2012.txt**.

After successful execution, the output file will be written in XML to the same directory and with the same name as the input file but with an *.xml* extension added. For each input file, the Stanford CoreNLP generates, by default, an XML file with all the relevant annotations. An example, the output for the jasist_2012.txt file (Fig. 10.2) shows each of the tokens processed by Stanford CoreNLP: (1) word, (2) lemma, (3) part of speech (POS), and (4) named entity recognition (NER). The word token identifies the word, which is delimited by whitespaces in a sentence. The Lemma token identifies the lemma, or canonical form of the word. For example, *retrieves, retrieved*, and *retrieving* are forms of the same lexeme, and have the same lemma, *retrieve*. The POS token is the linguistic category for the word, which is generally defined by its syntactic or morphological behavior in the sentence such as, noun or verb. The NER token identifies the word as an entity of a predefined category such as, the name of a person, organization, or location. The typical results of NER are a set of entity names for meaningful words extracted from each sentence.

After this, the XML output can then be rendered by the CoreNLP-to-HTML.xsl file into an easily readable HTML file. This is accomplished by running the

```
<?xml-stylesheet href="CoreNLP-to-HTML.xsl" type="text/xsl"?>
<root>
  <document>
    <sentences>
      <sentence id="1">
        <tokens>
          <token id="1">
            <word>Rapid</word>
            <lemma>rapid</lemma>
            <CharacterOffsetBegin>0</CharacterOffsetBegin>
            <CharacterOffsetEnd>5</CharacterOffsetEnd>
            <POS>JJ</POS>
            <NER>O</NER>
          </token>
          <token id="2">
            <word>understanding</word>
            <lemma>understanding</lemma>
            <CharacterOffsetBegin>6</CharacterOffsetBegin>
            <CharacterOffsetEnd>19</CharacterOffsetEnd>
            <POS>NN</POS>
            <NER>O</NER>
          </token>
          <token id="3">
            <word>of</word>
            <lemma>of</lemma>
            <CharacterOffsetBegin>20</CharacterOffsetBegin>
            <CharacterOffsetEnd>22</CharacterOffsetEnd>
            <POS>IN</POS>
            <NER>O</NER>
          </token>
          <token id="4">
```

Fig. 10.2 Example XML output results from the Stanford CoreNLP parser for the jasist_2012.txt input file

following command (2) from the command window in a similar way as the previous command.

(2) org.apache.xalan.xslt.Process -in jasist_2012.txt.xml -xsl CoreNLP-to-HTML.
    xsl -out jasist_2012.html

Note that in command (2) the **– in** command calls the previously created file (jasist_2012.txt.xml) and the **– out** command names the new html file (jasist_2012. html). The xsl style sheet used for this process comes standard with the Stanford CoreNLP package and enables a human-readable display of the XML content. Figure 10.3 shows an example of the results for jasist_2012 file presented in HTML format.

MetricsConversionHandler Process: To use these results for bibliometric study, we used a program called the MetricsConversionHandler, which converts the XML output from the Stanford CoreNLP parser. The MetricsConversionHandler program first reads the output XML (jasist_2012.txt.xml) using a SAX processing technique, then selects the terms that are nouns and in the form of a lemma as the input for the co-word analysis by POS and lemma in the XML

Stanford CoreNLP XML Output

Document

Sentences

Sentence #1

Tokens

Id	Word	Lemma	Char begin	Char end	POS	NER	Normalized NER
1	Rapid	rapid	0	5	JJ	O	
2	understanding	understanding	6	19	NN	O	
3	of	of	20	22	IN	O	
4	scientific	scientific	23	33	JJ	O	
5	paper	paper	34	39	NN	O	
6	collections	collection	40	51	NNS	O	
7	:	:	51	52	:	O	
8	Integrating	integrate	53	64	VBG	O	
9	statistics	statistics	65	75	NNS	O	
10	,	,	75	76	,	O	
11	text	text	77	81	NN	O	
12	analytics	analytic	82	91	NNS	O	
13	,	,	91	92	,	O	
14	and	and	93	96	CC	O	
15	visualization	visualization	97	110	NN	O	

Fig. 10.3 An example of the HTML output as converted from the XML for the jasist_2012 file

output. After which the program counts each pair of words per sentence and creates a GraphML file from the word pairs. An example output, using the jasist_2012.txt. xml file is shown in Fig. 10.4.

The MetricsConversionHandler program can be downloaded from http://infor matics.yonsei.ac.kr/stanford_metrics/stanford_metrics.zip. After unzipping the file, one will need to run the following command (3) from the command line to produce the corresponding GraphML file:

(3) java -Xms64m -Xmx2550m -cp .:./build:./bin:./lib/* edu.yonsei.metrics. MetricsConversionHandler jasist_2012.txt.xml jasist_2012.graphml

Note again that the MetricsConversionHandler program calls the XML file originally created by the Stanford CoreNLP (jasist_2012.txt.xml) to produce a GraphML file by the same name. The resulting GraphML file (in this case jasist_2012.graphml) can then be used as input in the Gephi visualization program (https://gephi.org/) for further analysis and visualization of the co-word network.

```
<?xml version="1.0" encoding="UTF-8"?>
<graphml xmlns="http://graphml.graphdrawing.org/xmlns"
 xmlns:xsi="http://www.w3.org/2001/XMLSchema-instance"
xsi:schemaLocation="http://graphml.graphdrawing.org/xmlns
http://graphml.graphdrawing.org/xmlns/1.0/graphml.xsd">
<key id="label" for="node" attr.name="label" attr.type="string"/>
<key id="id" for="node" attr.name="id" attr.type="string"/>
<key id="d1" for="edge" attr.name="weight" attr.type="double"/>
<graph edgedefault="undirected">
<node id="0">
<data key="label">information</data>
</node>
<node id="1">
<data key="label">study</data>
</node>
<node id="2">
<data key="label">research</data>
</node>
<node id="3">
<data key="label">article</data>
</node>
<node id="4">
<data key="label">analysis</data>
</node>
<node id="5">
<data key="label">search</data>
</node>
<node id="6">
<data key="label">journal</data>
</node>
<node id="7">
```

Fig. 10.4 The GraphML Result of MetricsConversionHandler Program

10.6 Results

Results of the analysis of the co-word network produced from the described process are presented in Table 10.2 and visualized in Fig. 10.5. The results show that the concepts of **system, citation, web, search,** and **knowledge** have the highest degrees within the network. This indicates that these concepts were the most presented that year in JASIST and that they are the concepts most associated with other concepts that year. Given that these concepts serve as information hubs in JASIST that year, they could reasonably be attributed to reflecting the general content of the journal that year. Additionally, we note that **system** and **citation** also have the highest betweenness scores which indicate that these concepts serve as bridges between other sub-concepts in the given network.

Table 10.2 Top 25 nodes listed by degree and including closeness and betweenness centrality measures

Node	Degree[a]	Closeness centrality[b]	Betweenness centrality[c]
system	72	1.37931	249.9775
citation	70	1.396552	252.0744
web	70	1.396552	212.4804
search	69	1.405172	213.9665
knowledge	67	1.422414	178.2214
method	64	1.448276	150.9842
task	60	1.482759	143.0883
user	59	1.491379	187.5244
model	58	1.5	155.9161
impact	57	1.508621	129.9933
term	57	1.508621	113.7171
publication	56	1.517241	159.7068
network	56	1.517241	148.5817
behavior	55	1.525862	104.3287
document	55	1.525862	99.88509
evidence	50	1.568966	81.63788
use	49	1.577586	86.14172
process	48	1.586207	82.94188
query	46	1.603448	81.10262
source	46	1.603448	72.36623
researcher	45	1.612069	72.14743
framework	45	1.612069	61.90369
evaluation	42	1.637931	71.98185
measure	41	1.646552	70.18668
retrieval	41	1.646552	61.44948

[a]The degree of a node is the number of links that node has with other nodes. Nodes with higher degrees act as information hubs in the network
[b]The closeness centrality of a node indicates the extent of its influence over the entire network
[c]The betweenness centrality of a node indicates the number of shortest paths that pass through that node. Nodes with high Betweenness centrality serve as bridges in a network that connect different sub groups

The visualization using Gephi shows similar results but detail the interaction between the nodes in more depth. As such we can see that outstanding themes of JASIST in 2012 were information retrieval, information network, and informetrics in the context of the Web. Specifically speaking, two major research themes are outstanding. First, information retrieval related studies interact with information network related studies. In addition, information retrieval is studied in the context of the Web. Second, citation analysis is heavily studied. Studies on citation analysis are carried out in bibliometrics, as well as, information networks such as the Web.

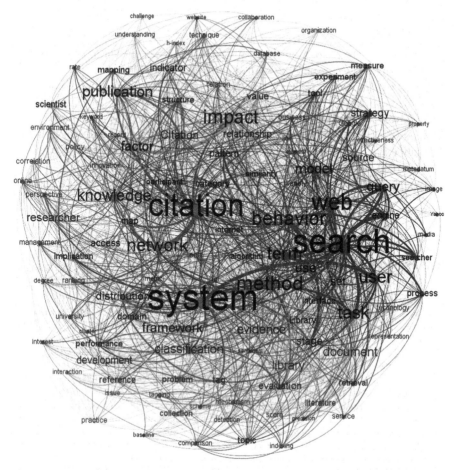

Fig. 10.5 Gephi Visualization of the Co-word Network based on JASIST Papers 2012

Conclusions

As a bibliometric tool, text mining offers the ability for large-scale topical analysis. The example provided here shows that processing large amounts of text data can easily be accomplished through the use of various programs and that the results can give researchers new insight into the topical offerings of a journal. Although there are other text mining tools available, we chose to use the Stanford CoreNLP due to its extensibility and enriched functionalities which can be applied to bibliometric research. Through the example

(continued)

(continued)

presented in the chapter, we have demonstrated how the Stanford CoreNLP
can easily be used for the co-word analysis of a set of abstracts.

Tseng, Lin, and Lin (2007) correctly summarized the text mining process
in their explanation of text mining in patent analysis: "the text mining process
involves a series of user interactions with the text mining tools to explore the
repository to find such patterns. After, supplemented with additional information and interpreted by experienced experts, these patterns can become
important intelligence for decision-making" (p. 1219). To this end, we recommend the Stanford CoreNLP processing suite and encourage interested
researchers to apply the demonstrated technique to their text mining-driven
bibliometric research.

References

Aggarwal, C. C., & Zhai, C. (2012). *Mining text data*. New York, NY: Springer.

Aggarwal, C. C., Zhao, Y., & Yu, P. S. (2012). On text clustering with side information. In *Proceedings from the 28th International Conference on Data Engineering (ICDE), 2012 IEEE* (pp. 894–904).

Bar-Ilan, J. (2008). Informetrics at the beginning of the 21st century—A review. *Journal of Informetrics, 2*, 1–52. doi:10.1016/j.joi2007.11.001.

Bekkerman, R., El-Yaniv, R., Tishby, N., & Winter, Y. (2001). On feature distributional clustering for text categorization. In *Proceedings of the 24th Annual International ACM SIGIR Conference on Research and Development in Information Retrieval (SIGIR'01)* (pp. 146–153).

Berger, A. L., Pietra, V. J. D., & Pietra, S. A. D. (1996). A maximum entropy approach to natural language processing. *Computational Linguistics, 22*(1), 39–71.

Bhattacharya, S., Kretschmer, H., & Meyer, M. (2003). Characterizing intellectual spaces between science and technology. *Scientometrics, 58*(2), 369–390. doi:10.1023/A:1026244828759.

Brody, S., & Elhadad, N. (2010). An unsupervised aspect-sentiment model for online reviews. In *Proceedings of the 2010 Annual Conference of the North American Chapter of the Association for Computational Linguistics (HLT'10: Human Language Technologies)* (pp. 804–812).

Callon, M., Courtial, J. P., Turner, W. A., & Bauin, S. (1983). From translations to problematic networks: An introduction to co-word analysis. *Social Science Information, 22*(2), 191–235. doi:10.1177/053901883022002003.

Callon, M., Courtial, J. P., & Laville, F. (1991). Co-word analysis as a tool for describing the network of interactions between basic and technological research: The case of polymer chemistry. *Scientometrics, 22*(1), 155–205. doi:10.1007/BF02019280.

Cambria, E., Rajagopal, D., Olsher, D., & Das, D. (2013). Big social data analysis. In R. Akerkar (Ed.), *Big data computing* (pp. 401–414). Boca Raton, FL: Taylor & Francis.

Cambria, E., & White, B. (2014). Jumping NLP curves: A review of natural language processing research. *IEEE Computational Intelligence Magazine, 9*(2), 48–57.

Carvalho, V. R., & Cohen, W. W. (2005). On the collective classification of email "speech acts." In *Proceedings of the 28th Annual International ACM SIGIR Conference on Research and Development in Information Retrieval (SIGIR'05)* (pp. 345–352). doi:10.1145/1076034.1076094

Cui, B., Mondal, A., Shen, J., Cong, G., & Tan, K. (2005). On effective e-mail classification via neural networks. In K. V. Andersen, J. Debenham, & R. Wagner (Eds.), *Database and Expert*

Systems Applications: 16th International Conference, DEXA 2005, Copenhagen, Denmark, August 22–26, 2005. Proceedings (pp. 85–94). Berlin: Springer. doi:10.1007/11546924_9.

Cutting, D., Karger, D., & Pederson, J. (1993). Constant interaction-time scatter/gather browsing of large document collections. In *Proceedings of the 16th Annual International ACM SIGIR Conference on Research and Development in Information Retrieval* (pp. 126–134). ACM.

De Looze, M., & Lemarie, J. (1997). Corpus relevance through co-word analysis: An application to plant proteins. *Scientometrics, 39*(3), 267–280.

De Marneff, M. C., MacCartney, B., & Manning, C. D. (2006). Generating typed dependency parses from phrase structure parses. In *Proceedings of LREC* (Vol. 6, pp. 449–454).

Ding, Y., Chowdhury, C. C., & Foo, S. (1999). Bibliometic cartography of information retrieval research by using co-word analysis. *Information Processing & Management, 37*(6), 817–842.

Ding, X., Liu, B., & Zhang, L. (2009). Entity discovery and assignment for opinion mining applications. In *Proceedings of the 15th ACM SIGKDD International Conference on Knowledge Discovery and Data Mining (KDD '09)* (pp. 1125–1134). doi:10.1145/1557019.1557141

Ding, Y., Song, M., Han, J., Yu, Q., Yan, E., Lin, L., & Chambers, T. (2013). Entitymetrics: Measuring the impact of entities. *PLoS One, 8*(8), 1–14. doi:10.1371/journal.pone.0071416

Du, R., Safavi-Naini, R., & Susilon, W. (2003). Web filtering using text classification. In *Proceedings of the 11th IEEE International Conference on Networks, 28 September–1 October, 2003* (pp. 352–330).

Feldman, R., & Dagan, I. (1995). Knowledge discovery in textual databases (KDT). In *Proceedings of the Workshop in Knowledge Discovery, ECML-95* (pp. 112–117).

Feldman, R., Klösgen, W., & Ziberstein, A. (1997). Document explorer: Discovering knowledge in document collections. In Z. W. Raś & A. Skowron (Eds.), *Proceedings of the Foundations of Intelligent Systems: 10th International Symposium, ISMIS '97 Charlotte, North Carolina, USA October 15–18, 1997* (pp. 137–146). doi:10.1007/3-540-63614-5_13

Feldman, R., & Sanger, J. (2007). Introduction to text mining. In *The text mining handbook: Advanced approaches to analyzing unstructured data* (pp. 1–10). New York, NY: Cambridge University Press.

Finkel, J. R., Grenager, T., & Manning, C. D. (2005). Incorporating non-local information into information extraction systems by Gibbs sampling. In *Proceedings of the 43rd Annual Meeting of the Association for Computational Linguistics (ACL 2005)* (pp. 363–370). doi:10.3115/1219840.1219885

Glenisson, P., Glänzel, W., Janssens, F., & De Moor, B. (2005). Combining full text and bibliometric information in mapping scientific disciplines. *Information Processing & Management, 41*, 1548–1572.

Glenisson, P., Glänzel, W., & Persson, O. (2005). Combining full text analysis and bibliometric indicators: A pilot study. *Scientometrics, 63*(1), 163–180.

Gunes, E., & Radev, D. (2004). Lexrank: Graph-based lexical centrality as salience in text summerication. *Journal of Artificial Intelligence Research, 22*(1), 457–479.

Hepple, M., Ireson, N., Allegrini, P., Marchi, S., Monemagni, S., & Hidalgo, J. M. G. (2004). NLP-enhanced content filtering within the POESIA project. In *Proceedings of the International Conference on Language Resources and Evaluation*.

Hotho, A., Nürnberger, A., & Paaß, G. (2005). A brief survey of text mining. *LDV Forum, 20*(1), 19–26.

Janssens, F., Glänzel, W., & De Moor, B. (2008). A hybrid mapping of information science. *Scientometrics, 75*(3), 607–631.

Janssens, F., Leta, J., Glänzel, W., & De Moor, B. (2006). Towards mapping library and information science. *Information Processing & Management, 42*(6), 1614–1642. doi:10.1016/j.ipm.2006.03.025.

Joachims, T. (2002). *Learning to classify text using support vector machines: Methods, theory and algorithms.* Boston, MA: Kluwer Academic Publishers.

Kim, H., & Lee, J. Y. (2008). Exploring the emerging intellectual structure of archival studies using text mining: 2001–2004. *Journal of Information Science, 34*(2), 356–369.

Kim, H., & Lee, J. Y. (2009). Archiving research trends in LIS domain using profiling analysis. *Scientometrics, 80*(1), 75–90.

Klein, D., & Manning, C. D. (2003a). Accurate unlexicalized Parsing. In *Proceedings of the 41st Meeting of the Association for Computational Linguistics* (pp. 423–430). doi:10.3115/1075096.1075150

Klein, D., & Manning, C. D. (2003). Fast exact inference with a factored model for natural language parsing. In *Advances in neural information processing systems 15 (NIPS 2002)* (pp. 3–10). Cambridge, MA: MIT Press.

Kostoff, R. N., del Rio, J. A., Cortés, H. D., Smith, C., Smith, A., Wagner, C., . . . Tshiteya, R. (2007). Clustering methodologies for identifying country core competencies. *Journal of Information Science, 33*(1), 21–40. doi:10.1177/0165551506067124

Kostoff, R. N., del Río, J. A., Humenik, J. A., García, E. O., & Ramírez, A. M. (2001). Citation mining: Integrating text mining and bibliometrics for research user profiling. *Journal of the American Society for Information Science and Technology, 52*(13), 1148–1156. doi:10.1002/asi.1181.

Kostoff, R. N., Eberhart, H. J., Toothman, D. R., & Pallenbarg, R. (2006). Database tomography for technical intelligence: Comparative roadmaps of research impact assessment literature and the journal of the American Chemical Society. *Scientometrics, 40*(1), 103–138.

Kostoff, R. N., Eberhart, H. J., & Toothman, D. R. (1998). Database tomography for technical intelligence: A roadmap of the near-earth space science and technology literature. *Information Processing & Management, 34*(1), 69–85.

Kostoff, R. N., Green, K. A., Toothman, D. R., & Humenik, J. A. (2000). Database tomography applied to an aircraft science and technology investment strategy. *Journal of Aircraft, 37*(4), 727–730.

Kostoff, R. N., Miles, D. L., & Eberhart, H. J. (1995). System and method for database tomography (No. PAT-APPL-9967 341). Washingtion, DC.

Kostoff, R. N., Toothman, D. R., Eberhart, H. J., & Humenik, J. A. (2001). Text mining using database tomography and bibliometrics: A review. *Technological Forecasting and Social Change, 68*(3), 223–253.

Lafferty, J., McCallum, A., & Pereira, F. (2001). Conditional random fields: Probablistic models for segmenting and labeling sequence data. In *Proceedings of the 18th International Conference on Machine Learning 2001 (ICML 2001)* (pp. 282–289).

Lang, K. (1995). Newsweeder: Learning to filter netnews. In *Proceedings of the Twelfth International Conference on Machine Learning*.

Lawson, M., Kemp, N., Lynch, M. F., & Chowdhury, G. G. (1996). Automatic extraction of citations from the text of English-language patents—An example of template mining. *Journal of Information Science, 22*(6), 423–436.

Lee, J. Y., Kim, H., & Kim, P. J. (2010). Domain analysis with text mining: Analysis of digital library research trends using profiling methods. *Journal of Information Science, 36*(2), 144–161.

Lee, H., Peirsman, Y., Chang, A., Chambers, N., Surdeanu, M., & Jurafsky, D. (2011). Stanford's multi-pass sieve coreference resolution system at the CoNLL-2011 shared task. In *Proceedings of the Fifteenth Conference on Computational Natural Language Learning: Shared Task* (pp. 28–34). Association for Computational Linguistics.

Lent, B., Agrawal, R., & Srikant, R. (1997). Discovering trends in text databases. In *Proceedings of the Fourth International Conference on Knowledge Discovery and Data Mining (KDD-97)* (pp. 227–230).

Leydesdorff, L., & Hellsten, I. (2005). Metaphors and diaphors in science communication: Mapping the case of stem cell research. *Science Communication, 27*(1), 64–99. doi:10.1177/1075547005278346.

Li, R., Chambers, T., Ding, Y., Zhang, G., & Meng, L. (2014). Patent citation analysis: Calculating science linkage motivation. *Journal of the Association for Information Science and Technology*. doi:10.1002/asi.23054.

Lin, J., & Demner-Fushman, D. (2007). Semanic clustering of answers to clinical questions. In *Proceedings of the Annual Symposium of the American Medical Informatic Association (AMIA 2007), Chicago* (pp. 458–462).

Liu, X., Yu, S., Janssens, F., Glänzel, W., Moreau, Y., & De Moor, B. (2010). Weighted hybrid clustering by combing text mining and bibliometrics on a large-scale journal database. *Journal of the American Society for Information Science and Technology, 61*(6), 1105–1119.

Liu, X., Zhang, J., & Guo, C. (2012). Full-text citation analysis: enhancing bibliometric and scientific publication ranking. In *Proceedings of the 21st ACM International Conference on Information and Knowledge Management, ACM, 2012.* (pp. 1975–1979). doi:10.1145/2396761.2398555

Marcus, M. P., Marcinkiewicz, M. A., & Santorini, B. (1993). Building a large annotated corpus of English□: The Penn Treebank. In *Proceedings of the Computational Intelligence in Security for Information Systems: CISIS'09, 2nd International Workshop Burgos, Spain, September 2009* (Vol. 19, pp. 313–330).

Marcus, M. P., Santorini, B. & Marcinkiewicz, M. A. (1993). Building a large annotated corpus of English: The penn Treebank. *Computational Linguistics, 19*: 313–330.

Mihalcea, R., & Tarau, P. (2004). TextRank: Bringing order into texts. *Proceedings of EMNLP, 4*(4), 404–411. doi:10.3115/1219044.1219064.

Ming, Z., Wang, K., & Chua, T. S. (2010). Prototype hierarchy-based clustering for the categorization and navigation of web collections. In *Proceedings of the 33rd International ACM SIGIR Conference on Research and Development in Information Retrieval* (pp. 2–9).

Nigam, K., McCallum, A., Thrun, S., & Mitchell, T. (2000). Text classification from labeled and unlabeled documents using EM. *Machine Learning, 39*(2–3), 103–134.

Onyancha, O. B., & Ocholla, D. N. (2005). An informetric investigation of the relatedness of opportunistic infections to HIV/AIDS. *Information Processing & Management, 41*(6), 1573–1588. doi:10.1016/j.ipm.2005.03.015.

Page, L., Brin, S., Motwani, R., & Winograd, T. (1999). *The pagerank citation ranking: Bringing order to the web.* Working paper, Department of computer science, Stanford University (1999).

Porter, A. L., Kongthon, A., & Lu, J. (2002). Research profiling: Improving the literature review. *Scientometrics, 53*(3), 351–370. doi:10.1023/A:1014873029258.

Rajman, M., & Vesely, M. (2004). From text to knowledge: Document processing and visualization: A text mining approach. In S. Sirmakessis (Ed.), *Text mining and its applications: Results of the NEMIS Launch Conference* (pp. 7–24). Berlin: Springer. doi:10.1007/978-3-540-45219-5_2.

Sebastiani, F. (2002). Machine learning in automated text categorization. *ACM Computing Surveys (CSUR), 34*(1), 1–47.

Socher, R., Perelygin, A., Wu, J. Y., Chuang, J., Manning, C. D., Ng, A. Y., & Potts, C. (2013). Recursive deep models for semantic compositionality over a sentiment Treebank. In *Proceedings of the Conference on Empirical Methods in Natural Language Processing (EMNLP)* (pp. 1631–1642).

Song, M., Han, N. G., Kim, Y. H., Ding, Y., & Chambers, T. (2013). Discovering implicit entity relation with the gene-citation-gene network. *PLoS One, 8*(12), e84639. doi:10.1371/journal.pone.0084639.

Song, M., & Kim, S. Y. (2013). Detecting the knowledge structure of bioinformatics by mining full-text collections. *Scientometrics, 96*, 183–201. doi:10.1007/s11192-012-0900-9.

Suchanek, F. M., Kasneci, G., & Weikum, G. (2007). Yago: A core of semantic knowledge. In *Proceedings of the 16th International Conference of World Wide Web (WWW'07)* (pp. 697–706).

The Stanford Natural Language Processing Group. (2013). *Stanford CoreNLP.* Stanford University. Retrieved from http://nlp.stanford.edu/downloads/corenlp.shtml

Toutanova, K., Klein, D., Manning, C., & Singer, Y. (2003). Feature-rich part-of-speech tagging with a cyclic dependency network. In *Proceedings of the NLT-NAACL 2003* (pp. 252–259). Association for Computational Linguistics. doi:10.3115/1073445.1073478

Tseng, Y. H., Lin, C. J., & Lin, Y. I. (2007). Text mining techniques for patent analysis. *Information Processing & Management, 43*(5), 1216–1247. Retrieved from http://www. sciencedirect.com/science/article/pii/S0306457306002020

Tseng, Y. H., Wang, Y. M., Lin, Y. I., Lin, C. J., & Juang, D. W. (2007). Patent surrogate extraction and evaluation in the context of patent mapping. *Journal of Information Science, 33* (6), 718–736. doi:10.1177/0165551507077406.

Van Raan, A. F. J., & Tijssen, R. J. W. (1993). The neural net of neural network research. *Scientometrics, 26*(1), 169–192. doi:10.1007/BF02016799.

Wang, B. B., McKay, R. I., Abbass, H. A., & Barlow, M. (2002). Learning text classifier using the domain concept hierarchy. In *Proceedings of the International Conference on Communications, Circuits, and Systems, China.*

Zitt, M. (1991). A simple method for dynamic scientometrics using lexical analysis. *Scientometrics, 2*(1), 229–252.

Zitt, M., & Bassecoulard, E. (1994). Development of a method for detection and trend analysis of research fronts built by lexicoal or cocitation analysis. *Scientometrics, 30*(1), 333–351.

Chapter 11
Topic Modeling: Measuring Scholarly Impact Using a Topical Lens

Min Song and Ying Ding

Abstract Topic modeling is a well-received, unsupervised method that learns thematic structures from large document collections. Numerous algorithms for topic modeling have been proposed, and the results of those algorithms have been used to summarize, visualize, and explore the target document collections. In general, a topic modeling algorithm takes a document collection as input. It then discovers a set of salient themes that are discussed in the collection and the degree to which each document exhibits those topics. Scholarly communication has been an attractive application domain for topic modeling to complement existing methods for comparing entities of interest. In this chapter, we explain how to apply an open source topic modeling tool to conduct topic analysis on a set of scholarly publications. We also demonstrate how to use the results of topic modeling for bibliometric analysis.

11.1 Introduction

Clustering algorithms have been widely used to study scholarly communication. Most clustering methods group words together based on their similarity, characterized as "distance." Topic modeling is the next level of clustering; it groups words based on hidden topics. It can discover hidden topics in a collection of articles based on the assumption that, given that a document is about a certain topic, particular words related to this topic will appear in this article with higher frequency than in articles that are not about that topic. For example, "rain" and "snow" will appear often in documents talking about weather, while "apple" and "grape" will appear often in documents discussing fruit.

M. Song (✉)
Department of Library and Information Science, Yonsei University, Seoul, South Korea
e-mail: min.song@yonsei.ac.kr

Y. Ding
Department of Information and Library Science, School of Informatics and Computing, Indiana University, Bloomington, IN, USA
e-mail: dingying@indiana.edu

© Springer International Publishing Switzerland 2014
Y. Ding et al. (eds.), *Measuring Scholarly Impact*,
DOI 10.1007/978-3-319-10377-8_11

Topic modeling methods have proved useful for analyzing and summarizing large-scale textual data. They can handle streaming data, and have been applied in biomedical data, images, videos, and social media (Blei, 2012). The goal of topic modeling is to group sets of words that co-occur within texts as topics by assigning a high probability to words about the same topics. The most useful aspect of topic modeling is that it does not require any pre-annotated datasets, which often demand tremendous manual effort in annotating or labeling, and make output quality heavily dependent on training datasets.

In the family of topic modeling algorithms, Latent Dirichlet Allocation (LDA) is, in our opinion, the simplest. The concept behind LDA is that one document contains multiple topics, and each topic requires specific words to describe it. For example, a paper is entitled "Topics in dynamic research communities: An exploratory study for the field of information retrieval." This document deals with topic modeling, community detection, scholarly communication, and information retrieval. So terms such as "LDA," "author-conference topic modeling," and "statistical methods" are used for the topic, "topic modeling"; "Newman's method," "community detection," "clustering," and "graph partition" are used to describe the topic of "community detection"; "co-authorship network," "research topics," and "scientific collaboration" are used for the topic, "scholarly communication"; and "information retrieval model," "information retrieval method," "use case," and "search" are used for the topic, "information retrieval."

LDA is a generative model, like Naïve Bayes, that is a full probabilistic model of all the variables. In generative modeling, data is derived from a generative process, which defines a joint probability distribution of observed and hidden variables. It stands in contrast to discriminative modeling (e.g., linear regression), which only models the conditional probability of unobserved variables on the observed variables. In LDA, the observed variables are words in the documents, and the hidden variables are topics. This follows the assumption that authors first decide a number of topics for an article, and then pick up words related to these topics to write the article. So in LDA, all documents in the corpus cover the same set of topics, but each document contains different proportions of those topics (Blei, 2012).

Topic modeling algorithms aim to capture topics from a corpus automatically by using the observed words in documents to infer the hidden topic structure (e.g., document topic distribution and word topic distribution). The number of topics, usually decided by *perplexity*, can be heuristically set in a range from 20 to 300 (Blei, 2012). Perplexity is usually applied to measure how a probability distribution fits a set of data. It equals the inverse of the geometric mean per-word likelihood, and is used to evaluate models. A lower perplexity indicates a model that can achieve enhanced generalization performance (Blei, Ng, & Jordan, 2003). The inference mechanics in topic models are independent of language and content. They capture the statistical structure of language used to represent thematic content. LDA approximates its posterior distribution by using inference (e.g., Gibbs sampling) or optimization (e.g., variational methods) (Asuncion, Welling, Smyth, & Teh, 2009).

This chapter is organized as follows: Section 11.2 introduces several widely used topic models. Section 11.3 provides an overview of how topic models have been applied to study scholarly communication. Section 11.4 provides a use case with detailed guidelines on how to apply TMT (i.e., the topic-model software developed by Stanford University) to conduct analysis on 2,434 papers published in the *Journal of the American Society for Information Science (and Technology)* (JASIS(T)) between 1990 and 2013. Section 11.5 concludes the chapter with a brief summary.

11.2 Topic Models

Topics can be automatically extracted from a set of documents by utilizing different statistical methods. Figure 11.1 shows the plate notation for the major topic models, with gray and white circles indicating observed and latent variables, respectively. An arrow indicates a conditional dependency between variables and plates (Buntine, 1994). Here, d is a document, w is a word, a_d is a set of co-authors, x is an author, and z is a topic. α, β, and μ are hyperparameters, and θ, ϕ, and ψ are multinomial distributions over topics, words, and publication venues, respectively. Table 11.1 lists notations for these formulas.

11.2.1 Language Model (LM)

The language model is an early effort to model topics in natural language processing and information retrieval. There is no latent variable in this model (see Fig. 11.1). For a given query q, the probability between a document and a query word is calculated as (Ponte & Croft, 1998)

$$P(w|d) = \frac{N_d}{N_d + \lambda} \times \frac{tf(w,d)}{N_d} + \left(1 - \frac{N_d}{N_d + \lambda}\right) \times \frac{tf(w,D)}{N_D} \qquad (11.1)$$

where $tf(w,d)$ is the word frequency of a word w in a document d, N_d is the number of words in the current document, N_D is the number of words in the entire collection, $tf(w,D)$ is the frequency of a word w in the collection D, and λ is the Dirichlet smoothing factor that is usually set equal to the average document length in the collection (Zhai & Lafferty, 2001).

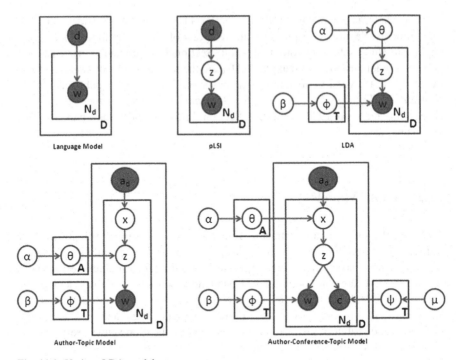

Fig. 11.1 Various LDA models

11.2.2 *Probabilistic Latent Semantic Indexing (pLSI)*

Hofmann (1999) proposed the probabilistic latent semantic indexing (pLSI) model by introducing a latent topic layer z between words and documents (see Fig. 11.1). In this model, the probability of generating a word w from a document d is based on the latent topic layer as

$$P(w|d) = \sum_{z=1}^{T} P(w|z)P(z|d) \qquad (11.2)$$

where pLSI does not provide a mathematical grounding for this latent topic layer and is thus susceptible to severe overfitting (Blei et al., 2003).

11.2.3 *Latent Dirichlet Allocation (LDA)*

Latent Dirichlet allocation (LDA) provides a probabilistic model for the latent topic layer (Blei et al., 2003). For each document d, a multinomial distribution θ_d over topics is sampled from a Dirichlet distribution with parameter α. For each word w_{di}, a topic z_{di} is chosen from the topic distribution. A word w_{di} is generated from a

Table 11.1 Notations for various LDA formulas

Notations	Meaning
d	Document
w	Word
x	Author
z	Topic
c	Publication venue
N_d	The number of words in the current document
N_D	The number of words in the entire collection of documents
a_d	The set of co-authors
α	The hyperparameter for generating Θ from Dirichlet distribution
β	The hyperparameter for generating φ from Dirichlet distribution
μ	The hyperparameter for generating Ψ from Dirichlet distribution
Θ	A multinomial distribution over topics
φ	A multinomial distribution over words
Ψ	A multinomial distribution over publication venues
D	Collection of documents
A	Collection of authors
T	Collection of topics
C_{mj}^{WT}	The number of times the mth word in a lexicon is assigned to topic j
C_{dj}^{DT}	The number of times the dth document is assigned to topic j
C_{aj}^{AT}	The number of times the ath author is assigned to topic j
C_{cj}^{CT}	The number of times the cth conference is assigned to topic j
z_{-di}	All word-topic assignment does not include current situation (assign word i in document d to a random topic in current instance)
x_{-di}	All word-author assignment does not include current situation (assign word i in document d to a random author in current instance)
m_{xz}	The number of times topic z is assigned to author x
n_{zv}	The number of times word v is assigned to topic z
n_{zc}	The number of times conference c is assigned to topic z

topic-specific multinomial distribution $\phi_{z_{di}}$. The probability of generating a word w from a document d is

$$P\left(w|d, \theta, \phi\right) = \sum_{z \in T} P\left(w|z, \phi_z\right) P\left(z|d, \theta_d\right) \tag{11.3}$$

Therefore, the likelihood of a document collection D is defined as

$$P\left(Z, W|\Theta, \Phi\right) = \prod_{d \in D} \prod_{z \in T} \theta_{dz}^{n_{dz}} \times \prod_{z \in T} \prod_{v \in V} \phi_{zv}^{n_{zv}} \tag{11.4}$$

where n_{dz} is the number of times that a topic z has been associated with a document d, and n_{zv} is the number of times that a word w_v has been generated by a topic z. The model can be explained as follows: an author first decides on topics and then, to write a paper, uses words that have a high probability of being associated with these topics.

11.2.4   Author-Topic Model

Rosen-Zvi, Griffiths, Steyvers, and Smyth (2004) proposed the author-topic model to represent both document content and author interests. In this model, an author is chosen randomly when a group of authors a_d decide to write a document d containing several topics. A word w is generated from a distribution of topics specific to a particular author. There are two latent variables, z and x. The formula to calculate these variables is

$$P\left(z_i, x_i | z_{-i}, x_{-i}, w, a_d, \alpha, \beta\right) \propto \frac{C_{mj}^{wT} + \beta}{\sum_{m'}\left(C_{m'j}^{wT} + V\beta\right)} \times \frac{C_{kj}^{AT} + \alpha}{\sum_{j'}\left(C_{kj'}^{AT} + T\alpha\right)} \quad (11.5)$$

where z_i and x_i represent the assignments of the ith word in a document to a topic j and an author k, respectively, w represents the observation that the ith word is the mth word in the lexicon, z_{-i} and x_{-i} represent all topic and author assignments not including the ith word, and C_{kj}^{AT} is the number of times an author k is assigned to a topic j, not including the current instance. The random variables ϕ (the probability of a word given a topic) and θ (the probability of a topic given an author) can be calculated as

$$\phi_{mj} = \frac{C_{mj}^{wT} + \beta}{\sum_{m'}\left(C_{m'j}^{wT} + V\beta\right)} \quad (11.6)$$

$$\theta_{kj} = \frac{C_{kj}^{AT} + \alpha}{\sum_{j'}\left(C_{kj'}^{AT} + T\alpha\right)} \quad (11.7)$$

This model can be used to recommend reviewers for peer-reviewed journals. The outcome of this model is a list of topics, each of which is associated with the top-ranked authors and words. Top-ranked authors are not necessarily the most highly cited authors in that area, but are those productive authors who use the most words for a given topic (Steyvers, Smyth, & Griffiths, 2004). Top-ranked words of a topic are those having a high probability of being selected when an author writes a paper on that particular topic.

11.2.5   Author-Conference-Topic Model

Tang, Jin, and Zhang (2008) proposed the author-conference-topic (ACT) model, an extended LDA used to model papers, authors, and publication venues simultaneously. Conference represents a general publication venue (e.g., journal, workshop, or organization). The ACT model can be interpreted as: Co-authors determine the topics for a paper, and each topic generates words and determines a publication venue. The ACT model calculates the probability of a topic for a given author, the probability of a word for a given topic, and the probability of a conference for a given topic. Gibbs sampling is used for inference, and the hyperparameters α, β, and μ are set at fixed values ($\alpha = 50/T$, $\beta = 0.01$, and $\mu = 0.1$). The posterior distribution is estimated on x and z, and the results are used to infer θ, φ, and ψ. The posterior probability is calculated as

$$P\left(z_{di},\, x_{di} \middle| z_{-di},\, x_{-di},\, w,\, c,\, \alpha,\, \beta,\, \mu\right) \propto \frac{m_{x_{di}z_{di}}^{-di} + \alpha_{z_{di}}}{\sum_{z}\left(m_{x_{di}z}^{-di} + \alpha_{z}\right)}$$

$$\times \frac{n_{z_{di}w_{di}}^{-di} + \beta_{w_{di}}}{\sum_{w_v}\left(n_{z_{di}w_v}^{-di} + \beta_{w_v}\right)}$$

$$\times \frac{n_{z_{di}c_d}^{-d} + \mu_{c_d}}{\sum_{c}\left(n_{z_{di}c}^{-d} + \mu_{c}\right)} \qquad (11.8)$$

After Gibbs sampling, the probability of a word given a topic φ, probability of a conference given a topic ψ, and probability of a topic given an author θ, can be estimated as

$$\phi_{zw_{di}} = \frac{n_{zw_{di}} + \beta_{w_{di}}}{\sum_{w_v}\left(n_{zw_v} + \beta_{w_v}\right)} \qquad (11.9)$$

$$\psi_{zc_d} = \frac{n_{zc_d} + \mu_{c_d}}{\sum_{c}(n_{zc} + \mu_{c})} \qquad (11.10)$$

$$\theta_{xz} = \frac{m_{xz} + \alpha_z}{\sum_{z'}\left(m_{xz'} + \alpha_{z'}\right)} \qquad (11.11)$$

A paper d is a vector w_d of N_d words, in which each w_{di} is chosen from a vocabulary of size V. A vector a_d of A_d authors is chosen from a set of authors of size A, and c_d represents a publication venue. A collection of papers D is defined by $D = \{(w_1, a_1, c_1), \ldots (w_D, a_D, c_D)\}$. The number of topics is denoted as T.

11.2.6 Hierarchical Latent Dirichlet Allocation (Hierarchical LDA)

Learning a topic hierarchy from a corpus is a challenge. Blei, Griffiths, and Jordan (2010) presented a stochastic process to assign probability distributions to form infinitely deep branching trees. LDA assumes that topics are flat with no hierarchical relationship between two topics; therefore, it fails to identify different levels of abstraction (e.g., relationships among topics). Blei et al. (2010) proposed a nested Chinese restaurant process (nCRP) as a hierarchical topic-modeling approach and applied Bayesian nonparametric inference to approximate the posterior distribution of topic hierarchies. Hierarchical LDA data treatment is different from hierarchical clustering. Hierarchical clustering initially treats every datum (i.e., word) as a leaf in a tree, and then merges similar data points until no word is left over—a process that finally forms a tree. Therefore, the upper nodes in the tree summarize their child nodes, which indicate that upper nodes share high probability with their children. In hierarchical topic modeling, a node in the tree is a topic that consists of a distribution of a set of words. The upper nodes do not summarize their child nodes, but instead reflect the shared distribution of words of their child nodes assigned to the same paths with them.

11.2.7 Citation LDA

Scientific documents are linked using citations. While common practices in graph mining focus on the link structure of a network (e.g., Getoor & Diehl, 2005), they ignore the topical features of nodes in that network. Erosheva, Fienberg, and Lafferty (2004) proposed the link-LDA as the mixed-membership model that groups publications into different topics by considering abstracts and their bibliographic references. Link-LDA models a document as a bag of words and a bag of citations. Chang and Blei (2010) proposed a relational topic model by considering both link structures and node attributes. This model can be used to suggest citations for new articles, and predict keywords from citations of articles. Nallapati, Ahmed, Xing, and Cohen (2008) proposed pairwise-link-LDA and link-LDA-PLSA models to address the issue of joint modeling of articles and their citations in the topic-modeling framework. The pairwise-link-LDA models the presence or absence of citations in each pair of documents, and is computationally expensive; Link-PLSA-LDA solves this issue by assuming that the link structure is a bipartite graph, and combines PLSA and LDA into one single graph model. Their experiments on CiteSeer show that their models outperform the baseline models and capture the topic similarity between contents of cited and citing articles. The link-PLSA-LDA performs better on citation prediction and is also highly scalable.

11.2.8 Entity LDA

LDA usually does not distinguish between different categories or concepts, but rather treats them equally as text or strings. But with the significant increase of available information, there exists a great need to organize, summarize, and visualize information based on different concepts or categories. For example, news articles emphasize information about who (e.g., entity person), when (e.g., entity time), where (e.g., entity location), and what (entity topic). In the biomedical domain, for example, genes, drugs, diseases, and proteins are major entities for studies and clinical trials. Newman, Chemudugunta, and Smyth (2006) proposed a statistical entity-topic method to model entities and make predictions about entities based on learning on entities and words. Traditional LDA assumes that each document contains one or more topics, and each topic is a distribution over words, while Newman's entity-topic models relate entities, topics, and words altogether. The conditionally independent LDA model (CI-LDA) makes a priori distinctions between words and entities during learning. SwitchLDA includes an additional binominal distribution to control the fraction of topic entities. But the word topics and entity topics generated by CI-LDA and SwitchLDA can be decoupled. CorrLDA1 enforces the connection between word topics and entity topics by first generating word topics for a document, and then generating entity topics based on the existing word topics in a document. This results in a direct correlation between entities and words. CorrLDA2 improves CorrLDA1 by allowing different numbers of word topics and entity topics. These entity-topic models can be used to compute the likelihood of a pair of entities co-occurring together in future documents. Kim, Sun, Hockenmaier, and Han (2012) proposed an entity topic model (ETM) to model the generative process of a term, given its topic and entity information, and the correlation between entity word distributions and topic word distributions.

11.3 Applying Topic Modeling Methods in Scholarly Communication

Mann, Mimno, and McCallum (2006) applied topic modeling methods to 300,000 computer science publications, to provide a topic-based impact analysis. They extended journal impact factor measures to topics, and introduced three topic impact measures: topical diversity (i.e., ranking papers based on citations from different topics), topical transfer (i.e., ranking papers based on citations from outside of their own topics), and topical precedence (i.e., ranking papers based on whether they are among the first to create a topic). They developed the topical N-Grams LDA, using phrases rather than words to represent topics. Gerrish and Blei (2010) proposed the document influence model (DIM) based on the dynamic LDA model to identify influential articles without using citations. Their hypothesis is that the influence of an article in the future is corroborated by how the language of its field changes subsequent to its publication. Thus, an article with words that contribute to the word frequency change will have a high influence score. They applied their model to three large corpora, and found that their influence measurement significantly correlates with an article's citation counts.

Liu, Zhang, and Guo (2012) applied labelled LDA to full-text citation analysis, to enhance traditional bibliometric analysis. Ding (2011a) combined topic-modeling and pathfinding algorithms to study scientific collaboration and endorsement in the field of information retrieval. The results show that productive authors tend to directly coauthor with and closely cite colleagues sharing the same research topics, but they do not generally collaborate directly with colleagues working on different research topics. Ding (2011b) proposed topic-dependent ranks based on the combination of a topic model and a weighted PageRank algorithm. She applied the author-conference Topic (ACT) model to extract the topic distribution for individual authors and conferences, and added this as a weighted vector to the PageRank algorithm. The results demonstrated that this method can identify representative authors with different topics over different time spans. Later, Ding (2011c) applied the author-topic model to detect communities of authors, and compared this with traditional community detection methods, which are usually topology-based graph partitions of co-author networks. The results showed that communities detected by the topology-based community detection approach tend to contain different topics within each community, and communities detected by the author-topic model tend to contain topologically diverse sub-communities within each community. Natale, Fiore, and Hofherr (2012) examined the aquaculture literature using bibliometrics and computational semantic methods, including latent semantic analysis, topic modeling, and co-citation analysis, to identify main themes and trends. Song, Kim, Zhang, Ding, and Chambers (2014) adopted the Dirichlet multinomial regression (DMR)-based topic modeling method to analyze the overall trends of bioinformatics publications during the period between 2003 and 2011. They found that the field of bioinformatics has undergone a significant shift, to coevolve with other biomedical disciplines.

11.4   Topic Modeling Tool: Case Study

In this section, we introduce an open-source tool for topic modeling and provide a concrete example of how to apply this tool to conduct topic analysis on a set of publications.

The Stanford Topic Modeling Toolbox (TMT) is a Java-based topic modeling tool (http://nlp.stanford.edu/software/tmt/tmt-0.4/), and a subset of the Stanford Natural Language Processing software. The current version of TMT is 0.4, and the tool is intended to be used by non-technical personnel who want to apply topic models to their own datasets. TMT accepts tab-separated and comma-separated values, and is seamlessly integrated with spreadsheet programs, such as Microsoft

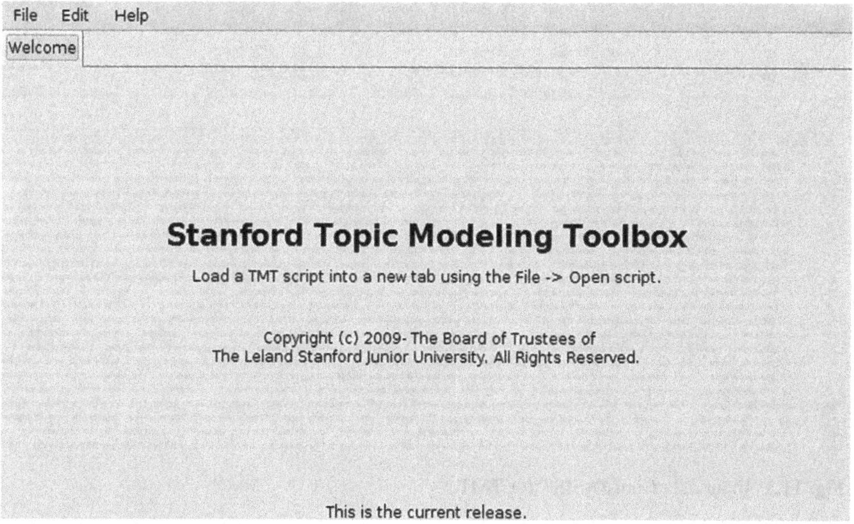

File   Edit   Help

Welcome

Stanford Topic Modeling Toolbox

Load a TMT script into a new tab using the File -> Open script.

This is the current release.

Fig. 11.2 Welcome page of TMT

Excel. While TMT provides several topic models—such as LDA, labeled LDA, and PLDA—it unfortunately cannot support the author-topic model or author-conference topic. Mallet (http://mallet.cs.umass.edu/) provides a toolkit for LDA, Pachinoko LDA, and hierarchical LDA. At David Blei's homepage (http://www.cs.princeton.edu/~blei/topicmodeling.html), there are codes for a variety of LDA models.

To run TMT, the following software needs to be pre-installed:

1. Any text editor, such as NotePad, for creating TMT processing scripts; and
2. Java 6SE, or a higher version.

Once the prerequisite software is in place, the TMT executable program needs to be downloaded from the TMT homepage (http://nlp.stanford.edu/software/tmt/tmt-0.4/tmt-0.4.0.jar). The simple GUI of TMT can be seen by either double-clicking the file to open the toolbox, or running java -jar tmt-0.4.0.jar from the command line (Fig. 11.2).

Once the GUI is displayed, there is an option for designating a CSV or tab-delimited input file. To demonstrate how topic modeling via TMT can be applied to analyze scientific publication datasets, we downloaded 2,534 records published in the *Journal of the American Society for Information Science (and Technology)* (JASIS(T)) between 1990 and 2013 from Web of Science. We made the dataset publicly available at http://informatics.yonsei.ac.kr/stanford_metrics/jasist_2012.txt.

Figure 11.3 shows the JASIST input data, opened in Microsoft Excel. To load the dataset into TMT, select "Open script …" from the file menu of the TMT GUI.

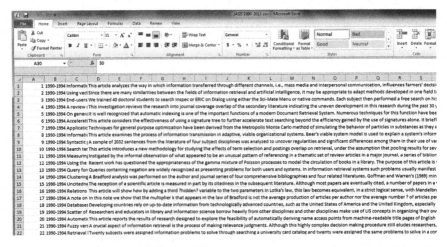

Fig. 11.3 Input data from JASIST for TMT

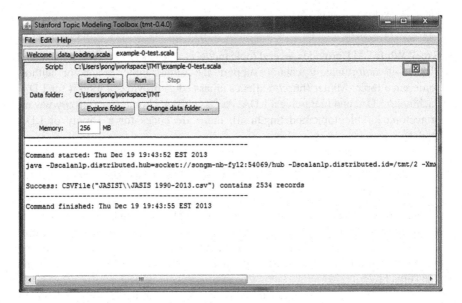

Fig. 11.4 Execution result of loading the example dataset

If the dataset is successfully loaded, as shown in Fig. 11.4, below, the message is "Success: CSVFile ("JASIST-oa-subset.csv") contains 2534 records" is displayed.

First, prepare the input data. As explained earlier, the dataset can be imported from a CSV file. Once the dataset is loaded into TMT, a simple Scala script that comes with TMT will convert a column of text from a file into a sequence of words. To this end, the script that comes with TMT must be executed; a basic understanding of the script is required to do this. Figure 11.5 shows a snippet of the script.

```
 1 // tells Scala where to find the TMT classes
 2 import scalanlp.io._;
 3 import scalanlp.stage._;
 4 import scalanlp.stage.text._;
 5 import scalanlp.text.tokenize._;
 6 import scalanlp.pipes.Pipes.global._;
 7
 8 import edu.stanford.nlp.tmt.stage._;
 9 import edu.stanford.nlp.tmt.model.lda._;
10 import edu.stanford.nlp.tmt.model.llda._;
11
12 val source = CSVFile("JASIST/JASIS 1990-2013.csv") ~> IDColumn(1);
13
14 val tokenizer = {
15   SimpleEnglishTokenizer() ~>              // tokenize on space and punctuation
16   CaseFolder() ~>                          // lowercase everything
17   WordsAndNumbersOnlyFilter() ~>           // ignore non-words and non-numbers
18   MinimumLengthFilter(3)                   // take terms with >=3 characters
19 }
20
21 val text = {
22   source ~>                                // read from the source file
23   Columns(3,4) ~> Join(" ") ~>             // select column containing text
24   TokenizeWith(tokenizer) ~>               // tokenize with tokenizer above
25   TermCounter() ~>                         // collect counts (needed below)
26   TermMinimumDocumentCountFilter(4) ~>     // filter terms in <4 docs
27   TermDynamicStopListFilter(30) ~>         // filter out 30 most common terms
28   DocumentMinimumLengthFilter(5)           // take only docs with >=5 terms
29 }
```

Fig. 11.5 Snippet of the Scala code for converting text into a sequence of words

In line 12, TMT is instructed to use the value in column 1 as the record ID, a unique identifier for each record in the file. If you have record IDs in a different column, change the 1 in line 12 to the right column number.

After identifying the record id, tokenization must be applied (lines 14–19 in Fig. 11.5). The SimpleEnglishTokenizer class (line 15) is used to remove punctuation from the ends of words and then split up the input text by white-space characters. The CaseFolder (line 16) is then used to lower-case each word. Next, the WordsAndNumbersOnlyFilter (line 17) is used to remove words that are entirely punctuation and other non-word or non-number characters. Finally, the MinimumLengthFilter class (line 18) is used to remove terms that are shorter than three characters.

After defining the tokenizer (line 14–19), the tokenizer is used to extract text from the appropriate column(s) in the CSV file. If your text data is in a single column (for example, the text is in the fourth column), this procedure is coded in line 21–29: source ~ > Column (3,4) ~ > TokenizeWith(tokenizer). After that, the function of lines 25–29 is to retain only meaningful words. The code above removes terms appearing in fewer than four documents (line 26), and the list of the 30 most common words in the corpus (line 27). The DocumentMinimumLengthFilter (5) class removes all documents shorter than length 5.

```
31 // turn the text into a dataset ready to be used with LDA
32 val dataset = LDADataset(text);
33
34 // define the model parameters
35 val params = LDAModelParams(numTopics = 10, dataset = dataset,
36   topicSmoothing = 0.01, termSmoothing = 0.01);
37
38 // Name of the output model folder to generate
39 val modelPath = file("lda-"+dataset.signature+"-"+params.signature);
40
41 // Trains the model: the model (and intermediate models) are written to the
42 // output folder.  If a partially trained model with the same dataset and
43 // parameters exists in that folder, training will be resumed.
44 TrainCVB0LDA(params, dataset, output=modelPath, maxIterations=1000);
```

Fig. 11.6 Snippet of the code of learning topic models

The next step is to select parameters for training an LDA model (line 37–47).

First, the number of topics needs to be pre-defined, as in the K-means clustering algorithm. In the code snippet above (Fig. 11.6), besides the number of topics, LDA model parameters for a smoothing term and topic need to be pre-defined to build topic models. Those parameters are shown in the LDAModelParams constructor on lines 35 and 36: termSmoothing is 0.01 and topicSmoothing is set to 0.01. The second step is to train the model to fit the documents. TMT supports several inference techniques on most topic models, including the possibility to use a collapsed Gibbs sampler (Griffiths & Steyvers, 2004) or the collapsed variational Bayes approximation to the LDA objective (Asuncion et al., 2009). In the example above (Fig. 11.6), the collapsed variational Bayes approximation is used (line 44).

To learn the topic model, the script "example-2-lda-learn.scala" is run by using the TMT GUI. The topic model outputs status messages as it trains, and writes the generated model into a folder in the current directory named, in this case, "lda-59ea15c7-30-75faccf7," as shown in Fig. 11.7. This process may take a few minutes, depending on the size of the dataset.

After the learning of topic models is successfully done, the model output folder "lda-59ea15c7-30-75faccf7" is generated. As shown in Fig. 11.8, the folder contains the following files that are required to analyze the learning process and to load the model back in from disk: description.txt, document-topic-distributions.csv.gz, tokenizer.txt, summary.txt, term-index.txt, and topic-term-distributions.csv.gz. Description.txt contains a description of the model saved in this folder, while document-topic-distributions.csv.gz is a csv file containing the per-document topic distribution for each document in the dataset. Tokenizer.txt contains a tokenizer that is employed to tokenize text for use with this model. Summary.txt provides the human-readable summary of the topic model, with the top 20 terms per topic. Term-index.txt maps terms in the corpus to ID numbers, and topic-term-distributions.csv.gz contains the probability of each term for each topic.

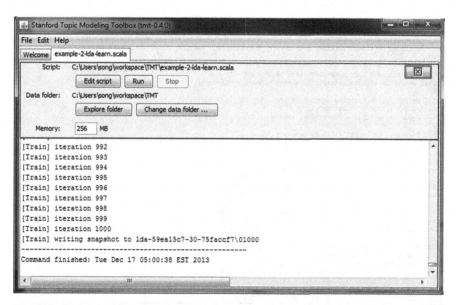

Fig. 11.7 Output message of running the script for learning topics

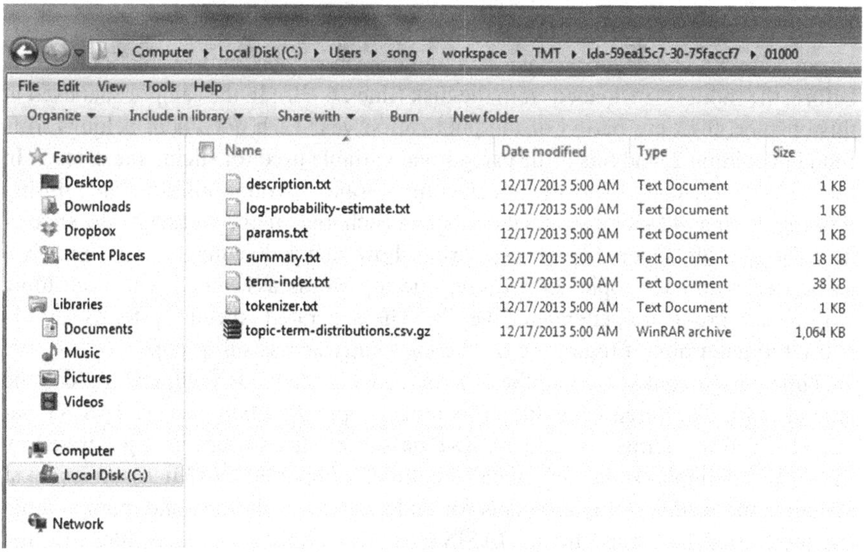

Fig. 11.8 Output folder as a result of learning topic models

The next code snippet shows how to slice the results of the topic model (Fig. 11.9). The code snippet shown in Fig. 11.9 is from the script "example-4-lda-slice.scala." The technique embodied in this snippet is helpful in examining how a topic is used in each slice of the data, where the slice is the subset of data

```
26 val text = {
27   source ~>                              // read from the source file
28   Columns(3,4) ~> Join(" ") ~>                   // select column containing text
29   TokenizeWith(model.tokenizer.get)      // tokenize with existing model's tokenizer
30 }
31
32 // turn the text into a dataset ready to be used with LDA
33 val dataset = LDADataset(text, termIndex = model.termIndex);
34
35 // define fields from the dataset we are going to slice against
36 val slice = source ~> Column(2);
37 // could be multiple columns with: source ~> Columns(2,7,8)
38
39 // Base name of output files to generate
40 val output = file(modelPath, source.meta[java.io.File].getName.replaceAll(".csv",""));
41
42 println("Loading document distributions");
43 val perDocTopicDistributions = LoadLDADocumentTopicDistributions(
44   CSVFile(modelPath,"document-topic-distributions.csv"));
45 // This could be InferDocumentTopicDistributions(model, dataset)
46 // for a new inference dataset. Here we load the training output.
47
48 println("Writing topic usage to "+output+"-sliced-usage.csv");
49 val usage = QueryTopicUsage(model, dataset, perDocTopicDistributions, grouping=slice);
50 CSVFile(output+"-sliced-usage.csv").write(usage);
51
52 println("Estimating per-doc per-word topic distributions");
53 val perDocWordTopicDistributions = EstimatePerWordTopicDistributions(
54   model, dataset, perDocTopicDistributions);
55 println("Writing top terms to "+output+"-sliced-top-terms.csv");
56
57 val topTerms = QueryTopTerms(model, dataset, perDocWordTopicDistributions, numTopTerms=50, grouping=slice);
58 CSVFile(output+"-sliced-top-terms.csv").write(usage);
```

Fig. 11.9 Snippet of the code for slicing the topic model's output

associated with one or more meta-data items, such as year, author, and journal. As before, the model is re-loaded from the disk (line 26–30). In the sample data used in this chapter, the time period of the publication year each document belongs to is found in column 2, and this is the categorical variable used for slicing the dataset. In lines 32–37, the code loads the per-document topic distributions generated during training. In lines 42–58, it shows the usage of each topic in the dataset by the slice of data. In line 49, QueryTopicUsage prints how many documents and words are associated with each topic. In addition, the top words associated with each topic within each group are generated (line 57). The generated -sliced-top-terms.csv file is used to determine if topics are used consistently across sub-groups.

Time period is indicated on the X-axis, count is the value field, and topic is the legend field. The three CSV files (document-topic-distributions.csv, JASIST-oa-subset-sliced-top-terms.csv, and JASIST-oa-subset-sliced-usage.csv) generated by the script "example-4-lda-slice.scala" are directly imported into Microsoft Excel to visualize the results of topic models for understanding, plotting, and manipulating the topic model outputs. In the JASIST-oa-subset-sliced-usage.csv file, the first column is the topic id, the second column is the group which is year, the third column contains the total number of documents associated with each topic within each slice, and the fourth column contains the total number of words associated with each topic within each slice.

Figure 11.10 shows several interesting results. First, there are topics showing a consistent increase in topic trends (topics 1, 3, 4, 7, 8, and 9). These topics are information resource, informetrics, information network, information science—

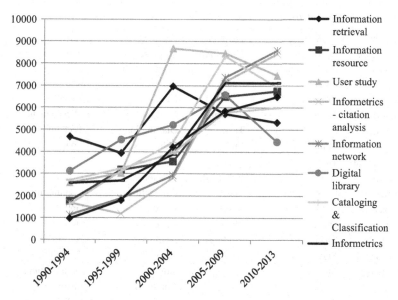

Fig. 11.10 Graphic representation of slicing results

general, and information structure. Second, the topic information retrieval shows a fluctuating pattern (i.e., a decrease in period 1995–1999, an increase in period 2000–2004, and then another decrease). User study (topic 2), has a big increase between the period 1995–1999 and the period 2000–2004, and then a mild decrease. Informetrics (topic 3) and information network (topic 4) show a similar pattern. With the advent of the Internet, informetrics and information network have become trendy topics published in JASIS(T). Digital library (topic 5) shows an increasing pattern until 2004, then decreases.

Table 11.2 shows 20 topical terms per topic for 10 topics. These topical terms are generated by TMT and stored in the summary.txt file.

The results of topic modeling indicate which salient topics were covered in JASIS(T). In addition, the results show that informetrics is the dominant topic studied by papers published in JASIS(T) for the past two decades.

The major limitations of TMT when applied to bibliometric research are as follows: First, it is not quite clear what common stop-word list is used by TMT. In the Scala script provided by TMT, the option for applying the stop-word list is TermDynamicStopListFilter(30), which removes the most common 30 terms. However, it is not clear what those 30 terms are and how to change them. Second, this filter is not adequate for processing a huge amount of data. To generate topic models from millions of records, TMT needs to be extended to the MapReduce platform. Third, the front end of TMT, written in Scala, does not provide as rich a set of functionalities as those of back-end components written in Java.

Table 11.2 20 topical terms for 10 topics

Topic 00	Topic 01	Topic 02	Topic 03	Topic 04
Information retrieval	*Information resource*	*User study*	*Informetrics-citation analysis*	*Information network*
Query	Internet	Users	Impact	Knowledge
Documents	Online	User	Number	Social
Document	About	Searching	Citation	Management
Relevance	Not	Task	Distribution	Technology
Term	Content	Image	Indicators	Communication
Performance	They	Design	Papers	Behavior
Method	Source	Behavior	Citations	Factors
Queries	Electronic	They	Publications	Community
Terms	Sites	Systems	Between	Group
Used	Other	Tasks	Different	How
Systems	Than	Findings	Authors	Network
Approach	May	Students	Also	Technologies
Effectiveness	Most	Participants	Not	Organizational
Space	Such	Used	Countries	Learning
Based	Personal	Different	Bibliometric	Collaborative
Methods	But	Two	Distributions	Through
Not	Site	Process	Than	Tagging
Our	Health	Interface	Index	Between
Database	Social	Terms	Law	Perceived
Models	Quality	Cognitive	One	Environment
Topic 05	Topic 06	Topic 07	Topic 08	Topic 09
Digital library	*Cataloging and classification*	*Informetrics*	*Information science*	*Information structure and analysis*
Digital	Text	Citation	Science	structure
Library	Indexing	Journals	Systems	Measures
Development	Approach	Journal	Knowledge	Network
Access	Language	Science	Theory	Clustering
Design	Words	Articles	Work	Two
Systems	Documents	Citations	What	Based
Libraries	Classification	Scientific	Approach	Between
Metadata	Chinese	Literature	Concepts	Relationships
Resources	Based	Impact	View	Different
Has	Automatic	Scholarly	Understanding	New
Services	Semantic	Cited	Its	Clusters
Electronic	Techniques	Published	How	Used
Tools	Our	Publication	Between	Similarity
Project	Terms	Authors	Concept	Map
Collection	Method	Between	Not	Networks
New	Has	Sciences	Nature	Pages
How	Been	Disciplines	New	Measure
Technology	Word	Databases	Framework	Mapping
Support	Algorithm	Than	Role	Categories
Been	Features	Subject	Process	Author

Conclusion

Topic modeling represents a recent surge in text-mining applications for analyzing large amounts of unstructured text data. Among a number of topic modeling algorithms, LDA is the best-received topic modeling method. LDA and its variations—such as hierarchical LDA and labeled LDA—are used in different research domains, such as Physics, Computer Science, Information Science, Education, and Life Sciences. In an effort to help bibliometric researchers adapt LDA to their own research problems, the present chapter provides an overview of topic modeling techniques, and walks readers through the steps needed to perform analysis on datasets for a bibliometric study.

To this end, we demonstrate how the topic-modeling technique can be applied to real-world research problems by using the Stanford Topic Modeling Tool (TMT). Stanford TMT allows for (1) importing and manipulating data from Microsoft Excel, (2) training topic models with summaries of the input data, (3) selecting parameters (such as the number of topics) with several easy steps, and (4) generating CSV-style outputs to track word usage across topics and time.

LDA can be applied in any field where texts are the main data format. There are several challenges for LDA, however. First, the labeling of topics can be done in different ways (Mei, Shen, & Zhai, 2007), usually using the top-ranked keywords with high probabilities from each topic to label that topic. But such labels can be hard to interpret, and they are sometimes contradictory. Since LDA uses soft clustering, one keyword can appear in more than one topic, and some topics can have very similar labels. How to provide a meaningful label for each topic automatically remains a challenge. Evaluating LDA is another challenge (Chang, Gerrish, Wang, Boyd-Graber, & Blei, 2009) because LDA is an unsupervised probabilistic model, and the generated latent topics are not always semantically meaningful. LDA assumes that each document can be described as a set of latent topics, which are multinomial distributions of words. Chang et al. (2009) found that models achieving better perplexity often generate less interpretable latent topics. By using the Amazon Mechanical Turk, they found that people appreciate the semantic coherence of topics, and they therefore recommended incorporating human judgments into the model-fitting process as a way to increase the thematic meanings of topics.

In the present chapter, we use the Stanford TMT to demonstrate how topic-modeling techniques can help bibliometric studies. As described earlier, Stanford TMT provides the following features: (1) Imports text datasets from cells in Microsoft Excel's CSV spreadsheets; (2) uses LDA modeling to create summaries of the text datasets; (3) selects parameters for training LDA models, such as the number of topics, the number of top words in each

(continued)

(continued)

topic, the filtering of most common words, and the selection of columns containing the text datasets; and (4) slices the LDA topic-model output and converts it into rich Microsoft-Excel-compatible outputs for tracking word usage across topics and respondent categories. As a case study, we collected 2,534 records published in the *Journal of the American Society for Information Science (and Technology)* (JASIS(T)) between year 1990 and year 2013 from Web of Science. With topic modeling, we discover hidden topical patterns that pervade the collection through statistical regularities and use them for bibliometric analysis. In the future we plan to explore how to combine a citation network with topic modeling, which will map out topical similarities between a cited article and its citing articles.

Appendix: Normalization, Mapping, and Clustering Techniques Used by VOSviewer

In this appendix, we provide a more detailed description of the normalization, mapping, and clustering techniques used by VOSviewer.

Normalization

We first discuss the association strength normalization (Van Eck & Waltman, 2009) used by VOSviewer to normalize for differences between nodes in the number of edges they have to other nodes. Let aij denote the weight of the edge between nodes i and j, where aij = 0 if there is no edge between the two nodes. Since VOSviewer treats all networks as undirected, we always have aij = aji. The association strength normalization constructs a normalized network in which the weight of the edge between nodes i and j is given by

$$s_{ij} = \frac{2ma_{ij}}{k_i k_j},$$ (11.12)

where k_i (k_j) denotes the total weight of all edges of node i (node j) and m denotes the total weight of all edges in the network. In mathematical terms,

$$k_i = \sum_j a_{ij} \quad \text{and} \quad m = \frac{1}{2}\sum_i k_i.$$ (11.13)

We sometimes refer to s_{ij} as the similarity of nodes i and j. For an extensive discussion of the rationale of the association strength normalization, we refer to Van Eck and Waltman (2009).

Mapping

We now consider the VOS mapping technique used by VOSviewer to position the nodes in the network in a two-dimensional space. The VOS mapping technique minimizes the function

$$V(\mathbf{x}_1, \ldots, \mathbf{x}_n) = \sum_{i<j} s_{ij} \|\mathbf{x}_i - \mathbf{x}_j\|^2 \qquad (11.14)$$

subject to the constraint

$$\frac{2}{n(n-1)} \sum_{i<j} \|\mathbf{x}_i - \mathbf{x}_j\| = 1, \qquad (11.15)$$

where n denotes the number of nodes in a network, $\mathbf{x}_i$ denotes the location of node i in a two-dimensional space, and $\|\mathbf{x}_i - \mathbf{x}_j\|$ denotes the Euclidean distances between nodes i and j. VOSviewer uses a variant of the SMACOF algorithm (e.g., Borg & Groenen, 2005) to minimize (11.14) subject to (11.15). We refer to Van Eck et al. (2010) for a more extensive discussion of the VOS mapping technique, including a comparison with multidimensional scaling.

Clustering

Finally, we discuss the clustering technique used by VOSviewer. Nodes are assigned to clusters by maximizing the function

$$V(c_1, \ldots, c_n) = \sum_{i<j} \delta(c_i, c_j)(s_{ij} - \gamma) \qquad (11.16)$$

where c_i denotes the cluster to which node i is assigned, $\delta(c_i, c_j)$ denotes a function that equals 1 if $c_i = c_j$ and 0 otherwise, and γ denotes a resolution parameter that determines the level of detail of the clustering. The higher the value of γ, the larger the number of clusters that will be obtained. The function in (11.16) is a variant of the modularity function introduced by Newman and Girvan (2004) and Newman (2005) for clustering the nodes in a network. There is also an interesting mathematical relationship between on the one hand the problem of minimizing (11.14) subject to (11.15) and on the other hand the problem of maximizing (11.16). Because of this relationship, the mapping and clustering techniques used by VOSviewer constitute a unified approach to mapping and clustering the nodes in a network. We refer to Waltman et al. (2010) for more details. We further note that VOSviewer uses the recently introduced smart local moving algorithm (Waltman & Van Eck, 2013) to maximize (11.16).

References

Asuncion, A., Welling, M., Smyth, P., & Teh, Y. (2009). On smoothing and inference for topic models. *Proceedings of the Conference on Uncertainty in Artificial Intelligence*, Montreal, Canada, 18–21 June.

Blei, D. M. (2012). Probabilistic topic models. *Communications of the ACM, 55*(4), 77–84.

Blei, D. M., Griffiths, T. L., & Jordan, M. (2010). The nested Chinese restaurant process and Bayesian nonparametric inference of topic hierarchies. *Journal of the ACM, 57*(2), 1–30.

Blei, D. M., Ng, A. Y., & Jordan, M. I. (2003). Latent dirichlet allocation. *Journal of Machine Learning Research, 3*, 993–1022.

Borg, I., & Groenen, P. J. F. (2005). *Modern multidimensional scaling* (2nd ed.). New York: Springer.

Buntine, W. L. (1994). Operations for learning with graphical models. *Journal of Artificial Intelligence Research, 2*, 159–225.

Chang, J., & Blei, D. M. (2010). Hierarchical relational models for document networks. *The Annals of Applied Statistics, 4*(1), 124–150.

Chang, J., Gerrish, S., Wang, C., Boyd-Graber, J. L., & Blei, D. M. (2009). Reading tea leaves: How humans interpret topic models. *Proceedings of 23rd Advances in Neural Information Processing Systems*, Vancouver, Canada, 7–12 December.

Ding, Y. (2011a). Scientific collaboration and endorsement: Network analysis of coauthorship and citation networks. *Journal of Informetrics, 5*(1), 187–203.

Ding, Y. (2011b). Topic-based PageRank on author co-citation networks. *Journal of the American Society for Information Science and Technology, 62*(3), 449–466.

Ding, Y. (2011c). Community detection: Topological vs. topical. *Journal of Informetrics, 5*(4), 498–514.

Erosheva, E., Fienberg, S., & Lafferty, J. (2004). Mixed-membership models of scientific publications. *Proceedings of the National Academy of Sciences, 101*(1), 5220–5227.

Gerrish, S., & Blei, D. M. (2010). A language-based approach to measuring scholarly impact. *Proceedings of the 26th International Conference on Machine Learning*, Haifa, Israel, 21–24 June.

Getoor, L., & Diehl, C. P. (2005). Link mining: A survey. *ACM SIGKDD Explorations Newsletter, 7*(2), 3–12.

Griffiths, T. L., & Steyvers, M. (2004). Finding scientific topics. *Proceedings of the National Academy of Sciences, 101*, 5228–5235.

Hofmann, T. (1999, August 15–19). Probabilistic latent semantic indexing. *Proceedings of the 22nd Annual International ACM SIGIR Conference on Research and Development in Information Retrieval* (pp. 50–57), Berkeley, CA, USA.

Kim, H., Sun, Y., Hockenmaier, J., & Han, J. (2012). ETM: Entity topic models for mining documents associated with entities. *2012 I.E. 12th International Conference on Data Mining* (pp. 349–358). IEEE.

Liu, X., Zhang, J., & Guo, C. (2012). Full-text citation analysis: Enhancing bibliometric and scientific publication ranking. *Proceedings of the 21st ACM International Conference on Information and Knowledge Management* (pp. 1975–1979), Brussels, Belgium. ACM.

Mann, G. S., Mimno, D., & McCallum, A. (2006). Bibliometric impact measures leveraging topic analysis. *The ACM Joint Conference on Digital Libraries*, Chapel Hill, North Carolina, USA, 11–15 June.

Mei, Q., Shen, X., & Zhai, C. (2007). Automatic labeling of multinomial topic models. *Proceedings of Knowledge Discovery and Data Mining Conference* (pp. 490–499).

Nallapati, R., Ahmed, A., Xing, E. P., & Cohen, W. W. (2008). Joint latent topic models for text and citations. *Proceedings of the 14th ACM SIGKDD International Conference on Knowledge Discovery and Data Mining*, Las Vegas, Nevada, USA, 24–27 August.

Natale, F., Fiore, G., & Hofherr, J. (2012). Mapping the research on aquaculture. A bibliometric analysis of aquaculture literature. *Scientometrics, 90*(3), 983–999.

Newman, D., Chemudugunta, C., & Smyth, P. (2006). Statistical entity-topic models. *Proceedings of 12th ACM SIGKDD International Conference on Knowledge Discovery and Data Mining*, Philadelphia, Pennsylvania, USA, 20–23 August.

Newman, M., & Girvan, M., (2004). Finding and evaluating community structure in networks. *Physical Review E: Statistical, Nonlinear, and Soft Matter Physics, 69*, 026113

Newman, M. E. J. (2005), Power laws, Pareto distributions and Zipf's law. *Contemporary Physics, 46*(5), 323–351

Ponte, J. M., & Croft, W. B. (1998, August 24–28). A language modeling approach to information retrieval. *Proceedings of the 21st Annual International ACM SIGIR Conference on Research and Development in Information Retrieval*, Melbourne, Australia (pp. 275–281).

Rosen-Zvi, M., Griffiths, T., Steyvers, M., & Smyth, P. (2004). The author-topic model for authors and documents. *Proceedings of the 20th Conference on Uncertainty in Artificial Intelligence*, Banff, Canada (pp. 487–494).

Song, M., Kim, S. Y., Zhang, G., Ding, Y., & Chambers, T. (2014). Productivity and influence in bioinformatics: A bibliometric analysis using PubMed central. *Journal of the American Society for Information Science and Technology, 65*(2), 352–371.

Steyvers, M., Smyth, P., & Griffiths, T. (2004 August 22–25). Probabilistic author-topic models for information discovery. *Proceeding of the 10th ACM SIGKDD Conference on Knowledge Discovery and Data Mining* (pp. 306–315), Seattle, Washington, USA.

Tang, J., Jin, R., & Zhang, J. (2008, December 15–19). A topic modeling approach and its integration into the random walk framework for academic search. *Proceedings of 2008 I.E. International Conference on Data Mining (ICDM2008)* (pp. 1055–1060), Pisa, Italy.

Van Eck, N.J., & Waltman, L. (2009). How to normalizecooccurance data? An analysis of some well-known similarity measures. *Journal of the American Society for Information Science and Technology, 60*(8), 1635–1651.

Van Eck, N. J., Waltman, L., Noyons, E. C. M., & Butter, R.K. (2010). Automatic term identification for bibliometric mapping. *Sceientometrics, 82*(3), 581–596.

Zhai, C., & Lafferty, J. (2001, September 9–13). A study of smoothing methods for language models applied to ad hoc information retrieval. *Proceedings of the 24th Annual International ACM SIGIR Conference on Research and Development in Information Retrieval* (pp. 334–342), New Orleans, LA, USA.

Chapter 12
The Substantive and Practical Significance of Citation Impact Differences Between Institutions: Guidelines for the Analysis of Percentiles Using Effect Sizes and Confidence Intervals

Richard Williams and Lutz Bornmann

Abstract In this chapter we address the statistical analysis of percentiles: How should the citation impact of institutions be compared? In educational and psychological testing, percentiles are already used widely as a standard to evaluate an individual's test scores—intelligence tests for example—by comparing them with the scores of a calibrated sample. Percentiles, or percentile rank classes, are also a very suitable method for bibliometrics to normalize citations of publications in terms of the subject category and the publication year and, unlike the mean-based indicators (the relative citation rates), percentiles are scarcely affected by skewed distributions of citations. The percentile of a certain publication provides information about the citation impact this publication has achieved in comparison to other similar publications in the same subject category and publication year. Analyses of percentiles, however, have not always been presented in the most effective and meaningful way. New APA guidelines (Association American Psychological, *Publication manual of the American Psychological Association* (6 ed.). Washington, DC: American Psychological Association (APA), 2010) suggest a lesser emphasis on significance tests and a greater emphasis on the substantive and practical significance of findings. Drawing on work by Cumming (*Understanding the new statistics: effect sizes, confidence intervals, and meta-analysis.* London: Routledge, 2012) we show how examinations of effect sizes (e.g., Cohen's d statistic) and confidence intervals can lead to a clear understanding of citation impact differences.

R. Williams (✉)
Department of Sociology, University of Notre Dame, 810 Flanner Hall,
Notre Dame, IN 46556, USA
e-mail: Richard.A.Williams.5@ND.Edu

L. Bornmann
Division for Science and Innovation Studies, Administrative Headquarters
of the Max Planck Society, Hofgartenstr. 8, 80539, Munich, Germany
e-mail: bornmann@gv.mpg.de

© Springer International Publishing Switzerland 2014
Y. Ding et al. (eds.), *Measuring Scholarly Impact*,
DOI 10.1007/978-3-319-10377-8_12

259

12.1   Introduction

Researchers in many fields, including bibliometricians, have typically focused on the statistical significance of results. Often, however, relatively little attention has been paid to substantive significance. Chuck Huber (2013, p. 1) provides an example of the possible fallacies of the typical approaches:

> What if I told you that I had developed a new weight-loss pill and that the difference between the average weight loss for people who took the pill and those who took a placebo was statistically significant? Would you buy my new pill? If you were overweight, you might reply, "Of course!" ... Now let me add that the average difference in weight loss was only one pound over the year. Still interested? My results may be statistically significant but they are not practically significant. Or what if I told you that the difference in weight loss was not statistically significant—the p-value was "only" 0.06—but the average difference over the year was 20 lb? You might very well be interested in that pill. The size of the effect tells us about the practical significance. P-values do not assess practical significance.

The American Psychological Association (APA) (2010) has recently called on researchers to pay greater attention to the practical significance of their findings. Geoff Cumming (2012) has taken up that challenge in his book, *Understanding the New Statistics: Effect Sizes, Confidence Intervals, and Meta-Analysis*. The need for the methods he outlines is clear: despite the serious flaws of approaches that only examine statistical significance, Tressoldi, Giofre, Sella, and Cumming (2013) find that "Null Hypothesis Significance Testing without any use of confidence intervals, effect size, prospective power and model estimation, is the prevalent statistical practice used in articles published in Nature, 89 %, followed by articles published in Science, 42 %. By contrast, in all other journals [The New England Journal of Medicine, The Lancet, Neuropsychology, Journal of Experimental Psychology-Applied, and the American Journal of Public Health], both with high and lower impact factors, most articles report confidence intervals and/or effect size measures." In bibliometrics, it has been also recommended to go beyond statistical significance testing (Bornmann & Leydesdorff, 2013; Schneider, 2012).

In this chapter we review some of the key methods outlined by Cumming (2012), and show how they can contribute to a meaningful statistical analysis of percentiles. The percentile of a certain publication provides information about the citation impact this publication has achieved in comparison to other similar publications in the same subject category and publication year. Following Cumming's (2012) lead, we explain what effect sizes and confidence intervals (CIs) are. We further explain how to assess the ways in which the percentile scores of individual institutions differ from some predicted values and from each other; and how the proportions of highly cited papers (i.e., the top 10 % most frequently cited papers) can be compared across institutions. Throughout, our emphasis will be in not only demonstrating whether or not statistically significant effects exist, but in assessing whether the effects are large enough to be of practical significance.

We begin by discussing the types of measures that bibliometricians will likely wish to focus on when doing their research. Specifically, we argue that percentile rankings for all papers, and the proportion of papers that are among the top 10 % most frequently cited, deserve special consideration.

12.2 Percentile Rankings

Percentiles used in bibliometrics provide information about the citation impact of a focal paper compared with other comparable papers in a reference set (all papers in the same research field and publication year). For normalizing a paper under study, its citation impact is evaluated by its rank in the citation distribution of similar papers in the corresponding reference set (Leydesdorff & Bornmann, 2011; Pudovkin & Garfield, 2009). For example, if a paper in question was published in 2009 and was categorized by Thomson Reuters into the subject category "physics, condensed matter", all papers published in the same year and subject category build up its reference set. Using the citation ranks of all papers in the reference set, percentiles are calculated which also lead to a corresponding percentile for the paper in question. This percentile expresses the paper's citation impact position relative to comparable papers.

This percentile-based approach arose from a debate in which it was argued that frequently used citation impact indicators based on using arithmetic averages for the normalization—e.g., "relative citation rates" (Glänzel, Thijs, Schubert, & Debackere, 2009; Schubert & Braun, 1986) and "crown indicators"(Moed, De Bruin, & Van Leeuwen, 1995; van Raan, van Leeuwen, Visser, van Eck, & Waltman, 2010)—had been both technically (Lundberg, 2007; Opthof & Leydesdorff, 2010) and conceptually (Bornmann & Mutz, 2013) flawed. Among their many advantages, percentile rankings limit the influence of extreme outliers. Otherwise, a few papers with an extremely large number of citations could have an immense impact on the test statistics and parameter estimates.

An example will help to illustrate this. The Leiden Ranking uses citation impact indicators based on using arithmetic averages for the normalization (the mean normalized citation score, MNCS) and based on percentiles ($PP_{top\ 10\ \%}$, which measures the proportion of papers among the 10 % most frequently cited papers in a subject category and publication year). For the University of Göttingen, an extreme outlier leads to a large ranking position difference between MNCS and $PP_{top\ 10\ \%}$:

This university is ranked 2nd based on the MNCS indicator, while it is ranked 238th based on the $PP_{top\ 10\ \%}$ indicator. The MNCS indicator for University of Göttingen turns out to have been strongly influenced by a single extremely highly cited publication. This publication ... was published in January 2008 and had been cited over 16,000 times by the end of 2010. Without this single publication, the MNCS indicator for University of Göttingen would have been equal to 1.09 instead of 2.04, and University of Göttingen would have been ranked 219th instead of 2nd. Unlike the MNCS indicator, the $PP_{top\ 10\ \%}$ indicator is hardly influenced by a single very highly cited publication. This is because the $PP_{top\ 10\ \%}$ indicator only takes into account whether a publication belongs to the top 10 % of its field or not. The indicator is insensitive to the exact number of citations of a publication (Waltman et al., 2012, p. 2425).

Since the percentile approach has been acknowledged in bibliometrics as a valuable alternative to the normalization of citation counts based on mean citation rates, some different percentile-based approaches have been developed (see an overview in Bornmann, Leydesdorff, & Mutz, 2013). More recently, one of these approaches ($PP_{top\ 10\ \%}$, also known as the Excellence Rate) has been prominently used in the Leiden Ranking (Waltman et al., 2012) and the SCImago institutions ranking (Bornmann, de Moya Anegon, & Leydesdorff, 2012) as evaluation tools.

Three steps are needed in order to calculate percentiles for a reference set: First, all papers in the set are ranked in ascending order of their numbers of citations. Second, each paper is assigned a percentile based on its rank (percentile rank). Percentiles can be calculated in different ways (Bornmann et al., 2013; Cox, 2005; Hyndman & Fan, 1996). The most commonly used formula is $(100 * (i - 1)/n)$, where n is the total number of papers, and i the rank number in ascending order. For example, the median value or the 50th percentile rank separates the top-half of the papers from the lower half. However, one can also calculate percentiles as $(100 * (i/n))$. This calculation is used, for example, by InCites (Thomson Reuters, see below). Third, the minimum or maximum of the percentile rank can be adjusted. Papers with zero citations can be assigned a rank of zero. By assigning the rank zero to the papers with zero citations, one ensures that the missing citation impact of these papers is reflected in the percentiles in the same way in every case. Different ranks for papers with zero citations would arise if percentiles are calculated without using a constant rank of zero at the bottom (Leydesdorff & Bornmann, 2012; Zhou & Zhong, 2012).

A technical issue in the case of using percentiles for research evaluation pertains to the handling of ties (e.g., Pudovkin & Garfield, 2009; Schreiber, 2013; Waltman & Schreiber, 2013). Imagine 50 papers with 61, 61, 61, 58, 58, 58, 58, 58, 58, 58 citations, with the rest (40 papers) each receiving 1 citation. For this fictitious reference set it is not possible to calculate exactly the top 10 % most frequently cited papers. You can take 3/50 (6 %) or 10/50 (20 %). Thus, the tying of the ranks at the threshold level generates an uncertainty (Leydesdorff, 2012). Schreiber (2012) and Waltman and Schreiber (2013) solved this problem by proposing fractional counting in order to attribute the set under study to percentile rank classes that are predefined (for example, $PP_{top10\ \%}$).

By proportional attribution of the fractions to the different sides of the threshold, the uncertainty can be removed from the resulting indicator. However, this approach can only be used to determine the exact proportion of $PP_{top\ x\ \%}$ (e.g., with x = 10) papers in a reference set, but cannot be used for the calculation of percentile ranks of the individual papers under study. Furthermore, the fractional attribution of percentile ranks is computationally intensive. Since individual papers are the units of analysis in many studies, the fractional attribution of percentile ranks is not functional in many situations.

12.3   Data and Statistical Software

Publications produced by three research institutions in German-speaking countries from 2001 and 2002 are used as data. Institutions 1, 2, and 3 have 268, 549, and 488 publications, respectively, for 1,305 publications altogether. The percentiles for the publications were obtained from InCites (Thomson Reuters, http://incites. thomsonreuters.com/), which is a Web-based research evaluation tool allowing the assessment of the productivity and citation impact of institutions. Percentiles are defined by Thomson Reuters as follows: "The percentile in which the paper ranks in its category and database year, based on total citations received by the paper. The higher the number of citations, the smaller the percentile number. The maximum percentile value is 100, indicating 0 cites received. Only article types *article*, *note*, and *review* are used to determine the percentile distribution, and only those same article types receive a percentile value. If a journal is classified into more than one subject area, the percentile is based on the subject area in which the paper performs best, i.e., the lowest value" (see http://incites.isiknowledge.com/common/help/h_ glossary.html). Since in a departure from convention low percentile values mean high citation impact (and vice versa), the percentiles received from InCites are inverted percentiles. To identify papers which belong to the 10 % most frequently cited papers within their subject category and publication year ($P_{top\ 10\ \%}$), publications from the universities with an inverted percentile smaller than or equal to 10 are coded as 1; publications with an inverted percentile greater than 10 are coded as 0.

For the calculation of the statistical procedures, we used Stata (StataCorp, 2013). However, many other statistical packages could also be used for these calculations (e.g., SAS or R).

12.4   Effect Sizes and related concepts

Cumming (2012, p. 34) defines an effect size as the amount of something that might be of interest. He offers several examples. In Table 12.1, we present measures of effect size that we think are of special interest to bibliometricians.

In isolation, however, effect sizes have only limited utility. First, because of sampling variability, estimated effect sizes will often be larger or smaller than the true effect is, i.e., just by chance alone an institution's performance could appear to be better or worse than it truly is; or apparent differences between institutions could seem larger or smaller than they actually are. Second, we need a criterion by which effect sizes can be evaluated. A common criterion is to look at statistical significance, e.g., are the differences between two institutions so large that they are unlikely to be due to chance alone? The APA, however, has called on researchers to go beyond statistical significance and assess substantive significance as well. This can be done both via theory (e.g., theory or past experience might say that a

Table 12.1 Examples of effect size measures for bibliometric analyses

Sample effect size	Example
Mean (M)	Mean percentile ranking, e.g., institution's average percentile ranking is 20
Difference between two means	Institution A's average percentile ranking is 40, Institution B's is 50, for a 10 % point difference
Cohen's d (both for individual institutions and for institutional comparisons)	The average effect of institution type (A or B) on percentile rankings is .25
Proportion[a]	20 % of the institution's publications are $PP_{top\ 10\ \%}$
Relative proportions and/or differences in proportions[a]	Institution B is twice as likely to have $PP_{top\ 10\ \%}$ as is institution A

[a]Cumming (2012) uses the terms risk and relative risk. His examples refer to accidents. But we can also think of "risk" as pertaining to other events that might happen, e.g., a published paper is "at risk" of becoming highly cited

Table 12.2 Effect sizes and significance tests using mean percentile rankings for individual institutions

Statistical measure	Institution 1	Institution 2	Institution 3
Mean	49.67	32.15	45.98
Standard deviation	30.66	27.49	29.40
Standard error of the mean	1.87	1.17	1.33
Lower bound of the 95 % CI	45.99	29.85	43.37
Upper bound of the 95 % CI	53.36	34.46	48.59
T (for test of $\mu = 50$)	−0.17	−15.21	−3.02
N	268	549	488
P value (two-tailed test)	.8613	.0000	.003
Cohen's d	−.011	−.649	−.137

5-point difference between institutions is substantively important while a 1-point difference is not) and via empirical means (using suggested guidelines for when effects should be considered small, moderate, or large).

Therefore, when discussing effect sizes, we present not only the measure of effect size itself, but the related measures that are needed to assess the statistical and substantive significance of the measure. We begin with the mean percentile ranking.

Table 12.2 presents the mean percentile rankings for the three institutions along with related measures.

The mean is one of the simplest and most obvious measures. It is simply the arithmetic average of the rankings for all the papers published for an institution. Because of the way percentile ranking is coded (see above), a lower score is a better score. There are obvious ways to use the mean to assess the citation impact of an institution. The population mean (50) is known. We can tell at a glance whether an institution is above the average or below it. However other criteria can also be used. An institution may wish to compare itself with the known values of its peer institutions or aspirational peers. Peer institutions might include other elite universities; schools located in the same general geographic area; or colleges that an institution competes against for students and grant money. Aspirational peers might include schools that are known to currently be stronger but who set a standard of excellence that the school is striving to achieve. Hence a university that considers itself among the best in the country or even the world might feel that its publications should average at least among the top 25 %. Conversely a school that sees itself as less research oriented might feel that an average ranking of 75 is sufficient if that is what its regional competitors are achieving.

Table 12.2 shows us that Institution 1 scores barely better than the population mean (49.67), Institution 3 is about 4 points better than the mean (45.98), while Institution 2 is nearly 18 points better than average (32.15).

However, the mean percentile of an institution should NOT be the only measure used to assess citation impact. The mean is merely a point estimate. Chance factors alone could increase or lower the value of the estimated mean. CIs (Confidence Intervals) therefore provide a more detailed way of assessing the importance of mean scores. Cumming (2012) and others discuss several ways in which CIs can be interpreted and used for assessment purposes. CIs provide a feel for the precision of measures. Put another way, they show the range that the true value of the mean may plausibly fall in. For example, if the observed mean was 40, the 95 % CI might range between 35 and 45. So, while 40 is our "best guess" as to what the mean truly is, values ranging between 35 and 45 are also plausible alternative values.

CIs also provide an approach to hypothesis testing. If the hypothesized value (e.g., the population mean of 50) falls within the CI, we do not reject the null hypothesis. Put another way, if the hypothesized value of 50 falls within the CI, then 50 is a plausible alternative value for the mean and hence cannot be ruled out as a possibility.[1]

Table 12.2 shows us the CIs for each of the three institutions, but Fig. 12.1 provides a graphical display of the same information that may be easier to understand. At a glance, we can see what the mean for each institution is and what the range of plausible values for the mean is. The horizontal line for mean = 50 makes it easy to see whether the CI does or does not include the value specified by the null hypothesis (the citation impact is equal to a medium impact, i.e., the population mean of 50). If the horizontal line passes through the CI we do not reject the null hypothesis; otherwise we do.

For Institution 1, the CI ranges between 45.99 (about 4 points better than average) to 53.36. Because the average value of 50 falls within that interval, we cannot rule out any of the possibilities that Institution 1 is below average, average,

---

[1] Cumming (2012) refers to the CI obtained from an analysis as "One from the dance." What he means is that it is NOT correct to say that there is a 95 % chance that the true value of the mean lies within the confidence interval. Either the true value falls within the interval or it doesn't. It is correct to say that, if this process were repeated an infinite number of times, then 95 % of the time the CI would include the true value of the mean while 5 % of the time it would not. Whether it does in the specific data we are analyzing, we don't know.

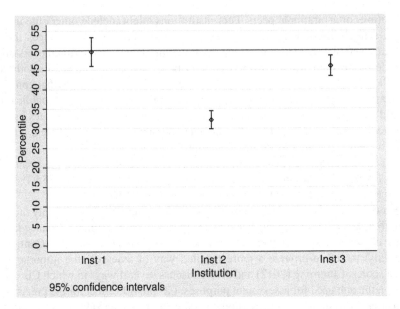

Fig. 12.1 Average percentile score by institution, with 95 % CIs

or above average. The CI for Institution 2 ranges from 29.85 to 34.46, suggesting that it is almost certainly well above average. For Institution 3 the CI range is 43.37–48.59, implying that it is probably at least a little better than average.

Significance tests (in this case, one-sample t-tests) can also be used to see whether the difference between the observed mean for an institution and the hypothesized mean is statistically significant. In our case we test whether the institutional mean differs from the known population mean of 50, but another criterion could be used if it was thought that a higher or lower criterion was appropriate. The formula for the one-sample t-test, along with the specific calculation for Institution 2, is

$$t = \frac{\bar{x} - \mu_0}{\frac{s}{\sqrt{N}}} = \frac{32.15 - 50}{\frac{27.49}{\sqrt{549}}} = \frac{-17.85}{1.1732} = -15.21 \qquad (12.1)$$

where $\bar{x} =$ the sample mean, μ_0 is the value for the mean specified under the null hypothesis (in this case 50), s is the standard deviation of x, and N is the sample size. $s/\sqrt{N}$ is also known as the standard error of the mean. Both the standard deviation of x and the standard error of the mean are reported in Table 12.2.

If the null hypothesis is true—in this case, if the population mean of Institution 2 really is 50—the t statistic will have a t distribution with $N-1$ degrees of freedom. The larger in magnitude the t statistic is (positive or negative), the less likely it is that the null hypothesis is true. The critical value for the t statistic, i.e., the value at which we conclude that the null hypothesis is unlikely to be true, depends on the sample size. In samples this large, the absolute value of the t statistic needs to be 1.96 or greater for us to conclude that observed deviations from the null hypothesis are probably not just due to chance factors alone. The *t*-test in Table 12.2 shows us that the difference between the mean for Institution 1 and the population mean is not statistically significant. The *t* values for Institutions 2 and 3 are statistically significant, but the *t* value for Institution 2 is more than five times as large as the *t* value for Institution 3.

Significance tests have their own limitations though. In particular, statistically significant results are not necessarily substantively meaningful. Significance tests are strongly affected by sample size. If the sample is large enough even trivial differences can be statistically significant. Conversely if the sample is small even a seemingly large difference may not achieve statistical significance. For example, if the sample is large enough, a mean score of 49 may statistically differ from the population mean of 50 at the .05 level of significance. Conversely if a sample is small a mean score of 40 might only be statistically significant at, say, the .06 level. As argued earlier, while significance tests can be helpful, their utility is also limited.

To make this important point clear, consider an example that is similar to our opening weight loss example: If you were told that an institution had scores that were statistically significantly above average, would you be impressed? Perhaps. But if you were also told that it was 1 point better than average and that this was statistically significant at the .04 level, would you still be impressed? Probably not. Conversely, if you were told that an institution's scores were not statistically significant from the average, would you be unimpressed? Perhaps. But if you were told that its observed score was 10 points better than average and that the difference was statistically significant at the .06 level, would you be impressed then? Probably in most cases more people would be impressed by the latter institution, even though its higher scores barely missed being statistically significant at the .05 level.

We may have enough of a theoretical or intuitive feel to decide whether an effect is large enough to care about, e.g., theory or intuition or past experience may tell us that a 1-point difference from the mean is not worth caring about while a 10-point difference is. However, in situations that are less clear, measures such as Cohen's d give us another way of assessing the substantive significance of effects.

12.5 Cohen's d (for Individual Institutions)

Table 12.2 also includes the Cohen's d statistic for each institution. Cohen's d, and related measures (e.g., Cohen's h, see below), try to illustrate the magnitude of an effect. Put another way, they try to illustrate the substantive importance, as opposed to the statistical significance, of results. Cohen's d and similar measures may be especially useful when it is not otherwise clear whether an effect is large or small. So, for example, if an institution scores 3 points above the mean, should that be considered a substantively large difference, a substantively small difference, or what?

Those with expertise in a well-understood field may be able to offer an informed opinion on that. But in less clear situations, Cohen's d provides a potentially superior means of assessing the substantive significance of results as opposed to simply noting what the observed differences are. As Cumming (2012) notes, Cohen's d "can help ES [effect size] communication to a wide range of readers, especially when the original units first used to measure the effect are not widely familiar" (p. 282). Cumming (2012) also notes that Cohen's d can be useful in meta-analysis where different researchers have measured key variables in different ways.

In the single sample case, Cohen's d equals the difference between the observed mean and the hypothesized mean (e.g., 50) divided by the sample standard deviation, i.e.,

$$\text{Cohen's d} = \frac{\bar{x} - \mu_0}{s} \tag{12.2}$$

Through simple algebra, it can also be shown that Cohen's d $= t/\sqrt{N}$. So, for example, the Cohen's d value for institution 2 is

$$\text{Cohen's d} = \frac{\bar{x} - \mu_0}{s} = \frac{32.15 - 50}{27.49} = \frac{-17.85}{27.49} = -.649 \tag{12.3}$$

As Cohen (1988) notes, Cohen's d is similar to a z-score transformation. For example, a Cohen's d value of .2 would mean that the mean in the sample was .2 standard deviations higher than the hypothesized mean. Cohen (1988) suggested that effect sizes of 0.2, 0.5, and 0.8 (or, if the coding is reversed, $-.2$, $-.5$ and $-.8$) correspond to small, medium, and large effects.

Put another way, the usefulness of Cohen's d depends, in part, on how obvious the meaning is of observed differences. If, for example, we knew that students in an experimental teaching program scored one grade level higher than their counterparts in traditional programs, such a difference might have a great deal of intuitive meaning to us. But if instead we knew that they scored 7 points higher on some standardized test, something like Cohen's d could help us to assess how large such a difference really is.[2]

Returning to table 12.2 and our three institutions, the Cohen's d statistic for Institution 1 is, not surprisingly, extremely small, almost 0. For Institution 3, even though the difference between the institution's sample mean and the population mean 50 is statistically significant, the Cohen's d statistic is only $-.137$. This falls below the value of $-.2$ that Cohen (1988) had suggested represented a small effect. For Institution 2, Cohen's d is equal to $-.649$. This falls almost exactly halfway between Cohen's suggested values of $-.5$ for medium and $-.8$ for large.

---

[2] Cumming (2012) notes various cautions about using Cohen's d (p. 283). For example, while it is common to use sample standard deviations as we do here, other "standardizers" are possible, e.g., you might use the standard deviation for a reference population, such as elite institutions. Researchers should be clear exactly how Cohen's d was computed.

Table 12.3 Effect sizes and significance tests for differences in percentile rankings across institutions

Statistical measure	Institution 1 vs. Institution 2	Institution 1 vs. Institution 3	Institution 3 vs. Institution 2
Difference between means	17.52	3.69	13.83
Standard deviation (pooled)	28.57	29.85	28.40
Standard error of the mean difference	2.13	2.27	1.77
Lower bound of the 95 % CI for the difference	13.34	−.76	10.36
Upper bound of the 95 % CI for the difference	21.70	8.15	17.30
T (for test of μs are equal)	8.23	1.63	7.83
P value (two-tailed test)	.0000	.1042	.0000
Cohen's d	.613	.124	.487

Therefore, if we had no other clear criteria for assessing the magnitude of effects, Cohen's d would lead us to conclude that differences between institutions 1 and 3 and the population mean of 50 are not substantively important, while the difference between Institution 2 and the population mean is moderately to highly important.

Before leaving Table 12.2, it should be noted that we did additional analyses to confirm the validity of our results. In Table 12.2 (and also in Table 12.3), we make heavy use of t-tests and related statistics. As Acock (2010) points out, t-tests assume that variables are normally distributed; and, when two groups are being compared, it is often assumed that the variances of the two groups are equal. Percentile rankings violate these assumptions in that they have a uniform, rather than normal, distribution. However, Acock (2010) also adds that t-tests are remarkably robust against violations of assumptions.

Nonetheless, to reassure ourselves that our results are valid, we double-checked our findings by using techniques that are known to work well when distributional assumptions are violated. In particular, for both Tables 12.2 and 12.3, we verified our findings using bootstrapping techniques. Bootstrapping is often used as an alternative to inference based on parametric assumptions when those assumptions are in doubt (Cameron & Trivedi, 2010). Bootstrapping resamples observations (with replacement) multiple times. Standard errors, confidence intervals, and significance tests can then be estimated from the multiple resamples. Bootstrapping produced significance tests and confidence intervals that were virtually identical to those reported in our tables, giving us confidence that our procedures are valid.

12.6   Mean Differences Between Institutions

Rather than use some absolute standards for assessment (e.g., is the average score for an institution above or below the population average?) we may wish to compare institutions against each other. For example, a school might wish to compare itself against a school that it considers its rival, or that competes for students in the same geographic area. Does one institution have average scores that are significantly higher than the others, or are their scores about the same? Alternatively, we might want to compare the same institution at two different points in time—have its average scores gotten better across time or have they gotten worse? Table 12.3 presents such comparisons for the institutions in our study.

As Table 12.3 shows, Institution 1 scores 17.52 points worse than Institution 2. Similarly, Institution 2 averages almost 14 points better than Institution 3. The mean difference between institutions 1 and 3 is a much more modest 3.69 points.

Again, simply comparing the means for two institutions is not adequate. Apparent differences may not be statistically significant; just by chance alone one institution could have scored higher than the other. And even if more than chance was likely involved in the differences, the substantive significance of differences still needs to be assessed.

CIs can again be useful. Referring back to Fig. 12.1, we can see whether the 95 % CIs for two institutions overlap. As Cumming (2012) notes, if they do, then the difference between the institutions is not statistically significant at the .01 level. A common error is to assume that if two 95 % CIs overlap then the difference in values is not statistically significant at the .05 level. This is wrong because it is unlikely that, by chance alone, one variable would have an atypically low observed value while the other would have a value that was atypically high.

Even more useful is that we can compute the CI for the difference between the scores of two institutions. If 0 falls within the 95 % CI, then the difference between the two groups is not statistically significant. Or, if the observed difference is 10 but the CI ranges between 5 and 15, then the actual difference could plausibly be as low as 5 points or as much as 15. Figure 12.2 provides a graphical and possibly clearer illustration of the information that is also contained in Table 12.3. If the horizontal line at $y = 0$ crosses the CI, we know that the difference between the two means is not statistically significant. The CIs show that the differences between 1 and 3 are modest or even nonexistent while the differences between 2 and the other institutions are large (10 points or more) even at the lower bounds of the CIs.

Significance tests (in this case an independent sample *t*-test) can again be used. Because two groups are being compared, the calculations are somewhat more complicated but still straightforward. It is often assumed that the two groups have the same variance.[3] But, in the samples there are separate estimates of the variance

---

[3] With independent samples there are two different types of t-tests that can be conducted. The first type, used here, assumes that the variances for each group are equal. The second approach allows the variances for the two groups to be different. In our examples, it makes little difference which

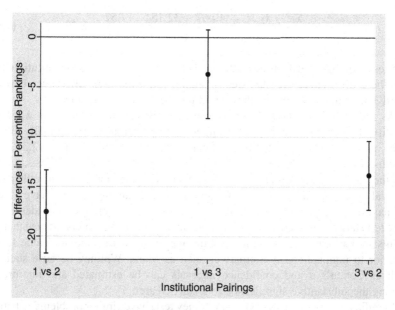

Fig. 12.2 Differences in mean percentile rankings with 95 % CIs

for each group. A pooled estimate for the variance of the two groups is therefore estimated as follows (again we show the general formula, and the specific calculation for institutions 1 and 2).

$$s_p = \sqrt{\frac{(n_1 - 1)s_1^2 + (n_2 - 1)s_2^2}{n_1 + n_2 - 2}} = \sqrt{\frac{(268 - 1)30.66^2 + (549 - 1)27.49^2}{815}}$$
$$= \sqrt{816.098} = 28.57 \tag{12.4}$$

The standard error of the difference (again, both in general, and specifically for institutions 1 and 2) is

$$se_p = \sqrt{s_p^2 \left(\frac{N_1 + N_2}{N_1 N_2}\right)} = \sqrt{28.57^2 \left(\frac{249 + 568}{249 * 568}\right)}$$
$$= \sqrt{28.57^2 \left(\frac{817}{141432}\right)} = 2.13 \tag{12.5}$$

The *t*-test (both in general and for institutions 1 and 2) is

---

approach is used, since, as Table 12.2 shows, the standard deviations for the three groups are similar. In cases where the variances do clearly differ the second approach should be used. Most, perhaps all, statistical software packages can compute either type of *t*-test easily.

$$t = \frac{\overline{x_1} - \overline{x_2}}{se_p} = \frac{49.67 - 32.15}{2.13} = \frac{17.52}{2.13} = 8.23 \qquad (12.6)$$

The tests confirm that Institution 2 does better than the other two institutions and that these differences are statistically significant, while the differences between 1 and 3 are small enough that they could just be due to chance factors.

As noted earlier, bootstrapping techniques, which are often used when the validity of parametric assumptions is in doubt (e.g., variables are normally distributed), produced results virtually identical to those reported in Table 12.3. As an additional check, for Table 12.3 we also computed Mann–Whitney tests. Mann–Whitney tests are appropriate when dependent variables have ordinal rather than interval measurement (Acock, 2010), and percentile rankings clearly have at least ordinal measurement. The Mann–Whitney test statistics were virtually identical to the t-test statistics we reported in the table, again increasing our confidence that our results are valid. Further, we think that the approach we are using for Table 12.3 is superior to nonparametric alternatives such as Mann–Whitney because statistics such as Cohen's d and confidence intervals can be estimated and interpreted, making the substantive significance of results clearer.

Significance tests (t-test or Mann–Whitney test) have similar problems as before. If sample sizes are large, even small differences between the two groups can be statistically significant, e.g., a difference of only 1 point could be statistically significant if the samples are large enough. Conversely, even much larger differences (e.g., 10 points) may fail to achieve significance at the .05 level if the samples are small.

To better assess substantive significance, Cohen's d can be calculated for the difference between means. The formula (both in general and for institutions 1 and 2 is)

$$\text{Cohen's d} = \frac{\overline{x_1} - \overline{x_2}}{s_p} = \frac{49.67 - 32.15}{28.57} = \frac{17.52}{28.57} = .613 \qquad (12.7)$$

The Cohen's d indicates that the differences between Institution 2 and Institution 1 (.613) and between Institution 2 and Institution 3 (.487) are at least moderately large. Conversely, the Cohen's d statistic of .124 for comparing institutions 1 and 3 falls below Cohen's suggested level for a small effect.

12.7 Proportions (Both for One Institution and for Comparisons Across Institutions)

As noted above, one way of evaluating institutions is to see how their average scores compare. However, it could be argued that evaluations should be made, not on average scores, but on how well an institution's most successful publications

Table 12.4 Effect sizes and significance tests for PP_{top} 10 %—individual institutions

Statistical measure	Institution 1	Institution 2	Institution 3
PP_{top} 10 %[a]	11.19	29.14	11.68
Standard error[a]	1.93	1.94	1.45
Lower bound of the 95 % CI[a]	7.42	25.34	8.83
Upper bound of the 95 % CI[a]	14.97	32.95	14.53
Z (for test of PP_{top} 10 % = .10)	.65	14.95	1.24
P value (two-tailed test)	.51	.0000	.22
Cohen's h	.039	.497	.054
N	268	549	488

[a]Numbers are multiplied by 100 to convert them into percentages

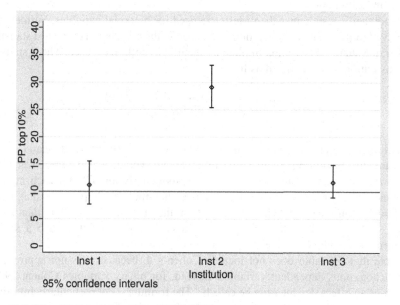

Fig. 12.3 PP_{top} 10 % by institution, with 95 % CIs

do. In particular, what proportion of an institution's publications rank among the 10 % most frequently cited papers?

Again, there is an obvious criterion: overall we know that 10 % of all papers rank among the top 10 % of those most cited. We use that criterion here, but the criterion could be made higher or lower as deemed appropriate for the type of institution.

There are important differences in how statistics and significance tests are computed for binary outcomes. Binary variables do not have a normal distribution, nor are their means and variances independent of each other. If the mean of $Y = P$, then $V(Y) = P(1 - P)$, e.g., if there is a .3 probability that $Y = 1$, then $V(Y) = .3 * 7 = .21$. As the Stata 13 Reference Manual (2013) points out, several different formulas have been proposed for computing confidence intervals (e.g., the

Wilson, Agresti, Klopper-Pearson, and Jeffries methods) and several other statistics. We use the large sample methods used by Stata's prtest command, but researchers, especially those with smaller samples, may wish to explore other options.

But, other than that, the arguments are largely the same as earlier. Significance tests and CIs have the same strengths and weaknesses as before. Visual means of presenting results are also quite similar. Effect size measures help to provide indicators of the substantive significance of findings. In short, the most critical difference from before is that a different criterion is being used for the assessment of impact.

Table 12.4 presents the effect sizes and related measures for $PP_{top\ 10\ \%}$ for each institution separately. Figure 12.3 illustrates how the proportions and their CIs can be graphically depicted.

Note that Z-tests rather than t-tests are used for binomial variables. Further, the observed sample standard deviation is not used in the calculation of the test statistic; rather, the standard deviation implied by the null hypothesis is used. The Z statistic for Institution 2 is calculated as follows:

$$z = \frac{p - p_0}{\sqrt{p_0(1 - p_0)/N}} = \frac{.2914 - .10}{\sqrt{.1 * .9/549}} = \frac{.1914}{.0128} = 14.95 \qquad (12.8)$$

The results are very consistent with what we saw when we analyzed mean percentile rankings. Institutions 1 and 3 are slightly above average in that a little over 11 % of their papers rank in the $PP_{top\ 10\ \%}$. Correspondingly, the CIs for each include 10, and the significance tests also indicate that the null hypothesis that $PP_{top\ 10\ \%} = 10$ cannot be rejected. Institution 2, on the other hand, has more than 29 % $PP_{top\ 10\ \%}$. Both the CIs and the significance test indicate that such a strong performance is highly unlikely to be due to chance alone.

Note that the table does not include Cohen's d, because it is not appropriate for dichotomous dependent variables. Instead, for binary variables Cohen, 1988) proposes an effect size measure he calls h.[4] The formula is not particularly intuitive, but it has several desirable properties. h is calculated as follows:

$$\begin{aligned} o &= 2 * \arcsin\left(\sqrt{P}\right), \\ o_0 &= 2 * \arcsin\left(\sqrt{P_0}\right), \\ h &= o - o_0 \end{aligned} \qquad (12.9)$$

So, for example, for Institution 2, P = .2914 ($PP_{top\ 10\ \%} = 29.14$). Since we are using $P_0 = .10$, The h value for institution 2 is

---

[4] Nonetheless, as we found for other measures in our analysis, Cohen's d seems robust to violations of its assumptions. When we estimated Cohen's d using binary dependent variables, we got almost exactly the same numbers as we did for Cohen's h.

Table 12.5 Effect sizes and significance tests for differences in PP_{top} 10 % across institutions

Statistical measure	Institution 1 vs Institution 2	Institution 1 vs Institution 3	Institution 3 vs Institution 2
Difference between proportions	−17.95	−0.49	−17.47
Standard error	2.73	2.43	2.42
Lower bound of the 95 % CI for the difference	−23.31	−5.22	−22.21
Upper bound of the 95 % CI for the difference	−12.59	4.24	−12.71
Z (for test of PP_{top}10 % are equal)	−5.70	−0.20	−6.90
Cohen's h	−.458	−.015	−.443
P value (two-tailed test)	.0000	.8411	.0000

$$o = 2 * \arcsin(\sqrt{P}) = 2 * \arcsin(\sqrt{.2914}) = 1.1404341,$$
$$o_0 = 2 * \arcsin(\sqrt{P_0}) = 2 * \arcsin(\sqrt{.10}) = .64350111, \qquad (12.10)$$
$$h = (o - o_0) = (1.1404341 - .64350111) = .497$$

According to Cohen (1988) the suggested small, medium and large values of .2, .5, and .8, respectively, continue to be reasonable choices for the h statistic, at least when there is little guidance as to what constitutes a small, medium or large effect. He further notes that in the fairly common case of $P_0 = .5$, h will equal approximately $2 * (P - .5)$.

Table 12.5 presents the corresponding measures for institutional comparisons.

Again, the results are very consistent with before. The differences between institutions 1 and 3 are very slight and may be due to chance factors alone. Institution 2, on the other hand, has more than twice as many $PP_{top\ 10\ \%}$ as do institutions 1 and 3, and the differences are highly statistically significant. The calculation of Cohen's h is similar to before, except that P_2 is substituted for P_0, e.g., for institutions 1 and 2.

$$o_1 = 2 * \arcsin(\sqrt{P_1}) = 2 * \arcsin(\sqrt{.1119}) = .68218016,$$
$$o_2 = 2 * \arcsin(\sqrt{P_2}) = 2 * \arcsin(\sqrt{.2914}) = 1.1404341, \qquad (12.11)$$
$$h = (o_1 - o_2) = (.68218016 - 1.1404341) = -.458$$

Incidentally, several other measures of effect size have been proposed and are widely used for analyses of binary dependent variables. These include risk ratios, odds ratios, and marginal effects. For a discussion of some of these methods, see Williams (2012), Bornmann and Williams (2013), Deschacht and Engels (this book), and Long and Freese (2006).

Conclusions

The APA has called on researchers to employ techniques that illustrate both the statistical and the substantive significance of their findings. Similarly, in the statistics paragraph of the *Uniform Requirements for Manuscripts* (URM) of the International Committee of Medical Journal Editors (ICMJE, 2010) it is recommended to "describe statistical methods with enough detail to enable a knowledgeable reader with access to the original data to verify the reported results. When possible, quantify findings and present them with appropriate indicators of measurement error or uncertainty (such as confidence intervals). Avoid relying solely on statistical hypothesis testing, such as P values, which fail to convey important information about effect size."

In this chapter, we have shown that the analysis of effect sizes for both means and proportions are worthwhile, but must be accompanied by criteria with which the statistical and the substantive significance of effect sizes can be assessed. Measures of statistical significance are, in general, well known, but we have shown how they can be applied to bibliometric data. Assessment of substantive significance depends, in part, on theory or empirical means: how large does an effect size need to be in order to be considered important? But, when theory and empirical evidence are unclear, measures such as Cohen's d can provide guidelines for assessing effects. As we have seen, effects that are statistically significant may not have much substantive importance. Conversely there may be situations where effects fail to achieve statistical significance but may nonetheless have a great deal of substantive significance. Using tools presented in this paper and in Cumming's (2012) book, researchers can assess both the statistical and substantive significance of their findings.

For those who would like to replicate our findings or try similar analyses with their own data, the Appendix shows the Stata code for the analyses presented in this chapter and for the additional double-checks we did to verify the validity of our results.

Appendix: Stata Code Used for These Analyses

```
* Stata code for Williams & Bornmann book chapter on effect sizes.
* Be careful when running this code - make sure it doesn't
* overwrite existing files or graphs that use the same names.
version 13.1
use "http://www3.nd.edu/~rwilliam/statafiles/rwlbes", clear
gen inst12 = inst if inst!=3
gen inst13 = inst if inst!=2
gen inst23 = inst if inst!=1
```

```
gen top10 = perc <= 10
* Limit to 2001 & 2002; this can be changed
keep if py <=2002
¶
* Table 12.2
* Single group designs - pages 286-287 of Cumming
* For each institution, test whether percentile mu = 50
* Note that negative differences mean better than average performance
forval instnum = 1/3 {
  display
  display "Institution `instnum'"
  ttest perc = 50 if inst==`instnum'
  display
  display "Cohen's d = " r(t) / sqrt(r(N_1))
  * DOUBLE CHECK: Compares above CIs and t-tests with bootstrap
  * Results from the test command should be similar to the t-test
  * significance level
  bootstrap, reps(100): reg perc if inst==`instnum'
  test _cons = 50
}
¶
* Table 12.3
* Two group designs - Test whether two institutions
* differ from each other on mean percentile rating.
* Starts around p. 155
* Get both the t-tests and the ES stats, e.g. Cohen's d
* Note: you should flip the signs for the 3 vs 2 comparison
¶
foreach iv of varlist inst12 inst13 inst23 {
  display   "perc is dependent, `iv'"
¶
  ttest perc, by(`iv')
  scalar n1 = r(N_1)
  scalar n2 = r(N_2)
  scalar s1 = r(sd_1)
  scalar s2 = r(sd_2)
  display
  display "Pooled sd is " ///
    sqrt(((n1 - 1) * s1^2 + (n2 - 1) * s2^2 ) / (n1 + n2 - 2))
  display
  esize two perc, by(`iv') all
  display
  * DOUBLE CHECKS: Compare Mann-Whitney & bootstrap results with above
  * Mann-Whitney test
  ranksum perc, by(`iv')
```

```
  * Bootstrap
  bootstrap, rep(100): reg perc i.`iv'
}
¶
¶
* Table 12.4
* Proportions in Top 10, pp. 399-402
* Single institution tests
* Numbers in table are multiplied by 100
forval instnum = 1/3 {
  display
  display "Institution `instnum'"
  prtest top10 = .10 if inst==`instnum'
  display
  display
  scalar phi1 = 2 * asin(sqrt(r(P_1)))
  scalar phi2 = 2 * asin(sqrt(.10))
  di "h effect size = " phi1 - phi2
  display
}
¶
¶
* Table 12.5
* Proportions in Top 10 - pairwise comparisons of institutions
* Numbers in table are multiplied by 100
foreach instpair of varlist inst12 inst13 inst23 {
  display
  display "`instpair'"
  prtest top10, by (`instpair')
  display
  scalar phi1 = 2 * asin(sqrt(r(P_1)))
  scalar phi2 = 2 * asin(sqrt(r(P_2)))
  di "h effect size = " phi1 - phi2
  display
  * NOTE: Cohen's d provides very similar results to Cohen's h
  esize two top10, by (`instpair') all
  display
}
* Do graphs with Stata
* NOTE: Additional editing was done with the Stata Graph Editor
* Use ciplot for Univariate graphs
¶
* Figure 12.1 - Average percentile score by inst with CI
ciplot perc, by(inst) name(fig1, replace)
¶
```

```
* Figure 12.3
* Was edited to multiply by 100
ciplot top10, bin by(inst) name(fig3, replace)
¶
*** Save figures before running figure 12.2 code
¶
* Figure 12.2 - Differences in mean percentile rankings
* Use statsby and serrbar for tests of group differences
* Note: Data in memory is overwritten
gen inst32 = inst23 * -1 + 4
tab2 inst32 inst23
statsby _b _se, saving(xb12, replace) : reg perc i.inst12
statsby _b _se, saving(xb13, replace) : reg perc i.inst13
statsby _b _se, saving(xb32, replace) : reg perc i.inst32
clear all
append using xb12 xb13 xb32, gen(pairing)
label define pairing 1 "1 vs 2" 2 "1 vs 3" 3 "3 vs 2"
label values pairing pairing
serrbar _stat_2 _stat_5 pairing, scale(1.96) name(fig2, replace)
```

References

Acock, A. (2010). *A gentle introduction to Stata* (3rd ed.). College Station, TX: Stata Press.

Association American Psychological. (2010). *Publication manual of the American Psychological Association* (6th ed.). Washington, DC: American Psychological Association (APA).

Bornmann, L., & Leydesdorff, L. (2013). Statistical tests and research assessments: A comment on Schneider (2012). *Journal of the American Society for Information Science and Technology, 64* (6), 1306–1308. doi:10.1002/asi.22860.

Bornmann, L., Leydesdorff, L., & Mutz, R. (2013). The use of percentiles and percentile rank classes in the analysis of bibliometric data: opportunities and limits. *Journal of Informetrics, 7* (1), 158–165.

Bornmann, L., & Mutz, R. (2013). The advantage of the use of samples in evaluative bibliometric studies. *Journal of Informetrics, 7*(1), 89–90. doi:10.1016/j.joi.2012.08.002.

Bornmann, L., & Williams, R. (2013). How to calculate the practical significance of citation impact differences? An empirical example from evaluative institutional bibliometrics using adjusted predictions and marginal effects. *Journal of Informetrics, 7*(2), 562–574. doi:10.1016/j.joi.2013.02.005.

Bornmann, L., de Moya Anegon, F., & Leydesdorff, L. (2012). The new Excellence Indicator in the World Report of the SCImago Institutions Rankings 2011. *Journal of Informetrics, 6*(2), 333–335. doi: 10.1016/j.joi.2011.11.006.

Cameron, A. C. & Trivedi, P. K. (2010). *Microeconomics using Stata* (Revised ed.). College Station, TX: Stata Press.

Cohen, J. (1988). *Statistical power analysis for the behavioral sciences* (2nd ed.). Hillsdale, NJ: Lawrence Erlbaum Associates, Publishers.

Cox, N. J. (2005). Calculating percentile ranks or plotting positions. Retrieved May 30, from http://www.stata.com/support/faqs/stat/pcrank.html

Cumming, G. (2012). *Understanding the new statistics: Effect sizes, confidence intervals, and meta-analysis*. London: Routledge.

Glänzel, W., Thijs, B., Schubert, A., & Debackere, K. (2009). Subfield-specific normalized relative indicators and a new generation of relational charts: methodological foundations illustrated on the assessment of institutional research performance. *Scientometrics, 78*(1), 165–188.

Huber, C. (2013). Measures of effect size in Stata 13. *The Stata Blog*. Retrieved December 6, 2013, from http://blog.stata.com/2013/09/05/measures-of-effect-size-in-stata-13.

Hyndman, R. J., & Fan, Y. N. (1996). Sample quantiles in statistical packages. *American Statistician, 50*(4), 361–365.

International Committee of Medical Journal Editors. (2010). Uniform requirements for manuscripts submitted to biomedical journals: Writing and editing for biomedical publication. *Journal of Pharmacology and Pharmacotherapeutics, 1*(1), 42–58. Retrieved April 10, 2014 from http://www.ncbi.nlm.nih.gov/pmc/articles/PMC3142758/.

Leydesdorff, L. (2012). Accounting for the uncertainty in the evaluation of percentile ranks. *Journal of the American Society for Information Science and Technology, 63*(11), 2349–2350.

Leydesdorff, L., & Bornmann, L. (2011). Integrated impact indicators (I3) compared with impact factors (IFs): An alternative research design with policy implications. *Journal of the American Society of Information Science and Technology, 62*(11), 2133–2146.

Leydesdorff, L., & Bornmann, L. (2012). Percentile ranks and the integrated impact indicator (I3). *Journal of the American Society for Information Science and Technology, 63*(9), 1901–1902. doi:10.1002/asi.22641.

Long, S., & Freese, J. (2006). *Regression models for categorical dependent variables using Stata* (2nd ed.). College Station, TX: Stata Press.

Lundberg, J. (2007). Lifting the crown - citation z-score. *Journal of Informetrics, 1*(2), 145–154.

Moed, H. F., De Bruin, R. E., & Van Leeuwen, T. N. (1995). New bibliometric tools for the assessment of national research performance - database description, overview of indicators and first applications. *Scientometrics, 33*(3), 381–422.

Opthof, T., & Leydesdorff, L. (2010). Caveats for the journal and field normalizations in the CWTS ("Leiden") evaluations of research performance. *Journal of Informetrics, 4*(3), 423–430.

Pudovkin, A. I., & Garfield, E. (2009). *Percentile rank and author superiority indexes for evaluating individual journal articles and the author's overall citation performance*. Paper presented at the Fifth International Conference on Webometrics, Informetrics & Scientometrics (WIS).

Schneider, J., & Schneider, J. (2012). Testing university rankings statistically: Why this is not such a good idea after all. Some reflections on statistical power, effect sizes, random sampling and imaginary populations. In E. Archambault, Y. Gingras, & V. Lariviere (Eds.), *The 17th International Conference on Science and Technology Indicators* (pp. 719–732). Montreal, Canada: Repro-UQAM.

Schreiber, M. (2012). Inconsistencies of recently proposed citation impact indicators and how to avoid them. *Journal of the American Society for Information Science and Technology, 63*(10), 2062–2073. doi:10.1002/asi.22703.

Schreiber, M. (2013). Uncertainties and ambiguities in percentiles and how to avoid them. *Journal of the American Society for Information Science and Technology, 64*(3), 640–643. doi:10.1002/asi.22752.

Schubert, A., & Braun, T. (1986). Relative indicators and relational charts for comparative assessment of publication output and citation impact. *Scientometrics, 9*(5–6), 281–291.

StataCorp. (2013). *Stata statistical software: Release 13*. College Station, TX: Stata Corporation.

Tressoldi, P. E., Giofre, D., Sella, F., & Cumming, G. (2013). High impact = high statistical standards? not necessarily so. *PLoS One, 8*(2). doi: 10.1371/journal.pone.0056180.

van Raan, A. F. J., van Leeuwen, T. N., Visser, M. S., van Eck, N. J., & Waltman, L. (2010). Rivals for the crown: Reply to Opthof and Leydesdorff. *Journal of Informetrics, 4*, 431–435.

Waltman, L., Calero-Medina, C., Kosten, J., Noyons, E. C. M., Tijssen, R. J. W., van Eck, N. J., et al. (2012). The Leiden Ranking 2011/2012: Data collection, indicators, and interpretation. *Journal of the American Society for Information Science and Technology, 63*(12), 2419–2432.

Waltman, L., & Schreiber, M. (2013). On the calculation of percentile-based bibliometric indicators. *Journal of the American Society for Information Science and Technology, 64*(2), 372–379.

Williams, R. (2012). Using the margins command to estimate and interpret adjusted predictions and marginal effects. *The Stata Journal, 12*(2), 308–331.

Zhou, P., & Zhong, Y. (2012). The citation-based indicator and combined impact indicator—new options for measuring impact. *Journal of Informetrics, 6*(4), 631–638. doi:10.1016/j.joi.2012.05.004.

Part IV
Visualization

Chapter 13
Visualizing Bibliometric Networks

Nees Jan van Eck and Ludo Waltman

Abstract This chapter provides an introduction to the topic of visualizing bibliometric networks. First, the most commonly studied types of bibliometric networks (i.e., citation, co-citation, bibliographic coupling, keyword co-occurrence, and coauthorship networks) are discussed, and three popular visualization approaches (i.e., distance-based, graph-based, and timeline-based approaches) are distinguished. Next, an overview is given of a number of software tools that can be used for visualizing bibliometric networks. In the second part of the chapter, the focus is specifically on two software tools: VOSviewer and CitNetExplorer. The techniques used by these tools to construct, analyze, and visualize bibliometric networks are discussed. In addition, tutorials are offered that demonstrate in a step-by-step manner how both tools can be used. Finally, the chapter concludes with a discussion of the limitations and the proper use of bibliometric network visualizations and with a summary of some ongoing and future developments.

13.1 Introduction

The idea of visualizing bibliometric networks, often referred to as "science mapping," has received serious attention since the early days of bibliometric research. Visualization has turned out to be a powerful approach to analyze a large variety of bibliometric networks, ranging from networks of citation relations between publications or journals to networks of coauthorship relations between researchers or networks of co-occurrence relations between keywords. Over time, researchers have started to analyze larger and larger networks, leading to the need for more advanced visualization techniques and tools. At the same time, professional users of bibliometrics, for instance research institutions, funding agencies, and publishers, have become more and more interested in bibliometric network visualizations. To make bibliometric network visualizations available to a wider public, both inside

N.J. van Eck (✉) • L. Waltman
Centre for Science and Technology Studies, Leiden University, Leiden, The Netherlands
e-mail: ecknjpvan@cwts.leidenuniv.nl; waltmanlr@cwts.leidenuniv.nl

© Springer International Publishing Switzerland 2014
Y. Ding et al. (eds.), *Measuring Scholarly Impact*,
DOI 10.1007/978-3-319-10377-8_13

and outside the bibliometric research community, researchers have developed a number of software tools, most of which are freely available. In this chapter, we focus mainly on two of these tools: VOSviewer and CitNetExplorer. VOSviewer is a tool that we have developed over the past few years and that offers in a relatively easy way the basic functionality needed for visualizing bibliometric networks. CitNetExplorer is a more specialized tool that we developed recently for visualizing and analyzing citation networks of publications.

This chapter offers an introduction to the topic of visualizing bibliometric networks. It first provides an overview of the literature and of the main software tools that are available. It then focuses specifically on the VOSviewer and CitNetExplorer tools. The most important techniques used by these tools are discussed, and tutorials are offered that provide step-by-step instructions on how the tools can be used. The chapter concludes with a discussion of the limitations and the proper use of bibliometric network visualizations and with a summary of some ongoing and future developments.

13.2   Literature Review

We first provide a brief overview of the literature on visualizing bibliometric networks. We start by discussing the types of bibliometric networks that have received most attention in the literature. We then discuss a number of commonly used visualization approaches. We refer to Börner, Chen, and Boyack (2003) and Börner (2010) for more extensive overviews of the literature.

13.2.1   Types of Bibliometric Networks

A bibliometric network consists of nodes and edges. The nodes can be, for instance, publications, journals, researchers, or keywords. The edges indicate relations between pairs of nodes. The most commonly studied types of relations are citation relations, keyword co-occurrence relations, and coauthorship relations. In the case of citation relations, a further distinction can be made between direct citation relations, co-citation relations, and bibliographic coupling relations. Bibliometric networks are usually weighted networks. Hence, edges indicate not only whether there is a relation between two nodes or not but also the strength of the relation. Below, we discuss the different types of relations in bibliometric networks in more detail.

Two publications are co-cited if there is a third publication that cites both publications (Marshakova, 1973; Small, 1973). The larger the number of publications by which two publications are co-cited, the stronger the co-citation relation between the two publications. Small and colleagues proposed to use co-citations to analyze and visualize relations between publications (Griffith, Small, Stonehill, &

Dey, 1974; Small & Griffith, 1974). Later on, the use of co-citations to study relations between researchers and between journals was introduced by, respectively, White and Griffith (1981) and McCain (1991). A well-known co-citation-based analysis is the study by White and McCain (1998) of researchers in the field of information science.

Bibliographic coupling is the opposite of co-citation. Two publications are bibliographically coupled if there is a third publication that is cited by both publications (Kessler, 1963). In other words, bibliographic coupling is about the overlap in the reference lists of publications. The larger the number of references two publications have in common, the stronger the bibliographic coupling relation between the publications. Although bibliographic coupling was introduced earlier than co-citation, it initially received less attention in the literature on visualizing bibliometric networks. In more recent years, however, the popularity of bibliographic coupling increased considerably (e.g., Boyack & Klavans, 2010; Jarneving, 2007; Small, 1997; Zhao & Strotmann, 2008).

Compared with co-citation and bibliographic coupling, direct citations, sometimes referred to as cross citations, offer a more direct indication of the relatedness of publications. Nevertheless, in the literature on visualizing bibliometric networks, it is relatively uncommon to work with direct citations. This is probably because the use of direct citations often leads to very sparse networks (i.e., networks with only a very small number of edges). In spite of this issue, there seems to be an increasing interest in direct citations in the more recent literature (e.g., Boyack & Klavans, 2010; Persson, 2010; Small, 1997; Waltman & Van Eck, 2012). Direct citations also play an essential role in Eugene Garfield's work on algorithmic historiography (Garfield, Pudovkin, & Istomin, 2003). We will come back to Garfield's work in Sect. 13.3.

In addition to the above-discussed citation-based bibliometric networks, networks of co-occurrences of keywords have also been studied extensively. Keywords can be extracted from the title and abstract of a publication, or they can be taken from the author-supplied keyword list of a publication. In some cases, especially in the older literature, keywords are restricted to individual words, but in other cases they also include terms consisting of multiple words. The number of co-occurrences of two keywords is the number of publications in which both keywords occur together in the title, abstract, or keyword list. For examples of early work on keyword co-occurrence networks, we refer to Callon, Courtial, Turner, and Bauin (1983), Callon, Law, and Rip (1986), and Peters and Van Raan (1993).

Finally, we briefly mention bibliometric networks based on coauthorship. In these networks, researchers, research institutions, or countries are linked to each other based on the number of publications they have authored jointly. Coauthorship networks have been studied extensively, but relatively little attention has been paid to the visualization of these networks. At the level of countries, the visualization of coauthorship networks is discussed by Luukkonen, Tijssen, Persson, and Sivertsen (1993).

13.2.2 Visualization Approaches

Many different approaches have been proposed for visualizing bibliometric networks. Our focus in this chapter is on three popular approaches. We refer to these approaches as the distance-based approach, the graph-based approach, and the timeline-based approach. We emphasize that these three approaches are definitely not the only approaches available. Alternative approaches for instance include circular visualizations (e.g., Börner et al., 2012) and self-organizing maps (e.g., Skupin, Biberstine, & Börner, 2013).

In the distance-based approach, the nodes in a bibliometric network are positioned in such a way that the distance between two nodes approximately indicates the relatedness of the nodes. In general, the smaller the distance between two nodes, the higher their relatedness. Nodes are usually positioned in a two-dimensional space. Edges between nodes are normally not shown. The most commonly used technique for determining the locations of the nodes in a distance-based visualization is multidimensional scaling (e.g., Borg & Groenen, 2005). The use of this technique for visualizing bibliometric networks has a long tradition, going back to early work on the visualization of co-citation networks by Griffith et al. (1974) and White and Griffith (1981). More recently, some alternatives to multidimensional scaling were introduced in the literature. One alternative is the VOS technique (Van Eck, Waltman, Dekker, & Van den Berg, 2010), which is used in the VOSviewer software that we will discuss in detail in the second part of this chapter. Another alternative is the VxOrd technique, also known as DrL or OpenOrd (http://www.sandia.gov/~smartin/software.html). The VxOrd technique has been used to create distance-based visualizations of very large bibliometric networks (e.g., Boyack, Klavans, & Börner, 2005; Klavans & Boyack, 2006). An example of a distance-based visualization is presented in Fig. 13.1. This example has been taken from a well-known study by White and McCain (1998). The visualization displays a co-citation network of researchers in the field of information science.

In the graph-based approach, nodes are positioned in a two-dimensional space, just like in the distance-based approach. The difference between the two approaches is that in the graph-based approach edges are displayed to indicate the relatedness of nodes. The distance between two nodes need not directly reflect their relatedness. The graph-based approach is most suitable for visualizing relatively small networks. Visualizing larger networks using the graph-based approach often does not give good results because of the large number of edges that need to be displayed. The most commonly used technique for creating graph-based visualizations of bibliometric networks is the graph drawing algorithm of Kamada and Kawai (1989). An alternative technique is the algorithm of Fruchterman and Reingold (1991). Graph drawing algorithms are sometimes used in combination with the pathfinder network technique for graph pruning (Schvaneveldt, Dearholt, & Durso, 1988). For examples of graph-based visualizations of bibliometric networks, we refer to Chen (1999), De Moya-Anegón et al. (2007), Leydesdorff and Rafols (2009), and White (2003). Figure 13.2 shows an example of a graph-based

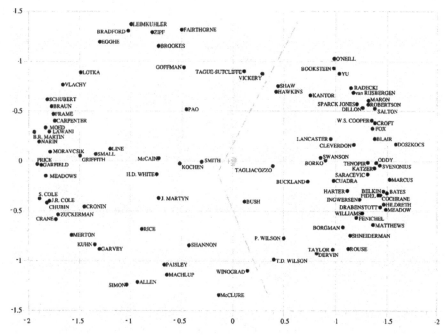

Fig. 13.1 Example of a distance-based visualization. Co-citation network of researchers in the field of information science. *Source*: White & McCain (1998, Fig. 4). Copyright 1998 by John Wiley & Sons, Inc. Reprinted with permission

visualization. This visualization displays the same researcher co-citation network as the visualization presented in Fig. 13.1, but instead of a distance-based approach a graph-based approach is used. The example shown in Fig. 13.2 has been taken from White (2003).

A third approach for visualizing bibliometric networks is the timeline-based approach. Unlike the distance-based and graph-based approaches, the timeline-based approach assumes that each node in a bibliometric network can be linked to a specific point in time. The timeline-based approach is especially suitable for visualizing networks of publications, since a publication can be easily linked to a specific point in time based on its publication date. In a timeline-based visualization, there are two dimensions, one of which is used to represent time. The other dimension can be used to represent the relatedness of nodes. The location of a node in the time-dimension is determined by the specific point in time to which the node is linked. The location of a node in the other dimension can be determined based on the relatedness of the node to other nodes. Timeline-based visualizations have for instance been used by Chen (2006), Garfield et al. (2003), and Morris, Yen, Wu, and Asnake (2003). They are also used in the CitNetExplorer software that will be discussed later on in this chapter. An example of a timeline-based visualization is provided in Fig. 13.3. The example, taken from Garfield (2004), displays a citation network of publications on the so-called small world phenomenon.

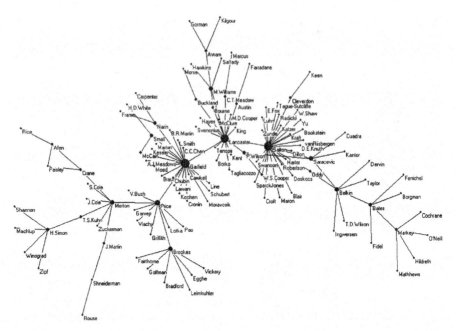

Fig. 13.2 Example of a graph-based visualization. Co-citation network of researchers in the field of information science. *Source*: White (2003, Fig. 1). Copyright 2003 by Wiley Periodicals, Inc. Reprinted with permission

13.3 Software Tools

There are many software tools that can be used for visualizing bibliometric networks. Some of these tools are general statistical or network analysis tools. Other tools have been developed specifically for visualizing bibliometric networks. In this section, we do not aim to give a comprehensive overview of all available tools. Instead, we have selected a number of tools that seem to be among the most important or most popular ones. We briefly discuss the main features of each of these tools. We refer to Cobo, López-Herrera, Herrera-Viedma, and Herrera (2011) for another recent overview of software tools for visualizing bibliometric networks. This overview includes a number of tools that are not considered in this section.

Below, we first discuss two general network analysis tools (Pajek and Gephi). We then discuss three tools (CiteSpace, Sci2, and VOSviewer) that have been developed specifically for analyzing and visualizing bibliometric networks. Finally, we consider two tools (HistCite and CitNetExplorer) that focus entirely on the analysis and visualization of one specific type of bibliometric network, namely citation networks of publications. The tools that we discuss are all freely available, at least for specific types of use. Some tools are open source, but most are not. Table 13.1 provides for each tool the URL of its website.

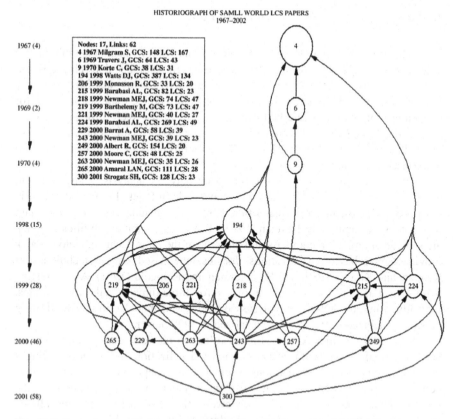

HISTORIOGRAPH OF SAMLL WORLD LCS PAPERS
1967–2002

Fig. 13.3 Example of a timeline-based visualization. Citation network of publications on the small world phenomenon. *Source*: Garfield (2004, Fig. 12). Copyright 2004 by SAGE Publications. Reprinted with permission

Table 13.1 URLs of the websites of a number of software tools that can be used for visualizing bibliometric networks

CitNetExplorer	http://www.citnetexplorer.nl
CiteSpace	http://cluster.cis.drexel.edu/~cchen/citespace/
Gephi	http://www.gephi.org
HistCite	http://www.histcite.com
Pajek	http://pajek.imfm.si
Sci2	https://sci2.cns.iu.edu
VOSviewer	http://www.vosviewer.com

13.3.1   General Network Analysis Tools

Two general network analysis tools that have been used for visualizing bibliometric networks are Pajek (De Nooy, Mrvar, & Batagelj, 2005) and Gephi (Bastian, Heymann, & Jacomy, 2009). Since these are general network analysis tools, they have no specific functionality for processing bibliographic data, for instance for extracting bibliometric networks from output files of the Web of Science bibliographic database. Other software tools are required to take care of this. In the case of Pajek, a tool called WoS2Pajek is available for this purpose. In the case of Gephi, Sci2 (discussed below) can be used.

Pajek is a popular software tool for social network analysis. Many different techniques for social network analysis are available in Pajek. From a bibliometric perspective, important techniques offered by Pajek include clustering and main path analysis (Hummon & Doreian, 1989). The visualization capabilities of Pajek are somewhat limited in comparison with some of the tools discussed below. Pajek provides graph-based visualizations, for instance based on the techniques of Kamada and Kawai (1989) and Fruchterman and Reingold (1991). An interesting feature of Pajek is the support it offers for three-dimensional visualizations. Recently, a link to VOSviewer (discussed below) was included in Pajek, allowing networks analyzed in Pajek to be visualized in VOSviewer.

Compared with Pajek, Gephi is focused less on network analysis and more on network visualization. Gephi offers extensive visualization capabilities, making it possible to customize visualizations in great detail. Gephi also supports various network analysis techniques, but not as many as Pajek. A comparison between the use of Gephi and the use of VOSviewer (discussed below) for visualizing bibliometric networks is made by Leydesdorff and Rafols (2012).

13.3.2   Tools for Analyzing and Visualizing Bibliometric Networks

We now discuss three software tools that have been developed specifically for analyzing and visualizing bibliometric networks: CiteSpace (Chen, 2004, 2006), Sci2, and our own VOSviewer (Van Eck & Waltman, 2010). All three tools can take Web of Science output files as input.

Unlike other software tools, CiteSpace has a strong focus on dynamic visualizations that show how bibliometric networks evolve over time. CiteSpace offers both graph-based and timeline-based visualizations. It also provides users with a large range of options. Because of the many options that are available, the user interface of CiteSpace may appear a bit overwhelming to new users, and it may take some effort to learn how to use the tool.

Sci2 is a general software tool for analyzing bibliometric networks. Many different analysis techniques are available in Sci2. Some of these techniques are implemented in external programs that have been linked to Sci2. To visualize a bibliometric network, Sci2 also relies on external software tools. By default, Sci2 uses a tool called GUESS (http://graphexploration.cond.org). Using this tool, graph-based visualizations of bibliometric networks are provided. Two other network visualization tools that can be used together with Sci2 are Gephi (discussed above) and Cytoscape (http://www.cytoscape.org).

Our aim with VOSviewer is to offer an easy-to-use software tool that is completely focused on the visualization of bibliometric networks. Unlike the above-discussed tools, VOSviewer provides distance-based visualizations of bibliometric networks. By default, VOSviewer therefore displays only the nodes in a bibliometric network and does not display the edges between the nodes. In the visualizations provided by VOSviewer, the distance between two nodes approximately indicates the relatedness of the nodes. By providing distance-based visualizations rather than graph-based ones, VOSviewer is especially suitable for visualizing larger networks. Because of its strong focus on visualization, VOSviewer offers less functionality for analyzing bibliometric networks than other tools. However, VOSviewer does have some special text mining features. VOSviewer will be discussed in more detail in the second part of this chapter.

13.3.3 Tools for Analyzing and Visualizing Publication Citation Networks

Finally, we consider two software tools that focus exclusively on the analysis and visualization of networks of direct citation relations between publications. These tools are Eugene Garfield's HistCite (Garfield et al., 2003) and our own recently developed CitNetExplorer. Both tools offer timeline-based visualizations of publication citation networks.

HistCite takes Web of Science output files as input. Based on the bibliographic data in these files, various statistics on publications, researchers, journals, etc. are presented. The bibliographic data can also be edited, for instance to correct errors and to add missing data elements. After the bibliographic data has been processed, the next step is to visualize the publication citation network. A visualization of a publication citation network is referred to as a historiograph in HistCite, and the algorithmic construction of such a visualization is called algorithmic historiography (Garfield et al., 2003). By default, HistCite includes the 30 most frequently cited publications in the visualization of a publication citation network. An example of a visualization produced by HistCite is shown in Fig. 13.3.

Compared with HistCite, CitNetExplorer offers more extensive analysis and visualization capabilities. Most importantly, CitNetExplorer provides functionality for drilling down into a publication citation network. This functionality makes it possible to explore a publication citation network in an interactive fashion. Initially, for instance, we may start with a large network that includes several millions of publications. We may then gradually drill down into this network until for instance we have reached a small subnetwork that includes no more than, say, 100 publications, all dealing with a specific topic that we are interested in. Like HistCite, CitNetExplorer is able to process Web of Science output files. CitNetExplorer will be discussed in more detail in the next two sections.

13.4 Techniques

In this section, we discuss the main techniques used in the two software tools on which we focus in the rest of this chapter: VOSviewer and CitNetExplorer. We first consider VOSviewer and then CitNetExplorer. The discussion relates to version 1.5.5 of VOSviewer and version 1.0.0 of CitNetExplorer. Readers who are interested mainly in the practical application of the two tools and not so much in the underlying techniques may skip this section and proceed to the next one.

13.4.1 VOSviewer

VOSviewer takes a distance-based approach to visualizing bibliometric networks. Any type of bibliometric network can be visualized. Directed networks, for instance networks based on direct citation relations, are treated as undirected. Networks may consist of several thousand nodes. Because of computational limitations and memory constraints, networks with more than 10,000 nodes tend to be difficult to handle in VOSviewer.

13.4.1.1 Normalization, Mapping, and Clustering

In a bibliometric network, there are often large differences between nodes in the number of edges they have to other nodes. Popular nodes, for instance representing highly cited publications or highly prolific researchers, may have several orders of magnitude more connections than their less popular counterparts. In the analysis of bibliometric networks, one usually performs a normalization for these differences between nodes. VOSviewer by default applies the association strength normalization. This normalization is discussed in detail by Van Eck and Waltman (2009).

After a normalized network has been constructed, the next step is to position the nodes in the network in a two-dimensional space in such a way that strongly related nodes are located close to each other while weakly related nodes are located far away from each other. For this purpose, VOSviewer uses the VOS mapping technique, where VOS stands for "visualization of similarities." A detailed discussion of the VOS mapping technique is provided by Van Eck et al. (2010).

VOSviewer by default also assigns the nodes in a network to clusters. A cluster is a set of closely related nodes. Each node in a network is assigned to exactly one cluster. The number of clusters is determined by a resolution parameter. The higher the value of this parameter, the larger the number of clusters. In the visualization of a bibliometric network, VOSviewer uses colors to indicate the cluster to which a node has been assigned. The clustering technique used by VOSviewer is discussed by Waltman, Van Eck, and Noyons (2010). The technique requires an algorithm for solving an optimization problem. For this purpose, VOSviewer uses the smart local moving algorithm introduced by Waltman and Van Eck (2013).

We refer to the Appendix for a technical summary of the normalization, mapping, and clustering techniques used by VOSviewer.

13.4.1.2 Displaying a Bibliometric Network

After the nodes in a bibliometric network have been positioned in a two-dimensional space and have been assigned to clusters, the network can be displayed. VOSviewer uses various techniques to optimize the way in which networks are displayed. In order to ensure that labels of nodes do not overlap each other, labels are displayed only for a selection of all nodes. This selection is determined in such a way that as many labels as possible are displayed while labels of more important nodes (i.e., nodes that have more edges) are given priority over labels of less important nodes. Like computer software for exploring geographical maps (e.g., Google Maps), VOSviewer offers zooming and panning (scrolling) functionality. This is especially useful for exploring larger networks consisting of hundreds or thousands of nodes. When zooming in, the selection of nodes for which labels are displayed is updated and labels that previously were not shown may become visible.

VOSviewer also supports overlay visualizations. In an overlay visualization, the color of a node indicates a certain property of the node. For instance, nodes may represent journals and the color of a node may indicate the number of times a journal has been cited. We refer to Van Eck, Waltman, Van Raan, Klautz, and Peul (2013) for an example of the use of overlay visualizations in VOSviewer. Another visualization supported by VOSviewer is the density visualization. In this visualization, colors indicate how nodes are distributed in the two-dimensional space underlying the visualization. The density visualization allows one to immediately identify dense areas in which many nodes are located close to each other. We refer to Van Eck and Waltman (2010) for a discussion of the technical details of the density visualization.

13.4.1.3 Fractional Counting Methodology

As will be demonstrated in Sect. 13.5, VOSviewer is able to process output files of the Web of Science bibliographic database. Based on these files, VOSviewer can construct co-citation and bibliographic coupling networks of publications, journals, and researchers. A special feature of VOSviewer is the possibility to construct co-citation and bibliographic coupling networks using a fractional counting methodology. To explain this methodology, we take bibliographic coupling networks of publications as an example, but the methodology works in a similar way for other types of networks.

Suppose we have 100 publications that all cite the same publication. Using an ordinary full counting methodology, each of the 100 publications has a bibliographic coupling relation with a weight of 1 with each of the 99 other publications. This yields a total bibliographic coupling weight of $99 \times 1 = 99$ for each of the 100 publications. Using the fractional counting methodology, on the other hand, each of the 100 publications has a bibliographic coupling relation with a weight of 1/99 with each of the 99 other publications, yielding a total bibliographic coupling weight of $99 \times (1/99) = 1$ for each publication. In other words, using the fractional counting methodology, giving a citation to a publication always results in a total bibliographic coupling weight of 1, irrespective of the number of other publications that also cite the same publication (although there must of course be at least one other citing publication). Hence, in the case of the fractional counting methodology, highly cited publications play a less important role in the construction of a bibliographic coupling network. In the same way, publications with a long reference list (e.g., review articles) play a less important role in the construction of a co-citation network. In general, we recommend the use of VOSviewer's fractional counting methodology instead of the ordinary full counting methodology.

13.4.1.4 Text Mining Techniques

Finally, we briefly mention VOSviewer's text mining functionality for constructing co-occurrence networks of terms extracted from English-language textual data, for instance from titles and abstracts of publications (Van Eck & Waltman, 2011). VOSviewer relies on the Apache OpenNLP toolkit (http://opennlp.apache.org) to perform part-of-speech tagging (i.e., to identify verbs, nouns, adjectives, and so on). It then uses a linguistic filter to identify noun phrases. The filter selects all word sequences that consist exclusively of nouns and adjectives and that end with a noun. Plural noun phrases are converted into singular ones. Some noun phrases (e.g., "conclusion," "interesting result," and "new method") are very general, and one usually does not want these noun phrases to be included in one's co-occurrence network. VOSviewer therefore calculates for each noun phrase a relevance score. Essentially, noun phrases have a low relevance score if their co-occurrences with other noun phrases follow a more or less random pattern, while they have a high relevance score if they co-occur mainly with a limited set of other noun phrases. Noun phrases with a low relevance score tend to be quite general, while noun phrases with a high relevance score typically have a more specific meaning. VOSviewer allows one to leave out noun phrases with a low relevance score. In this way, one gets rid of many general noun phrases. The remaining noun phrases usually represent relevant terms in the domain of interest. VOSviewer can be used to visualize the co-occurrence network of these terms.

13.4.2   CitNetExplorer

CitNetExplorer visualizes networks of direct citation relations between publications. A timeline-based approach is taken. CitNetExplorer supports very large networks. Networks may include millions of publications and citation relations. The maximum size of the networks that can be handled by CitNetExplorer depends mainly on the amount of computer memory that is available.

13.4.2.1   Constructing a Publication Citation Network

CitNetExplorer is able to construct a publication citation network based on Web of Science output files. Web of Science output files contain bibliographic data on publications, but they do not directly indicate the citation relations that exist between these publications. For each publication in a Web of Science output file, the list of cited references is given. To find out which publications cite which other publications, the cited references in a Web of Science file need to be matched with the publications in the file. For some cited references, it will not be possible to match them with a publication. This usually means that these cited references point to publications that are not included in the Web of Science file. For other cited references, matching will be possible. Cited references that can be matched with a publication indicate the existence of a citation relation between two publications in the Web of Science file.

Citation matching can be performed in different ways. CitNetExplorer first attempts to match based on DOI. However, DOI data often is not available. In that case, matching is done based on first author name (last name and first initial only), publication year, volume number, and page number. A perfect match is required for each of these data elements. Data on the title of the cited journal usually is available as well, but this data is not used by CitNetExplorer. This is because in many cases the title of a journal is not written consistently in the same way.

CitNetExplorer assumes publication citation networks to be acyclic. This for instance means that it is not allowed to have both a citation from publication A to publication B and a citation from publication B to publication A. Likewise, it is not allowed to have a citation from publication A to publication B, a citation from publication B to publication C, and a citation from publication C to publication A. In other words, when moving through a publication citation network by following citation relations from one publication to another, one should never get back again at a publication that has already been visited. In practice, publication citation networks are not always perfectly acyclic. There may for instance be publications in the same issue of a journal that mutually cite each other. CitNetExplorer solves this problem by checking whether a publication citation network is acyclic and by removing citation relations that cause the network to have cycles. CitNetExplorer also removes all citation relations that point forward in time, for instance a publication from 2013 citing a publication from 2014.

13.4.2.2 Displaying a Publication Citation Network

If a publication citation network consists of more than 50 or 100 publications, displaying all publications and all citation relations is typically of little or no use. There will usually be lots of citation relations, and when displaying these citation relations, many of them will inevitably cross each other, leading to a visualization that is hard to interpret. For this reason, in the case of a larger network, CitNetExplorer displays only a selection of all publications. By default, this selection consists of the 40 most frequently cited publications in the network, but the selection of publications to be displayed may be changed by the user. To keep things simple, in the discussion below, we assume that we are dealing with a small network and that all publications in the network are displayed. Larger networks are visualized in the same way as described below, with the exception that only a selection of all publications are included in the visualization.

In the timeline-based visualization of CitNetExplorer, the vertical dimension is used to represent time, with more recent years being located below older years. Publications are positioned in the vertical dimension based on the year in which they appeared. The vertical dimension is organized into layers, each of equal height. A year is represented by at least one layer, but some years may be represented by multiple layers. Multiple layers are used if there are citation relations between publications from the same year. The horizontal dimension in the timeline-based visualization of CitNetExplorer is used to provide an indication of the relatedness of publications. In general, publications that are strongly related to each other, based on citation relations, are positioned close to each other in the horizontal dimension.

Positioning the publications in a publication citation network in the horizontal and vertical dimensions of a timeline-based visualization is a hierarchical graph drawing problem. Following the literature on hierarchical graph drawing (e.g., Healy & Nikolov, 2013), CitNetExplorer first assigns each publication to a layer in the vertical dimension. This is done based on the year in which a publication appeared. In addition, publications are assigned to layers in such a way that citations always flow in an upward direction in the visualization. In other words, for any citation relation, the layer to which the citing publication is assigned must be located below the layer of the cited publication. After each publication has been assigned to a layer, CitNetExplorer positions the publications in the horizontal dimension.

Publications are assigned to layers in the vertical dimension in a year-by-year fashion. If the publications from a given year do not cite each other, they are all assigned to the same layer, unless the number of publications exceeds the maximum number of publications per layer. By default, the maximum number of publications per layer is 10. If the publications from a given year do cite each other, they are assigned to multiple layers in such a way that the layer of a citing publication is always located below the layer of the corresponding cited publication. This ensures that citations always flow in an upward direction. Assigning publications to multiple layers is done using a simple heuristic algorithm that aims to minimize the number of layers that are needed.

Publications are positioned in the horizontal dimension based on their citation relations with other publications. In addition to direct citation relations, various types of indirect citation relations (e.g., co-citation and bibliographic coupling relations) are taken into account as well. In general, the higher the relatedness of two publications, the closer the publications are located to each other in the horizontal dimension. However, publications that have been assigned to the same layer in the vertical dimension must have at least a certain minimum distance from each other in the horizontal dimension. This is done in order to minimize the problem of overlapping publications in the visualization. The technique used by CitNetExplorer to position publications in the horizontal dimension is fairly similar to the VOS mapping technique discussed in Sect. 13.4.1 and in the Appendix.

To optimize the way in which a publication citation network is displayed, CitNetExplorer uses similar techniques as VOSviewer. CitNetExplorer labels publications by the last name of the first author. To prevent labels from overlapping each other, labels may sometimes be displayed only for a selection of all publications. Like VOSviewer, CitNetExplorer offers zooming and panning functionality.

13.4.2.3 Analysis Techniques

CitNetExplorer offers various techniques for analyzing publication citation networks. At the moment, techniques are available for extracting connected components, for clustering publications, for identifying core publications, and for finding shortest and longest paths between publications. More techniques are expected to become available in future versions of CitNetExplorer.

Clustering of publications is accomplished following the methodology proposed by Waltman and Van Eck (2012). This methodology optimizes a variant of the modularity function of Newman and Girvan (2004) and Newman (2004). The level of detail of the clustering is determined by a resolution parameter. Like VOSviewer, CitNetExplorer uses the optimization algorithm introduced by Waltman and Van Eck (2013).

The identification of core publications is based on the idea of k-cores introduced by Seidman (1983). A core publication is a publication that has at least a certain minimum number of citation relations with other core publications. Both incoming and outgoing citation relations are counted. Hence, for the purpose of identifying core publications, a publication citation network is treated as undirected. The identification of core publications makes it possible to get rid of unimportant publications in the periphery of a publication citation network.

13.5 Tutorials

We now offer two tutorials, one for VOSviewer and one for CitNetExplorer. The aim of the tutorials is to provide a basic introduction to both software tools. For each tool, step-by-step instructions are given on how the tool can be used. The bibliometric networks that are analyzed and visualized in the two tutorials deal with the field of scientometrics and closely related fields. Analyses for other fields of science can be performed in a similar way. To run VOSviewer and CitNetExplorer, one needs to have a computer system that offers Java support. Java version 6 or higher needs to be installed.

13.5.1 Data Collection

Data was collected from the Web of Science bibliographic database produced by Thomson Reuters.[1] Bibliographic data was downloaded for all 25,242 publications in the 13 journals listed in Table 13.2. To select these journals, we started with *Scientometrics* and *Journal of Informetrics*, which we regard as the two core journals in the field of scientometrics. We then used the 2012 edition of Thomson Reuters' Journal Citation Reports to identify closely related journals. We took all journals listed among the five most closely related journals to either *Scientometrics* or *Journal of Informetrics*, excluding journals that seem to be mainly nationally oriented. For each of the selected journals, we determined whether it has any predecessors. These predecessors were included in the selection as well, provided that they are indexed in the Web of Science database. In this way, the list of 13 journals shown in Table 13.2 was obtained.

Bibliographic data for the 25,242 publications in the journals listed in Table 13.2 was downloaded from the Web of Science database. The database supports various file formats. We used the tab-delimited format. For each publication, the full record including cited references was obtained. In the Web of Science database, bibliographic data can be downloaded for at most 500 publications at a time. Downloading therefore took place in batches. We ended up with a large number of Web of Science output files, each containing bibliographic data for at most 500 publications.

As can be seen in Table 13.2, the publications included in the data collection cover the period 1945–2013. Bibliographic data was downloaded separately for publications from the period 1945–1999 and for publications from the period 2000–2013. In the VOSviewer tutorial, only publications from the latter period will be considered. In the CitNetExplorer tutorial, publications from both periods will be included in the analysis.

---

[1] The data collection took place on November 7, 2013.

Table 13.2 Journals included in the data collection

Journal	Time period	No. of pub.
American Documentation	1956–1969	796
ASLIB Proceedings	1956–2013	2,697
Information Processing &Management	1975–2013	3,036
Information Scientist	1968–1978	254
Information Storage and Retrieval	1963–1974	372
Journal of Documentation	1945–2013	3,778
Journal of Information Science	1979–2013	1,855
Journal of Informetrics	2007–2013	399
Journal of the American Society for Information Science	1970–2000	2,995
Journal of the American Society for Information Science and Technology	2001–2013	2,486
Research Evaluation	2000–2013	383
Research Policy	1974–2013	2,596
Scientometrics	1978–2013	3,595

13.5.2   VOSviewer

In this tutorial, we demonstrate the visualization of three bibliometric networks: A bibliographic coupling network of researchers, a co-citation network of journals, and a co-occurrence network of terms.

13.5.2.1   Bibliographic Coupling Network of Researchers

To construct and visualize a bibliographic coupling network of researchers, we take the following steps:

1. Launch VOSviewer. If VOSviewer has not yet been downloaded, then first download it from http://www.vosviewer.com.
2. Press the **Create** button on the **Action** tab to open the **Create Map** dialog box.
3. Select the **Create a map based on a network** option button and press the **Next** button.
4. Go to the **Web of Science** tab to open the **Select Web of Science File** dialog box. In this dialog box, the Web of Science output files that we want to work with can be selected. The CTRL key can be used to select multiple files. We select the files that we obtained in Sect. 13.5.1. We only include files containing bibliographic data for publications from the period 2000–2013. After the files have been selected, first press the **OK** button to close the **Select Web of Science File** dialog box and then press the **Next** button.
5. Select the **Bibliographic coupling of authors** and **Fractional counting** option buttons and press the **Next** button. By selecting the **Fractional counting** option

button, we indicate that we want to use the fractional counting methodology discussed in Sect. 13.4.1.

6. VOSviewer asks for the minimum number of publications a researcher must have in order to be included in the bibliographic coupling network. We choose the default value of five publications. In our data set, there turn out to be 792 researchers with at least five publications. To go to the next step, press the **Next** button.

7. Some researchers may have no or almost no bibliographic coupling relations with other researchers. It is usually best to exclude these researchers from a bibliographic coupling network. VOSviewer asks for the number of researchers to be included in the bibliographic coupling network. We choose to include 500 researchers. This means that the $792 - 500 = 292$ researchers with the smallest number of bibliographic coupling relations will be excluded. To go to the next step, press the **Next** button.

8. VOSviewer lists the 500 researchers included in the bibliographic coupling network and offers the possibility to remove individual researchers from the network. We choose not to remove anyone. To finish the construction of the bibliographic coupling network and to close the **Create Map** dialog box, press the **Finish** button.

VOSviewer now applies the normalization, mapping, and clustering techniques discussed in Sect. 13.4.1. Since we are working with a relatively small network, this will take at most a few seconds. VOSviewer then provides us with the visualization shown in Fig. 13.4.

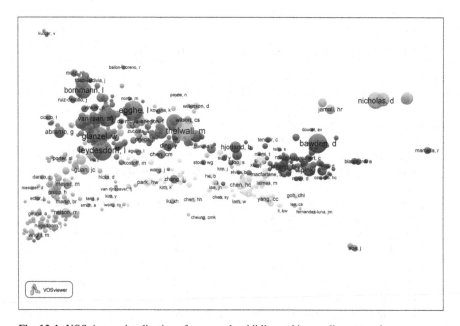

Fig. 13.4 VOSviewer visualization of a researcher bibliographic coupling network

In the visualization presented in Fig. 13.4, each circle represents a researcher. Large circles represent researchers that have many publications. Small circles represent researchers with only a few publications. In general, the closer two researchers are located to each other in the visualization, the more strongly they are related to each other based on bibliographic coupling. In other words, researchers that are located close to each other tend to cite the same publications, while researchers that are located far away from each other usually do not cite the same publications. Colors indicate clusters of researchers that are relatively strongly related to each other. There are 15 clusters, most of which are fairly small. It may be helpful to reduce the number of clusters a bit. This can be done as follows:

9. Go to the **Map** tab in the left part of the VOSviewer window.
10. In the **Clustering resolution** text box, decrease the value of the resolution parameter (see Sect. 13.4.1) from 1.00 to 0.50.
11. Select the **Clustering only** option button and press the **Run** button.

Instead of 15 clusters, we now have only six, of which three are very small. The three larger clusters consist of researchers in scientometrics (upper left area in the visualization), researchers in information science and information retrieval (right area), and researchers in technology and innovation studies (lower left area).

13.5.2.2 Co-citation Network of Journals

We now demonstrate the construction and visualization of a co-citation network of journals. The first four steps that we take are the same as described above for analyzing a bibliographic coupling network of researchers. After these four steps have been taken, we proceed as follows:

1. Select the **Co-citation of sources** and **Fractional counting** option buttons and press the **Next** button.
2. VOSviewer asks for the minimum number of citations a journal must have received in order to be included in the co-citation network. The default value is 20 citations, but we choose to require at least 50 citations. This means that a journal can be included in the co-citation network only if in our Web of Science output files there are at least 50 cited references that point to the journal. There turn out to be 619 journals that satisfy this requirement. To go to the next step, press the **Next** button.
3. VOSviewer asks for the number of journals to be included in the co-citation network. The journals with the smallest number of co-citation relations will be excluded. Since the 619 journals selected in the previous step have all been cited quite significantly (i.e., at least 50 times), we do not think there is a need to exclude any journals. We therefore simply choose to include all 619 journals. To go to the next step, press the **Next** button.

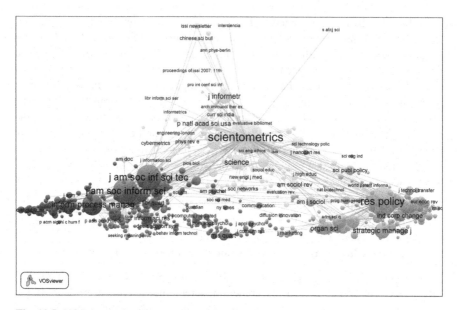

Fig. 13.5 VOSviewer visualization of a journal co-citation network

4. VOSviewer lists the 619 journals included in the co-citation network and offers
   the possibility to remove individual journals from the network. We do not
   remove any journal. To finish the construction of the co-citation network and
   to close the **Create Map** dialog box, press the **Finish** button.

The visualization that we obtain of our journal co-citation network is shown in
Fig. 13.5.[2]

Each circle in the visualization presented in Fig. 13.5 represents a journal. The
size of a circle reflects the number of citations a journal has received. Journals that
are located close to each other in the visualization tend to be more strongly related,
based on co-citations, than journals that are located far away from each other. Three
broad groups of journals can be distinguished: Journals that publish scientometrics
research in the upper area of the visualization, journals that publish information
science and information retrieval research in the lower left area, and journals that
publish technology and innovation studies research in the lower right area. These
three groups of journals can also be easily recognized in the density visualization
that is available on the **Density View** tab.

---

[2] To improve the visualization, the **Size variation** parameter in the **Options** dialog box has been
set to a value of 0.40. In addition, the **No. of lines** parameter has been set to a value of 500. This has
the effect that 500 lines, representing the 500 strongest co-citation relations between journals, are
displayed in the visualization.

VOSviewer has identified six clusters of journals.[3] These clusters are indicated using colors in the visualization shown in Fig. 13.5. Two clusters are very small and can be ignored. Of the four larger clusters, the yellow one consists mainly of journals publishing scientometrics research, while the green one covers journals focusing on technology and innovation studies research. The red and the blue cluster both consist of journals publishing information science and information retrieval research. By zooming in on the lower left area in the visualization, it can be seen that the journals in the blue cluster are more computer science oriented, focusing mainly on technical information retrieval research, while the journals in the red cluster are more social science oriented, focusing on general information science and library science research.

13.5.2.3   Co-occurrence Network of Terms

Finally, we consider the construction and visualization of a co-occurrence network of terms extracted from the titles and abstracts of publications. We take the following steps:

1. Launch VOSviewer.
2. Press the **Create** button on the **Action** tab to open the **Create Map** dialog box.
3. Select the **Create a map based on a text corpus** option button and press the **Next** button.
4. Go to the **Web of Science** tab, select the Web of Science output files to be used, and press the **Next** button. We use the same files as we did above in the analysis of researcher bibliographic coupling and journal co-citation networks.
5. Select the **Title and abstract fields** option button and press the **Next** button. VOSviewer will now extract noun phrases from the titles and abstracts of the publications in our Web of Science output files (see Sect. 13.4.1). This may take some time.
6. Select the **Binary counting** option button and press the **Next** button. The use of a binary counting methodology means that in the construction of a co-occurrence network the number of times a noun phrase occurs in the title and abstract of a publication plays no role. A noun phrase that occurs only once in the title and abstract of a publication is treated in the same way as a noun phrase that occurs for instance ten times.
7. VOSviewer asks for the minimum number of occurrences a noun phrase must have in order to be included in the co-occurrence network. We choose the default value of ten occurrences. There turn out to be 3,158 noun phrases that occur in the title or abstract of at least ten publications. To go to the next step, press the **Next** button.

---

[3] The resolution parameter of VOSviewer's clustering technique is set to its default value of 1.00, not to the value of 0.50 that was used in the case of the author bibliographic coupling network.

8. VOSviewer asks for the number of noun phrases to be included in the co-occurrence network. By default, VOSviewer suggests to include 60 % of the noun phrases selected in the previous step. We follow this suggestion and choose to include 1,894 noun phrases in the co-occurrence network. To determine the noun phrases to be excluded, VOSviewer will calculate for each noun phrase a relevance score (see Sect. 13.4.1). The noun phrases with the lowest relevance scores will be excluded. To go to the next step, press the **Next** button. VOSviewer will now perform the calculation of the relevance scores. This may take some time.

9. VOSviewer lists the 1,894 noun phrases included in the co-occurrence network and offers the possibility to remove individual noun phrases from the network. We order the noun phrases alphabetically and indicate that we want to remove all noun phrases starting with "Elsevier." These noun phrases result from copyright statements in the abstracts of publications in Elsevier journals. After removing six noun phrases that start with "Elsevier," we end up with a co-occurrence network of 1,888 noun phrases. We refer to these noun phrases as terms. To finish the construction of the co-occurrence network and to close the **Create Map** dialog box, press the **Finish** button.

Figure 13.6 shows the visualization of our term co-occurrence network.[4]

In the visualization presented in Fig. 13.6, each circle represents a term. The size of a circle indicates the number of publications that have the corresponding term in their title or abstract. Terms that co-occur a lot tend to be located close to each other in the visualization. VOSviewer has grouped the terms into six clusters, of which four are of significant size. The red cluster, located in the upper left area in the visualization, consists of scientometric terms. The green cluster, located in the lower left area, covers terms related to technology and innovation studies. In the right area in the visualization, the blue and yellow clusters consist of terms related to information science and information retrieval. Similar to what we have seen in the case of the journal co-citation network, one cluster (the blue one) is more computer science-oriented while the other (the yellow one) is more focused on the social sciences.

Comparing the visualizations shown in Figs. 13.4, 13.5, and 13.6, it can be concluded that we have obtained a quite consistent picture of the structure of the field of scientometrics and closely related fields. The three visualizations all suggest a similar division into subfields. The differences between the visualizations are fairly small and relate mainly to the positioning of the subfields relative to each other and to the level of detail that is provided.

---

[4] To improve the visualization, the **Size variation** parameter in the **Options** dialog box has been set to a value of 0.40. A few terms in the upper part of the visualization are not visible in Fig. 13.6.

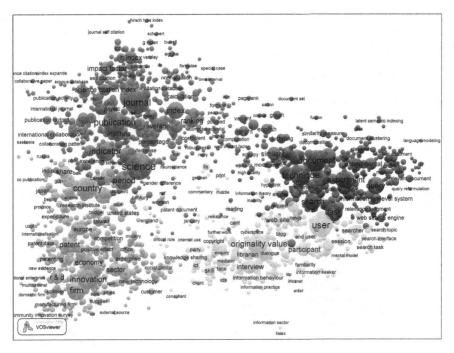

Fig. 13.6 VOSviewer visualization of a term co-occurrence network

13.5.3 CitNetExplorer

In this CitNetExplorer tutorial, we analyze and visualize the citation network of publications in the field of scientometrics.

We start by opening the network:

1. Launch CitNetExplorer. If CitNetExplorer has not yet been downloaded, then first download it from http://www.citnetexplorer.nl.
2. The **Open Citation Network** dialog box will appear. On the **Web of Science** tab, we select the Web of Science output files that we want to work with. These are the files that we obtained in Sect. 13.5.1. All files are included, covering the full period 1945–2013. By default, the **Include non-matching cited references** check box is checked. We leave it this way. If the **Include non-matching cited references** check box is not checked, only publications for which bibliographic data is available in our Web of Science output files will be included in the citation network. If the check box is checked, all publications with at least a certain minimum number of citations will be included, even if no bibliographic data (other than the data that can be extracted from cited references) is available. For the minimum number of citations, we choose the default value of ten.

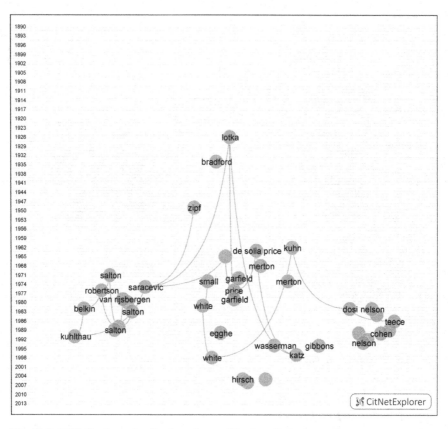

Fig. 13.7 CitNetExplorer visualization of our publication citation network

3. Press the **OK** button to close the **Open Citation Network** dialog box. The
   citation network will now be opened. CitNetExplorer will perform citation
   matching and will make sure that the network is acyclic (see Sect. 13.4.2).
   This may take some time.

The visualization of the citation network is shown in Fig. 13.7. Recall from
Sect. 13.4.2 that only the 40 most frequently cited publications in the network are
displayed. When the mouse is moved over a publication, some bibliographic
information (e.g., authors, title, and journal) is displayed in the left part of the
CitNetExplorer window. The visualization presented in Fig. 13.7 has a clear
structure. Publications on information science and information retrieval can be
found in the left part of the visualization, publications on scientometrics in the
middle part, and publications on technology and innovation studies in the right part.
The curved lines indicate citation relations between publications.

Suppose we are interested to learn more about the literature on visualizing
bibliometric networks. Among the 40 publications displayed in Fig. 13.7, we
recognize three publications on this topic: Small (1973), White and Griffith

(1981), and White and McCain (1998). All three publications were discussed in the literature review presented in Sect. 13.2. We use these publications to drill down into the citation network:

4. Click on Small (1973). This publication has now been marked. To indicate this, the publication is displayed using a square instead of a circle.
5. Click on White and McCain (1998). This publication has now been marked as well. Notice that Small (1973), White and McCain (1998), and a few other publications are displayed using a red border. This border indicates that these publications are selected. By default, a publication is selected if it has been marked or if it is on a citation path from one marked publication to another. For instance, White and Griffith (1981) is selected because it is on a citation path from White and McCain (1998) to Small (1973). White and Griffith (1981) is cited by White and McCain (1998) and is citing Small (1973). Publications that are not among the 40 displayed in the visualization may also be selected, but this is not directly visible. The number of selected publications is reported in the upper left part of the CitNetExplorer window. In our case, there turn out to be 261 selected publications. Each of these publications is on a citation path from White and McCain (1998) to Small (1973).
6. Press the **Drill down** button in the upper part of the CitNetExplorer window. By pressing this button, we drill down from the full citation network to a citation network that includes only the 261 selected publications. This network is now referred to as the current network.

After drilling down, we obtain the visualization shown in Fig. 13.8. Of the 261 publications, the 40 most frequently cited are displayed in the visualization.

Many of the publications displayed in Fig. 13.8 do not directly deal with the topic of visualizing bibliometric networks. We therefore drill down deeper into the citation network. We use a clustering technique (see Sect. 13.4.2) for this purpose:

7. Press the **Analysis** button in the upper part of the CitNetExplorer window. In the **Analysis** menu, choose the **Clustering** option. The **Clustering** dialog box will appear.
8. We use the default parameter values presented in the **Clustering** dialog box. Press the **OK** button to close the dialog box. The 261 publications in the current network will now be clustered.
9. A message box is displayed indicating that five clusters of publications have been identified. Press the **OK** button to close the message box. Publications now have colors in the visualization. The color of a publication indicates the group to which the publication belongs. There are five groups, each corresponding with one of the five clusters that have been identified.
10. Publications on the topic of visualizing bibliometric networks turn out to be concentrated in group 2. These publications have a green color. In the **Selection parameters** frame in the left part of the CitNetExplorer window, choose the **Based on groups** option in the **Selection** drop-down box. In the **Groups** list

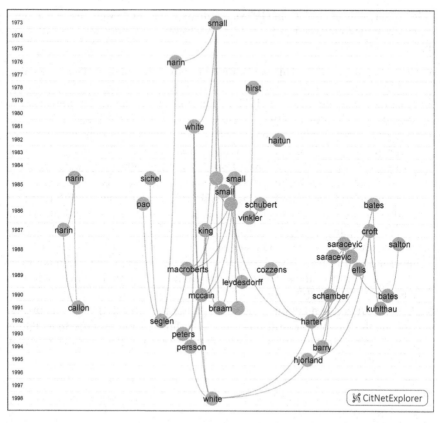

Fig. 13.8 CitNetExplorer visualization of our publication citation network after drilling down the first time

box, check **Group 2**. The 66 publications belonging to group 2 are now selected. These publications are displayed using a red border in the visualization.

11. Press the **Drill down** button.

The visualization obtained after drilling down is shown in Fig. 13.9.[5] The current network now includes 66 publications. The 40 most frequently cited ones are displayed in the visualization.

We now have an overview of an important part of the literature that appeared in the period 1973–1998 on the topic of visualizing bibliometric networks. However,

---

[5] Notice that in the visualization shown in Fig. 13.9, publications are displayed in green rather than in gray. This is because the publications included in the visualization all belong to the same cluster identified by CitNetExplorer's clustering technique.

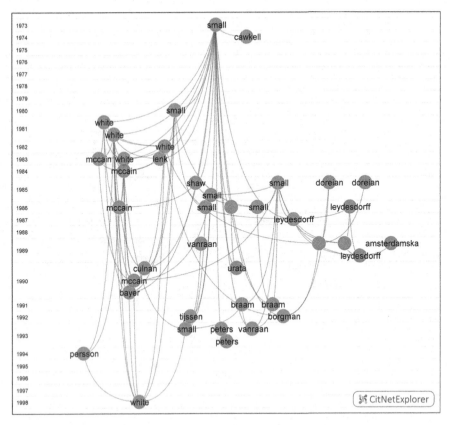

Fig. 13.9 CitNetExplorer visualization of our publication citation network after drilling down the second time

it may be that not all important publications from the period 1973–1998 are included in the current network. Moreover, publications from earlier and later periods are not included at all. To include a larger part of the literature, we expand the current network. We first look at publications that precede the 66 publications in the current network. We expand the current network with all publications cited by at least one of these 66 publications:

12. Press the **Expand** button in the upper part of the CitNetExplorer window. The **Expand Current Network** dialog box will appear.
13. In the **Publications** drop-down box, choose the **Predecessors** option. Do not change the values of the **Min. number of citation links** and **Max. distance** parameters, and do not check the **Add intermediate publications** check box. Press the **OK** button to close the **Expand Current Network** dialog box. The current network will now be expanded.

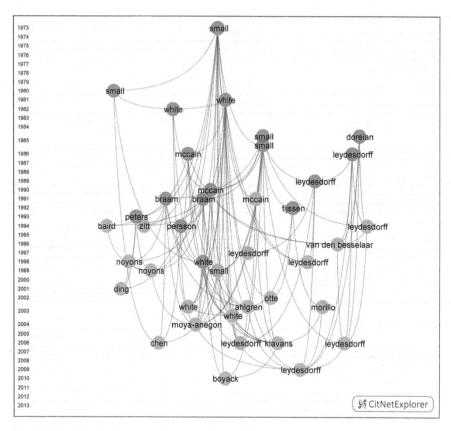

Fig. 13.10 CitNetExplorer visualization of our publication citation network after the second expansion

After expansion, the current network includes 342 publications. Among these publications, we for instance find classical works by Alfred Lotka, Eugene Garfield, and Derek de Solla Price.

Suppose we also want to see publications that are preceded by the above-mentioned 66 publications. In other words, we want to see publications by which the 66 publications are cited. The number of publications that cite at least one of our 66 publications will probably be quite large, and many of these publications may not be directly related to the topic of visualizing bibliometric networks. We therefore require publications to cite at least three of our 66 publications. We take the following steps:

14. Press the **Back** button in the upper part of the CitNetExplorer window. The expansion of the current network from 66 to 342 publications will be undone.
15. Press the **Expand** button to open the **Expand Current Network** dialog box.
16. In the **Publications** drop-down box, choose the **Successors** option. Set the **Min. number of citation links** parameter to a value of three. Do not change the

value of the **Max. distance** parameter, and do not check the **Add intermediate publications** check box. Press the **OK** button to close the **Expand Current Network** dialog box and to perform the expansion.

The visualization obtained after the expansion is shown in Fig. 13.10. The current network now includes 248 publications. Among the more recent publications, we find publications on the topic of visualizing bibliometric networks by for instance Kevin Boyack, Chaomei Chen, Richard Klavans, and Loet Leydesdorff.

The above tutorial has demonstrated the most essential elements of the user interface of CitNetExplorer. However, there are many features of the software that have not been discussed in this short tutorial. Some of these features are discussed in a more extensive tutorial available at http://www.citnetexplorer.nl/gettingstarted/.

Conclusion

In this chapter, we have provided an introduction into the topic of visualizing bibliometric networks. An overview has been given of the literature on this topic and of the main software tools that are available. Our focus has been mostly on two software tools: VOSviewer and CitNetExplorer. We have discussed the most important techniques used by these tools, and we have provided tutorials with step-by-step instructions on the use of the tools.

To conclude this chapter, we first discuss the limitations of bibliometric network visualizations and the proper use of these visualizations. We then summarize some ongoing and future developments in the visualization of bibliometric networks.

Limitations and Proper Use of Bibliometric Network Visualizations

The main idea of bibliometric network visualization is to allow large amounts of complex bibliographic data to be analyzed in a relatively easy way by visualizing core aspects of data. The strength of bibliometric network visualization is in the simplification it provides, but simplification comes at a cost. It typically implies a loss of information.

Loss of information takes place in reducing bibliographic data to a bibliometric network. For example, when textual data is reduced to a co-occurrence network of terms, information on the context in which terms co-occur is lost. Similarly, when we have constructed a citation network, we can see who is citing whom, but we can no longer see why someone may be citing someone else.

Loss of information also occurs in the visualization of a bibliometric network. In the case of a distance-based visualization, for instance, it is usually not possible to position the nodes in a two-dimensional space in such a way that for any pair of nodes the distance between the nodes reflects the relatedness of the nodes with perfect accuracy. Distances reflect relatedness only approximately, and we therefore lose information. In the case of graph-based and timeline-based visualizations, we may need to restrict

(continued)

(continued)

ourselves to visualizing a limited number of nodes, for instance, the nodes
with the highest degree in a network. This means that we lose information on
the other nodes in the network.

Loss of information is especially problematic because it is often difficult to
assess how much information is lost and to what extent this may affect the
conclusions that can be drawn from a bibliometric network visualization. For
instance, to what extent do the distances between the nodes in a distance-
based visualization accurately reflect the relatedness of the nodes? To what
extent does a visualization of a term co-occurrence network change if the
selection of terms included in the visualization is changed? Even if we are
aware that there may be inaccuracies in a bibliometric network visualization,
it remains difficult to assess the magnitude and the consequences of these
inaccuracies.

Related to this, it is often difficult to assess the sensitivity of a bibliometric
network visualization due to various technical choices. Would other technical
choices have resulted in a completely different visualization, or would the
differences have been minor? How strongly does a visualization depend on
the values of all kinds of technical parameters, and is it possible to justify the
choice of particular parameter values? Is a certain structure suggested by a
visualization a reflection of the underlying data, or is it merely an artifact of
the techniques used to produce the visualization? Researchers who regularly
work with bibliometric network visualizations develop an intuition that helps
them to give approximate answers to these types of questions, but most users
of bibliometric network visualizations lack such an intuition, making it
difficult for them to assess the accuracy of a visualization.

Given the above difficulties, our general recommendation is to use
bibliometric network visualizations as a complement rather than as a substitute to expert judgment. When expert judgment and bibliometric network
visualizations are in agreement with each other and point in the same direction, they strengthen each other. When they do not agree, this may be a reason
for experts to reconsider their opinion, it may also be a reason to ask for the
opinion of additional experts, or it may be a reason to check whether the
visualizations may be inaccurate because important information has been lost
or because of methodological issues. Bibliometric network visualizations are
most useful when they are interpreted in a careful manner and used in
combination with expert judgment. Also, visualization should be a means to
an end, not an end in itself. For instance, when dealing with only a small
amount of data, there often is no added value in the use of visualization. It
may be much better to simply study the data directly.

(continued)

(continued)

Ongoing and Future Developments in the Visualization of Bibliometric Networks

Finally, we discuss some ongoing and future developments in the visualization of bibliometric networks.

An important development, made possible by the enormous growth in computational resources that has taken place, is the increasing attention that is given to visualizing large bibliometric networks. Examples include the work by Boyack et al. (2005) on large journal networks, the work by Klavans and Boyack (2006) on large publication networks, and the work by Skupin et al. (2013) on large term networks. Interesting alternative data sources are being explored as well (e.g., Bollen et al., 2009).

A second and related development is the increasing use of interactive visualizations. The use of interactive visualizations is especially important given the above-mentioned trend toward visualizing increasingly large bibliometric networks. Static visualizations of large networks tend to be of limited use. It is rather difficult to show the detailed structure of a large network in a static visualization. Interactive visualizations allow large networks to be visualized and explored in a much more powerful way, for instance by allowing users to drill down from a general high-level overview to a very detailed low-level picture. The drill down functionality offered by CitNetExplorer can be seen as an example of this idea.

A third development is the increasing interest in dynamic visualizations that show how bibliometric networks have evolved over time. Traditionally, most attention has been paid to visualizations that offer a static picture showing the structure of a bibliometric network at a given point in time. Dynamic visualizations have received less attention, although some interesting work has been done, for instance in the CiteSpace tool (Chen, 2004, 2006) and in general network analysis tools such as Gephi and Visone (http://visone.info). Like visualizations of large networks, dynamic visualizations may benefit a lot from interactive visualization approaches.

In summary, we expect a trend toward more and more interactive and dynamic visualizations that involve increasingly large bibliometric networks. Clearly, an exciting and highly challenging research agenda lies ahead of us.

Appendix: Normalization, Mapping, and Clustering Techniques Used by VOSviewer

In this appendix, we provide a more detailed description of the normalization, mapping, and clustering techniques used by VOSviewer.

Normalization

We first discuss the association strength normalization (Van Eck & Waltman, 2009) used by VOSviewer to normalize for differences between nodes in the number of edges they have to other nodes. Let a_{ij} denote the weight of the edge between nodes i and j, where $a_{ij} = 0$ if there is no edge between the two nodes. Since VOSviewer treats all networks as undirected, we always have $a_{ij} = a_{ji}$. The association strength normalization constructs a normalized network in which the weight of the edge between nodes i and j is given by

$$s_{ij} = \frac{2ma_{ij}}{k_i k_j},\tag{1}$$

where k_i (k_j) denotes the total weight of all edges of node i (node j) and m denotes the total weight of all edges in the network. In mathematical terms,

$$k_i = \sum_j a_{ij} \quad \text{and} \quad m = \frac{1}{2}\sum_i k_i.\tag{2}$$

We sometimes refer to s_{ij} as the similarity of nodes i and j. For an extensive discussion of the rationale of the association strength normalization, we refer to Van Eck and Waltman (2009).

Mapping

We now consider the VOS mapping technique used by VOSviewer to position the nodes in the network in a two-dimensional space. The VOS mapping technique minimizes the function

$$V(\mathbf{x}_1,\ldots,\mathbf{x}_n) = \sum_{i<j} s_{ij}\|\mathbf{x}_i - \mathbf{x}_j\|^2\tag{3}$$

subject to the constraint

$$\frac{2}{n(n-1)}\sum_{i<j}\|\mathbf{x}_i - \mathbf{x}_j\| = 1,\tag{4}$$

where n denotes the number of nodes in a network, $\mathbf{x}_i$ denotes the location of node i in a two-dimensional space, and $\|\mathbf{x}_i - \mathbf{x}_j\|$ denotes the Euclidean distances between nodes i and j. VOSviewer uses a variant of the SMACOF algorithm (e.g., Borg & Groenen, 2005) to minimize (3) subject to (4). We refer to Van Eck et al. (2010) for a more extensive discussion of the VOS mapping technique, including a comparison with multidimensional scaling.

Clustering

Finally, we discuss the clustering technique used by VOSviewer. Nodes are assigned to clusters by maximizing the function

$$V(c_1,\ldots,c_n) = \sum_{i<j}\delta(c_i,c_j)(s_{ij} - \gamma),\tag{5}$$

where c_i denotes the cluster to which node i is assigned, $\delta(c_i, c_j)$ denotes a function that equals 1 if $c_i = c_j$ and 0 otherwise, and γ denotes a resolution parameter that determines the level of detail of the clustering. The higher the value of γ, the larger the number of clusters that will be obtained. The function in (5) is a variant of the modularity function introduced by Newman and Girvan (2004) and Newman (2004) for clustering the nodes in a network. There is also an interesting mathematical relationship between on the one hand the problem of minimizing (3) subject to (4) and on the other hand the problem of maximizing (5). Because of this relationship, the mapping and clustering techniques used by VOSviewer constitute a unified approach to mapping and clustering the nodes in a network. We refer to Waltman et al. (2010) for more details. We further note that VOSviewer uses the recently introduced smart local moving algorithm (Waltman & Van Eck, 2013) to maximize (5).

Acknowledgment We would like to thank Katy Börner and Ismael Rafols for their very helpful comments on an earlier draft of this chapter.

References

Bastian, M., Heymann, S., & Jacomy, M. (2009). Gephi: An open source software for exploring and manipulating networks. *International AAAI Conference on Weblogs and Social Media* [Online]

Bollen, J., Van de Sompel, H., Hagberg, A., Bettencourt, L., Chute, R., Rodriguez, M. A., et al. (2009). Clickstream data yields high-resolution maps of science. *PLoS ONE, 4*(3), e4803.

Borg, I., & Groenen, P. J. F. (2005). *Modern multidimensional scaling* (2nd ed.). New York, NY: Springer.

Börner, K. (2010). *Atlas of science: Visualizing what we know*. Cambridge, MA: MIT Press.

Börner, K., Chen, C., & Boyack, K. W. (2003). Visualizing knowledge domains. *Annual Review of Information Science and Technology, 37*(1), 179–255.

Börner, K., Klavans, R., Patek, M., Zoss, A. M., Biberstine, J. R., Light, R. P., et al. (2012). Design and update of a classification system: The UCSD map of science. *PLoS ONE, 7*(7), e39464.

Boyack, K. W., & Klavans, R. (2010). Co-citation analysis, bibliographic coupling, and direct citation. Which citation approach represents the research front most accurately? *Journal of the American Society for Information Science and Technology, 61*(12), 2389–2404.

Boyack, K. W., Klavans, R., & Börner, K. (2005). Mapping the backbone of science. *Scientometrics, 64*(3), 351–374.

Callon, M., Courtial, J. P., Turner, W. A., & Bauin, S. (1983). From translations to problematic networks: An introduction to co-word analysis. *Social Science Information, 22*(2), 191–235.

Callon, M., Law, J., & Rip, A. (Eds.). (1986). *Mapping the dynamics of science and technology*. London: MacMillan.

Chen, C. (1999). Visualising semantic spaces and author co-citation networks in digital libraries. *Information Processing & Management, 35*(3), 401–420.

Chen, C. (2004). Searching for intellectual turning points: Progressive knowledge domain visualization. *Proceedings of the National Academy of Sciences, 101*(suppl. 1), 5303–5310.

Chen, C. (2006). CiteSpace II: Detecting and visualizing emerging trends and transient patterns in scientific literature. *Journal of the American Society for Information Science and Technology, 57*(3), 359–377.

Cobo, M. J., López-Herrera, A. G., Herrera-Viedma, E., & Herrera, F. (2011). Science mapping software tools: Review, analysis, and cooperative study among tools. *Journal of the American Society for Information Science and Technology, 62*(7), 1382–1402.

De Moya-Anegón, F., Vargas-Quesada, B., Chinchilla-Rodríguez, Z., Corera-Álvarez, E., Munoz-Fernández, F. J., & Herrero-Solana, V. (2007). Visualizing the marrow of science. *Journal of the American Society for Information Science and Technology, 58*(14), 2167–2179.

De Nooy, W., Mrvar, A., & Batagelj, V. (2005). *Exploratory social network analysis with Pajek*. Cambridge: Cambridge University Press.

Fruchterman, T. M. J., & Reingold, E. M. (1991). Graph drawing by force-directed placement. *Software: Practice and Experience, 21*(11), 1129–1164.

Garfield, E. (2004). Historiographic mapping of knowledge domains literature. *Journal of Information Science, 30*(2), 119–145.

Garfield, E., Pudovkin, A. I., & Istomin, V. S. (2003). Why do we need algorithmic historiography? *Journal of the American Society for Information Science and Technology, 54*(5), 400–412.

Griffith, B. C., Small, H., Stonehill, J. A., & Dey, S. (1974). The structure of scientific literatures II: Toward a macro- and microstructure for science. *Science Studies, 4*(4), 339–365.

Healy, P., & Nikolov, N. S. (2013). Hierarchical drawing algorithms. In R. Tamassia (Ed.), *Handbook of graph drawing and visualization* (pp. 409–453). Boca Raton, FL: CRC Press.

Hummon, N. P., & Doreian, P. (1989). Connectivity in a citation network: The development of DNA theory. *Social Networks, 11*(1), 39–63.

Jarneving, J. (2007). Bibliographic coupling and its application to research-front and other core documents. *Journal of Informetrics, 1*(4), 287–307.

Kamada, T., & Kawai, S. (1989). An algorithm for drawing general undirected graphs. *Information Processing Letters, 31*(1), 7–15.

Kessler, M. M. (1963). Bibliographic coupling between scientific papers. *American Documentation, 14*(1), 10–25.

Klavans, R., & Boyack, K. W. (2006). Quantitative evaluation of large maps of science. *Scientometrics, 68*(3), 475–499.

Leydesdorff, L., & Rafols, I. (2009). A global map of science based on the ISI subject categories. *Journal of the American Society for Information Science and Technology, 60*(2), 348–362.

Leydesdorff, L., & Rafols, I. (2012). Interactive overlays: A new method for generating global journal maps from Web-of-Science data. *Journal of Informetrics, 6*(2), 318–332.

Luukkonen, T., Tijssen, R. J. W., Persson, O., & Sivertsen, G. (1993). The measurement of international scientific collaboration. *Scientometrics, 28*(1), 15–36.

Marshakova, I. (1973). System of documentation connections based on references (SCI). *Nauchno-TekhnicheskayaInformatsiya Seriya, 2*(6), 3–8.

McCain, K. W. (1991). Mapping economics through the journal literature: An experiment in journal cocitation analysis. *Journal of the American Society for Information Science, 42*(4), 290–296.

Morris, S. A., Yen, G., Wu, Z., & Asnake, B. (2003). Time line visualization of research fronts. *Journal of the American Society for Information Science and Technology, 54*(5), 413–422.

Newman, M. E. J. (2004). Fast algorithm for detecting community structure in networks. *Physical Review E, 69*(6), 066133.

Newman, M. E. J., & Girvan, M. (2004). Finding and evaluating community structure in networks. *Physical Review E, 69*(2), 026113.

Persson, O. (2010). Identifying research themes with weighted direct citation links. *Journal of Informetrics, 4*(3), 415–422.

Peters, H. P. F., & Van Raan, A. F. J. (1993). Co-word-based science maps of chemical engineering. Part I: Representations by direct multidimensional scaling. *Research Policy, 22* (1), 23–45.

Schvaneveldt, R. W., Dearholt, D. W., & Durso, F. T. (1988). Graph theoretic foundations of pathfinder networks. *Computers & Mathematics with Applications, 15*(4), 337–345.

Seidman, S. B. (1983). Network structure and minimum degree. *Social Networks, 5*(3), 269–287.

Skupin, A., Biberstine, J. R., & Börner, K. (2013). Visualizing the topical structure of the medical sciences: A self-organizing map approach. *PLoS ONE, 8*(3), e58779.

Small, H. (1973). Co-citation in the scientific literature: A new measure of the relationship between two documents. *Journal of the American Society for Information Science, 24*(4), 265–269.

Small, H. (1997). Update on science mapping: Creating large document spaces. *Scientometrics, 38* (2), 275–293.

Small, H., & Griffith, B. C. (1974). The structure of scientific literatures I: Identifying and graphing specialties. *Science Studies, 4*(1), 17–40.

Van Eck, N. J., & Waltman, L. (2009). How to normalize cooccurrence data? An analysis of some well-known similarity measures. *Journal of the American Society for Information Science and Technology, 60*(8), 1635–1651.

Van Eck, N. J., & Waltman, L. (2010). Software survey: VOSviewer, a computer program for bibliometric mapping. *Scientometrics, 84*(2), 523–538.

Van Eck, N. J., & Waltman, L. (2011). Text mining and visualization using VOSviewer. *ISSI Newsletter, 7*(3), 50–54.

Van Eck, N. J., Waltman, L., Dekker, R., & Van den Berg, J. (2010). A comparison of two techniques for bibliometric mapping: Multidimensional scaling and VOS. *Journal of the American Society for Information Science and Technology, 61*(12), 2405–2416.

Van Eck, N. J., Waltman, L., Van Raan, A. F. J., Klautz, R. J. M., & Peul, W. C. (2013). Citation analysis may severely underestimate the impact of clinical research as compared to basic research. *PLoS ONE, 8*(4), e62395.

Waltman, L., & Van Eck, N. J. (2012). A new methodology for constructing a publication-level classification system of science. *Journal of the American Society for Information Science and Technology, 63*(12), 2378–2392.

Waltman, L., & Van Eck, N. J. (2013). A smart local moving algorithm for large-scale modularity-based community detection. *European Physical Journal B, 86*, 471.

Waltman, L., Van Eck, N. J., & Noyons, E. C. M. (2010). A unified approach to mapping and clustering of bibliometric networks. *Journal of Informetrics, 4*(4), 629–635.

White, H. D. (2003). Pathfinder networks and author cocitation analysis: A remapping of paradigmatic information scientists. *Journal of the American Society for Information Science and Technology, 54*(5), 423–434.

White, H. D., & Griffith, B. C. (1981). Author cocitation: A literature measure of intellectual structure. *Journal of the American Society for Information Science, 32*(3), 163–171.

White, H. D., & McCain, K. W. (1998). Visualizing a discipline: An author co-citation analysis of information science, 1972–1995. *Journal of the American Society for Information Science, 49*(4), 327–355.

Zhao, D., & Strotmann, A. (2008). Evolution of research activities and intellectual influences in information science 1996–2005: Introducing author bibliographic-coupling analysis. *Journal of the American Society for Information Science and Technology, 59*(13), 2070–2086.

Chapter 14
Replicable Science of Science Studies

Katy Börner and David E. Polley

Abstract Much research in bibliometrics and scientometrics is conducted using proprietary datasets and tools making it hard if not impossible to replicate results. This chapter reviews free tools, software libraries, and online services that support science of science studies using common data formats. We then introduce plug-and-play macroscopes (Börner, Commun ACM 54(3):60–69, 2011) that use the OSGi industry standard to support modular software design, i.e., the plug-and-play of different data readers, preprocessing and analysis algorithms, but also visualization algorithms and tools. Exemplarily, we demonstrate how the open source Science of Science (Sci2) Tool can be used to answer temporal (when), geospatial (where), topical (what), and network questions (with whom) at different levels of analysis—from micro to macro. Using the Sci2 Tool, we provide hands-on instructions on how to run burst analysis (see Chapter 10 in this book), overlay data on geospatial maps (see Chapter 6 in this book), generate science map overlays, and calculate diverse network properties, e.g., weighted PageRank (see Chapter 4 in this book) or community detection (see Chapter 3 in this book), using data from Scopus, Web of Science or personal bibliography files, e.g., EndNote or BibTex. We exemplify tool usage by studying evolving research trajectories of a group of physicists over temporal, geospatial, and topic space as well as their evolving co-author networks. Last but not least, we show how plug-and-play macroscopes can be used to create bridges between existing tools, e.g., Sci2 and the VOSviewer clustering algorithm (see Chapter 13 in this book), so that they can be combined to execute more advanced analysis and visualization workflows.

K. Börner (✉)
Cyberinfrastructure for Network Science Center, School of Informatics and Computing,
Indiana University, 1320 E. Tenth Street, Bloomington, IN 47405, USA
e-mail: katy@indiana.edu

D.E. Polley
University Library, Indiana University-Purdue University Indianapolis,
755 West Michigan Street, Indianapolis, IN 46292, USA
e-mail: dapolley@iupui.edu

© Springer International Publishing Switzerland 2014
Y. Ding et al. (eds.), *Measuring Scholarly Impact*,
DOI 10.1007/978-3-319-10377-8_14

14.1   Open Tools for Science of Science Studies

Science of science studies seek to develop theoretical and empirical models of the scientific enterprise. Examples include qualitative and quantitative methods to estimate the impact of science (Cronin & Sugimoto, 2014) or models to understand the production of science (Scharnhorst, Börner, & van den Besselaar, 2012). There exist a variety of open-source tools that support different types of analysis and visualization. Typically, tools focus on a specific type of analysis and perform well at a certain level, e.g., at the micro (individual) level or meso level—using datasets containing several thousand records. What follows is a brief survey of some of the more commonly used tools that support four types of analyses and visualizations: temporal, geospatial, topical, and network.

Many tools have a temporal component, but few are dedicated solely to the interactive exploration of time-series data. The Time Searcher project is an excellent example of a tool that allows for interactive querying of time stamped data through the use of timeboxes, a graphical interface that allows users to build and manipulate queries (Hochheiser & Shneiderman, 2004). Now in its third iteration, the tool can handle more than 10,000 data points and offers data-driven forecasting through a Similarity-Based Forecasting (SBF) interface (Buono et al., 2007). The tool runs on Windows platforms and is freely available.

There are many tools available for advanced geospatial visualization and most require some knowledge of geographical information science. Exemplary tools include GeoDa, GeoVISTA, and CommonGIS. GeoDa is an open-source and cross-platform tool that facilitates common geospatial analysis functionality, such as spatial autocorrelation statistics, spatial regression functionality, full space-time data support, cartograms, and conditional plots (and maps) (Anselin, Syabri, & Kho, 2006). Another tool, GeoVISTA Studio, comes from the GeoVISTA Center at Penn State, which produces a variety of geospatial analysis tools. Studio offers an open-source graphical interface that allows users to build applications for geocomputation and visualization and allows for interactive querying, 3D rendering of complex graphics, and 2D mapping and statistical tools (Takatsuka & Gahegan, 2002). Finally, CommonGIS is a java-based geospatial analysis tool accessible via the web. This service allows for interactive exploration and analysis of geographically referenced statistical data through any web browser (Andrienko et al., 2002).

There are a variety of topical data analysis and visualization tools used in a variety of disciplines from bibliometrics to digital humanities and business. TexTrend is a freely available, cross-platform tool that aims to support decision-making in government and business. Specifically, the tool facilitates text mining and social network analysis, with an emphasis on dynamic information (Kampis, Gulyas, Szaszi, Szakolczi, & Soos, 2009). The aim of the tool is to extract trends and support predictions based on textual data. VOSviewer is another cross-platform and freely available topical analysis tool, designed specifically for bibliometric analysis (Van Eck & Waltman, 2014). The tool allows users to create maps of publications, authors, and journals based on co-citation networks or keywords, illuminating the topic coverage of a dataset. VOSviewer provides multiple ways to visualize data, including a label view, density view, cluster density view, and a scatter view.

Finally, there are many tools that are dedicated to network analysis and visualization. Some of the more prominent tools include Pajek and Gephi. Pajek has long been popular among social scientists. Originally designed for social network analysis, the program is not open-source but is freely available for noncommercial use on the Windows platform. Pajek is adept at handling large networks containing thousands of nodes (Nooy, Mrvar, & Batageli, 2011). The tool handles a variety of data objects, including networks, partitions, permutations, clusters, and hierarchies. Pajek offers a variety of network analysis and visualization algorithms and includes bridges to other programs, such as the ability to export to R for further analysis. Gephi is a more recent, and widely used network analysis and visualization program (Bastian, Heymann, & Jacomy, 2009). This tool is open-source and available for Windows, Mac OS, and Linux. Gephi is capable of handling large networks, and provides common network analysis algorithms, such as average network degree, graph density, and modularity. Gephi is also well suited for displaying dynamic and evolving networks and accepts a wide variety of input formats including NET and GRAPHML files.

14.2 The Science of Science (Sci2) Tool

The Science of Science (Sci2) Tool is a modular toolset specifically designed for the study of Science (Sci2 Team, 2009). Sci2 can be downloaded from http://sci2.cns.iu.edu and it can be freely used for research and teaching but also for commercial purposes (Apache 2.0 license). Extensive documentation is provided on the Sci2 Wiki (http://sci2.wiki.cns.iu.edu), in the Information Visualization MOOC (http://ivmooc.cns.iu.edu), and the *Visual Insights* textbook (Börner & Polley, 2014).

Instead of focusing on just one specific type of analysis, like many other tools, it supports the temporal, geospatial, topical, and network analysis and visualization of datasets at the micro (individual), meso (local), and macro (global) levels. The tool is built on the OSGi/CIShell framework that is widely used and supported by industry (http://osgi.org). It uses an approach known as "plug-and-play" in macroscope construction (Börner, 2011), allowing anyone to easily add new algorithms and tools using a wizard-supported process, and to customize the tool to suit their specific research needs. The tool is optimized for datasets of up to 100,000 records for most algorithms (Light, Polley, & Börner, 2014). This section reviews key functionality—from data reading to preprocessing, analysis, and visualization.

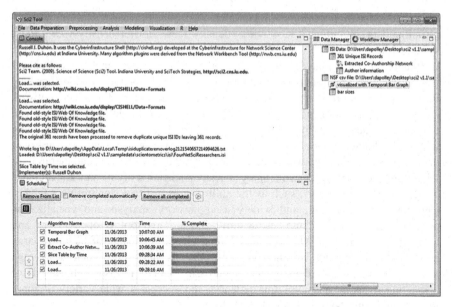

Fig. 14.1 Sci2 user interface with Menu on *top*, Console *below*, Scheduler in *lower left*, and Data Manager and Workflow Tracker on *right*

14.2.1 Workflow Design and Replication

The Sci2 Tool supports workflow design, i.e., the selection and parameterization of different data readers, data analysis, visualization and other algorithms, via a unified graphical interface, see Fig. 14.1.

Workflows are recorded and can be re-run to replicate results. They can be run with different parameter values to test and understand the sensitivity of results or to perform parameter sweeps. Workflows can also be run on other datasets to compare results. Last but not least, the same workflow can be run with different algorithms (e.g., clustering techniques) in support of algorithm comparisons. For a detailed documentation of workflow log formats and usage, please see our documentation wiki.[1]

All workflows discussed in this paper have been recorded and can be re-run. They can be downloaded from section 2.6, Sample Workflows, on the Sci2 wiki.[2]

---

[1] http://wiki.cns.iu.edu/display/CISHELL/Workflow+Tracker

[2] http://wiki.cns.iu.edu/display/SCI2TUTORIAL/2.6+Sample+Workflows

14.2.2 Data Readers

The Sci2 Tool reads a number of common file formats, including tabular formats (.CSV); output file formats from major data providers such as Thomson Reuter's Web of Science (.isi), Elsevier's Scopus (.scopus), Google Scholar, but also funding data from the U.S. National Science Foundation (.nsf) and the U.S. National Institutes of Health (using .CSV); output formats from personal bibliography management systems such as EndNote (.enw) and Bibtex (.bib). In addition, there exist data readers that retrieve data from Twitter, Flickr, and Facebook.[3] Last but not least, Sci2 was codeveloped with the Scholarly Database (SDB) (http://sdb.cns.iu.edu) that provides easy access to 27 million paper, patent, grant, clinical trials records. All datasets downloaded from SDB in tabular or network format, e.g., coauthor, coinventor, coinvestigator, patent–citation networks, are fully compatible with Sci2. File format descriptions and sample data files are provided at the Sci2 Wiki in section 4.2.[4]

14.2.3 Temporal Analysis (When)

14.2.3.1 Data Preprocessing

The "Slice Table by Time" algorithm[5] is a common data preprocessing step in many temporal visualization workflows. As an input, the algorithm takes a table with a date/time value associated with each record. Based on a user-specified time interval the algorithm divides the original table into a series of new tables. Depending on the parameters selected, these time slices are either cumulative or not, and aligned with the calendar or not. The intervals into which a table may be sliced include: milliseconds, seconds, minutes, hours, days, weeks, months, quarters, years, decades, and centuries.

14.2.3.2 Data Analysis

The "Burst Detection" algorithm[6] implemented in Sci2, adapted from Jon Kleinberg's (2002), identifies sudden increases or "bursts" in the frequency-of-use of character strings over time. It identifies topics, terms, or concepts important to the events being studied that increase in usage, are more active for a period of time, and then fade away. The input for the algorithm is time-stamped text, such as

---

[3] http://wiki.cns.iu.edu/display/SCI2TUTORIAL/3.1+Sci2+Algorithms+and+Tools

[4] http://wiki.cns.iu.edu/display/SCI2TUTORIAL/4.2+Data+Acquisition+and+Preparation

[5] http://wiki.cns.iu.edu/display/CISHELL/Slice+Table+by+Time

[6] http://wiki.cns.iu.edu/display/CISHELL/Burst+Detection

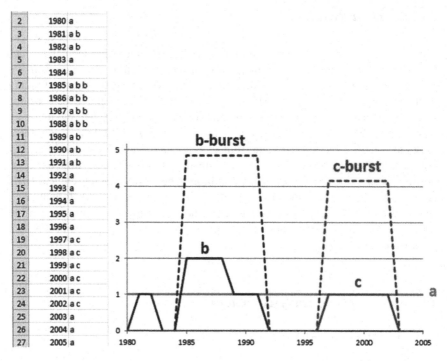

2	1980	a
3	1981	a b
4	1982	a b
5	1983	a
6	1984	a
7	1985	a b b
8	1986	a b b
9	1987	a b b
10	1988	a b b
11	1989	a b
12	1990	a b
13	1991	a b
14	1992	a
15	1993	a
16	1994	a
17	1995	a
18	1996	a
19	1997	a c
20	1998	a c
21	1999	a c
22	2000	a c
23	2001	a c
24	2002	a c
25	2003	a
26	2004	a
27	2005	a

Fig. 14.2 A burst analysis diagram for three letters (*right*) compared with the raw data (*left*)

documents with publication years. From titles, abstracts or other text, the algorithm generates a list of burst words, ranked according to burst weight, and the intervals of time in which these bursts occurred. Figure 14.2 shows a diagram of bursting letters (left) next to the raw data (right). The letter "b" (solid blue line plots frequency, dashed blue line plots burst) experienced a burst from just before 1985 to just after 1990. Similarly, the letter "c" (red line) experienced a burst starting just after 1995 and ending just before 2005. However, the letter "a" (green line) remains constant throughout this time series, i.e., there is no burst for that term.

14.2.3.3   Data Visualization

The "Temporal Bar Graph" visualizes numeric data over time, and is the only truly temporal visualization algorithm available in Sci2.[7] This algorithm accepts tabular (CSV) data, which must have start and end dates associated with each record. Records that are missing either start or end dates are ignored. The other input parameters include "Label," which corresponds to a text field and is used to label the bars; "Size By," which must be an integer and corresponds to the area of the

---

[7] http://wiki.cns.iu.edu/display/CISHELL/Temporal+Bar+Graph

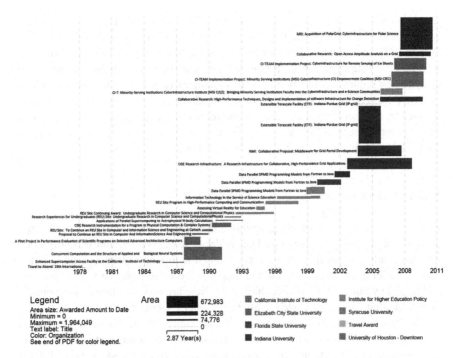

Fig. 14.3 Temporal Bar Graph showing the NSF funding profile for Dr. Geoffrey Fox

horizontal bars; "Date Format," which can either be in "Day-Month-Year Date Format (Europe, e.g. 31/10/2010)" or "Month-Day-Year (U.S., e.g. 10/31/2010)"; and "Category," which allows users to color code bars by an attribute of the data. For example, Fig. 14.3 shows the National Science Foundation (NSF) funding Profile for Dr. Geoffrey Fox, Associate Dean for Research and Distinguished Professor of Computer Science and Informatics at Indiana University, where each bar represents an NSF award on which Dr. Fox was an investigator. Each bar is labeled with the title of the award. The total area of the bars corresponds to the total amount awarded, and the bars are color-coded by the institution/organization affiliated with the award.

14.2.4 Geospatial Analysis (Where)

14.2.4.1 Data Preprocessing

The "Extract ZIP Code" algorithm[8] is a common preprocessing step in geospatial visualization workflows. This algorithm takes US addresses as input data and extracts the ZIP code from the address in either the standard five-digit short form (xxxxx) or the

---

[8] http://wiki.cns.iu.edu/display/CISHELL/Extract+ZIP+Code

standard nine-digit long form (xxxxx-xxxx). This feature facilitates quick spatial analysis and simplifies geocoding. However, the algorithm is limited to U.S. or U. S.-based ZIP code systems. Another useful preprocessing step for geospatial and other workflows is data aggregation. Redundant geo-identifiers are common in geospatial analysis and require aggregation prior to visualization. Sci2 provides basic aggregation with the "Aggregate Data" algorithm,[9] which groups together values in a column selected by the user. The other values in the records are aggregated as specified by the user. Currently, sum, difference, average, min, and max are available for numerical data. All text data are aggregated when a text delimiter is provided.

14.2.4.2 Data Analysis

Sci2 has a variety of geocoding options. The "Generic Geocoder"[10] is the most basic of these options. It converts U.S. addresses, U.S. states, and U.S. ZIP codes into longitude and latitude values. The input for this algorithm is a table with a geo-identifier for each record, and the output is the same table but with a longitude and latitude value appended to each record. There are no restrictions on the number of records that can be geocoded using the "Generic Geocoder."

The "Bing Geocoder"[11] expands the functionality of the "Generic Geocoder," allowing Sci2 to convert international addresses into longitude and latitude values. All coordinates are obtained by querying the Bing geocoder service and Internet access must be available while using this algorithm. Users must obtain an API key from Bing Maps in order to run this algorithm, and there is a limit of 50,000 records which can be geocoded in a 24 hour period. Finally, Sci2 provides a "Congressional District Geocoder,"[12] which converts nine-digit ZIP codes (five-digit ZIP codes can contain multiple districts) into congressional districts and geographic coordinates. The algorithm is available as an external plugin that can be downloaded from the Sci2 wiki.[13] The database that supports this algorithm is based on the 2012 ZIP Code Database for the 113th U.S. Congress and does not take into account any subsequent redistricting.

14.2.4.3 Data Visualization

The Sci2 Tool offers three geospatial visualization algorithms: a proportional symbol map, a choropleth map, or region-shaded map, and a network geomap overlay. The "Proportional Symbol Map" algorithm[14] takes a list of coordinates

---

[9] http://wiki.cns.iu.edu/display/CISHELL/Aggregate+Data

[10] http://wiki.cns.iu.edu/display/CISHELL/Geocoder

[11] http://wiki.cns.iu.edu/display/CISHELL/Bing+Geocoder

[12] http://wiki.cns.iu.edu/display/CISHELL/Congressional+District+Geocoder

[13] http://wiki.cns.iu.edu/display/SCI2TUTORIAL/3.2+Additional+Plugins

[14] http://wiki.cns.iu.edu/display/CISHELL/Proportional+Symbol+Map

Fig. 14.4 Proportional symbol map (*left*), choropleth map (*center*), and network layout overlaid on world map (*right*)

and at most three numeric attributes and visualizes them over a world or United States base map. The sizes and colors of the symbols are proportional to the numeric values. The "Choropleth Map" algorithm[15] allows users to color countries of the world or states of the United States in proportion to one numeric attribute. Finally, Sci2 has the ability to visualize networks overlaid on a geospatial basemap. As input, the "Geospatial Network Layout with Base Map" algorithm[16] requires a network file with latitude and longitude values associated with each node. The algorithm produces a network file and a PostScript base map. The network file is visualized in GUESS or Gephi, exported as a PDF, and overlaid on the PDF of the base map. Figure 14.4 shows an exemplary proportional symbol map (left), choropleth map (center), and the geospatial network layout with base map (right).

14.2.5 Topical Analysis (What)

14.2.5.1 Data Preprocessing

The topic or semantic coverage of a scholar, institution, country, paper, journal or area of research can be derived from the texts associated with it. In order to analyze and visualize topics, text must first be normalized. Sci2 provides basic text normalization with the "Lowercase, Tokenize, Stem, and Stopword Text" algorithm.[17] This algorithm requires tabular data with a text field as input and outputs a table with the specified text field normalized. Specifically, the algorithm makes all text lowercase, splits the individual words into tokens (delimited by a user-selected separator), stems each token (removing low content prefixes and suffixes), and removes stopwords, i.e., very common (and therefore dispensable) words or phrases such as "the" or "a". Sci2 provides a basic stopword list,[18] which can be edited to fit users' specific needs. The goal of text normalization is to facilitate the extraction of unique words or word profiles in order to identify topic coverage of bodies of text.

---

[15] http://wiki.cns.iu.edu/display/CISHELL/Choropleth+Map

[16] http://wiki.cns.iu.edu/display/CISHELL/Geospatial+Network+Layout+with+Base+Map

[17] http://wiki.cns.iu.edu/display/CISHELL/Lowercase%2C+Tokenize%2C+Stem%2C+and +Stopword+Text

[18] *yoursci2directory*/*configuration*/*stopwords.txt*

14.2.5.2 Data Analysis

Burst detection, previously discussed in the temporal section, is often used for identifying the topic coverage of a corpus of text. Since burst detection also involves a temporal component, it is ideal for demonstrating the evolution of scientific research topics as represented by bodies of text.

14.2.5.3 Data Visualization

Sci2 provides a map of science visualization algorithm to display the topical distributions, also called expertise profiles. The UCSD Map of Science (Börner et al., 2012) is a visual representation of 554 sub-disciplines within the 13 disciplines of science and their relationships to one another. There are two variations of this algorithm: "Map of Science via Journals"[19] and "Map of Science via 554 Fields".[20] The first works by matching journal titles to the underlying sub-disciplines, as specified in the UCSD Map of Science classification scheme.[21] The second works by directly matching the IDs for the 554 fields, integers 1–554, to the sub-disciplines. Both algorithms take tabular data as input, the first with a column of journal names and the second with a column of field IDs. It is recommended that users run the "Reconcile Journal Names" algorithm[22] prior to science mapping. Both map of science algorithms output a PostScript file and the "Map of Science via Journals" also outputs two tables: one for the journals located, and one for journals not located. Figure 14.5 shows the topic distribution of the *FourNetSciResearchers.isi* file, a dataset containing citations from four major network science researchers: Eugene Garfield, Stanley Wasserman, Alessandro Vespignani, and Albert-László Barabási with a total of 361 publication records.

14.2.6   Network Analysis (with Whom)

14.2.6.1   Data Preprocessing

Many datasets come in tabular format and a key step in most network visualization workflows involves extracting networks from these tables. Sci2 supports this process with a variety of network extraction algorithms. The "Extract Co-Occurrence Network" algorithm can be used to extract networks from columns

---

[19] http://wiki.cns.iu.edu/display/CISHELL/Map+of+Science+via+Journals

[20] http://wiki.cns.iu.edu/display/CISHELL/Map+of+Science+via+554+Fields

[21] http://sci.cns.iu.edu/ucsdmap

[22] http://wiki.cns.iu.edu/display/CISHELL/Reconcile+Journal+Names

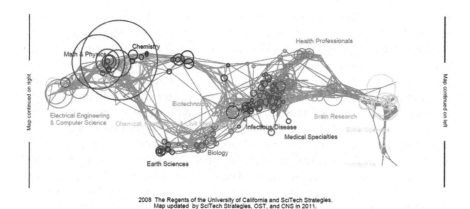

Fig. 14.5 Map of Science via Journals showing the topic coverage of the *FourNetSciResearchers. isi* file, see wiki for legend and additional information analogous to Fig. 14.8

that contain multiple values.[23] The "Extract a Directed Network" algorithm[24] will create a network between two columns with data of the same type. Sci2 makes extracting co-occurrence networks specific to bibliometric analysis even easier by providing algorithms such as "Extract Co-Author Network",[25] "Extract Word Co-Occurrence Network",[26] and "Extract Reference Co-Occurrence (Bibliographic Coupling) Network."[27] Finally, for two columns that contain different data types, the "Extract Bipartite Network" algorithm[28] can be used.

14.2.6.2   Data Analysis

Sci2 offers a large variety of network analysis algorithms for directed or undirected, and weighted or unweighted networks. A full list of all available network analysis algorithms can be found in section 3.1 Sci2 Algorithms and Tools of the online wiki.[29] Two of the more interesting algorithms for network analysis include the PageRank and Blondel Community Detection algorithms (also known as the Louvain algorithm). The PageRank algorithm was originally developed for the Google search engine to rank sites in the search result by relative importance, as measured

---

[23] http://wiki.cns.iu.edu/display/CISHELL/Extract+Word+Co-Occurrence+Network

[24] http://wiki.cns.iu.edu/display/CISHELL/Extract+Directed+Network

[25] http://wiki.cns.iu.edu/display/CISHELL/Extract+Co-Author+Network

[26] http://wiki.cns.iu.edu/display/CISHELL/Extract+Word+Co-Occurrence+Network

[27] http://wiki.cns.iu.edu/display/CISHELL/Extract+Reference+Co-Occurrence+%28Biblio graphic+Coupling%29+Network

[28] http://wiki.cns.iu.edu/display/CISHELL/Extract+Bipartite+Network

[29] http://wiki.cns.iu.edu/display/SCI2TUTORIAL/3.1+Sci2+Algorithms+and+Tools

by the number of links to a page (Brin & Page, 1998).[30] The same process can be used in directed networks to rank the relative importance of nodes. There are two versions of the "PageRank" algorithm in Sci2, one for directed and unweighted networks, which simply measures the importance of nodes based on the number of incoming edges, and one for directed and weighted networks, which measures the importance nodes based on incoming edges and takes into consideration the weight of those edges. Both algorithms are useful for identifying important nodes in very large networks. The "Blondel Community Detection" algorithm is a clustering algorithm for large networks.[31] The algorithm detects communities in weighted networks using an approach based on modularity optimization (Blondel, Guillaume, Lambiotte, & Lefebvre, 2008). The resulting network will be structurally the   same   but   each   node   will   have   an   attribute   labeled "blondel_community_level_x."

14.2.6.3   Data Visualization

The Sci2 Tool offers multiple ways to view networks. While the tool itself does not directly support network visualization, GUESS, a common network visualization program comes already bundled with the tool (Adar & Kim, 2007). For users who already have Gephi (Bastian et al., 2009) installed on their machines, Sci2 provides a bridge, allowing a user to select a network in the Data Manager, then run Gephi, and the tool will start with the selected network file loaded. Gephi is available for download at http://gephi.org. Cytoscape (Saito et al., 2012) was also made available as a plugin to the Sci2 Tool, further expanding network visualization options. The Cytoscape plugin is available for download from section 3.2 of the Sci2 wiki.[32]

14.3   Career Trajectories

This section demonstrates Sci2 Tool functionality for the analysis and visualization of career trajectories, i.e., the trajectory of people over time. In addition to studying movement over geospatial space, e.g., via the addresses of different institutions people might study and work at, the expertise profile of physicists are analyzed. Specifically, we use a dataset of authors in physics with a large number of institutional affiliations—assuming that those authors engaged in a large number of Postdocs. The original dataset of the top-10,000 authors in physics with the most affiliations was retrieved by Vincent Larivière. For the purposes of this study, we selected ten of top paper producing authors. Using the name-unification method

---

[30] http://wiki.cns.iu.edu/display/CISHELL/PageRank

[31] http://wiki.cns.iu.edu/display/CISHELL/Blondel+Community+Detection

[32] http://wiki.cns.iu.edu/display/SCI2TUTORIAL/3.2+Additional+Plugins

introduced by Kevin W. Boyack and Richard Klavans (2008), we consider names to refer to one person if the majority of his or her publications occur at one institution. Uniquely identifying people in this way is imperfect but practical. The resulting dataset contains ten authors that have more than 70 affiliation addresses in total and published in more than 100 mostly physics journals between 1988 and 2010. Subsequently, we show how Sci2 can be used to clean and aggregate data, provide simple statistics, analyze the geospatial and topical evolution of the different career trajectories over time and visualize results. We conclude this section with an interpretation of results and a discussion of related works.

14.3.1 Data Preparation Analysis

The list of the top 10,000 physicists was loaded into OpenRefine for cleaning.[33] The author names were normalized by placing them in uppercase and trimming leading and trailing white spaces. Then, the text facets feature of OpenRefine was used to identify groups of names. The count associated with each name corresponds to the number of papers associated with that name, which starts to give some idea which names uniquely identify a person, reducing homonymity. Once a name was identified as potentially uniquely identifying a person, the text facet was applied to this person's institutions, showing the number of papers produced at each institution. If the majority of the papers associated with a name occurred at one institution, the name was considered to uniquely identify one person in this dataset. Following this process for the highest producing authors in this dataset resulted in the final list of ten physicists, see Table 14.1.

Next, the addresses for each institution were obtained by searching the Web. The data for each of the ten authors was then saved in separate CSV files and loaded into Sci2. The "Bing Geocoder" algorithm[34] was used to geocode each institution. Then, the "Aggregate Data" algorithm[35] was used to aggregate the data for each author by institution, summing the number of papers produced at those institutions and summing the total citations for those papers. This aggregation was performed because many authors had affiliations with institutions far away from their home institutions for short periods of time, either due to sabbaticals or as visiting scholars.

In addition, the citation data for each of the ten physicists was downloaded from the Web of Science. The full records plus citations were exported as ISI-formatted text files. The extensions were changed from .txt to .isi and loaded into Sci2. Next,

---

[33] http://code.google.com/p/google-refine/

[34] http://wiki.cns.iu.edu/display/CISHELL/Bing+Geocoder

[35] http://wiki.cns.iu.edu/display/CISHELL/Aggregate+Data

Table 14.1 Top-ten
physicists with the most
publications plus the number
of their institutions, papers,
and citation counts

Name	Institutions	Papers	Citations
AGARWAL-GS	7	163	3,917
AMBJORN-J	3	185	3,750
BENDER-CM	6	118	3,962
BRODSKY-SJ	14	119	4,566
CHAICHIAN-M	7	123	2,725
ELIZALDE-E	16	135	2,151
GIANTURCO-FA	4	130	1,634
PERSSON-BNJ	6	100	3,472
YUKALOV-VI	8	150	1,772
ZHDANOV-VP	6	147	1,594

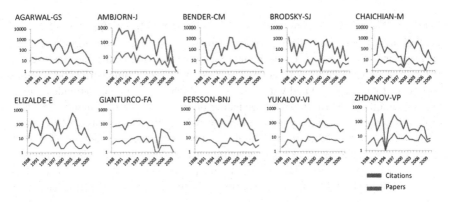

Fig. 14.6 Paper and citation counts over 23 years for each of the ten physicists

the "Reconcile Journal Names" algorithm[36] was run, which ensures all journal titles
are normalized and matched to the standard given in the UCSD Map of Science
Standard (Börner et al., 2012).

14.3.2 Data Visualization and Interpretation

Initially a comparison of all the authors, their paper production, and resulting
citation counts was created in MS Excel. The ten authors followed more or less
similar trajectories, with earlier works receiving more citations and newer works
fewer. Figure 14.6 shows the number of papers per publication year (blue line) and

---

[36] http://wiki.cns.iu.edu/display/CISHELL/Reconcile+Journal+Names

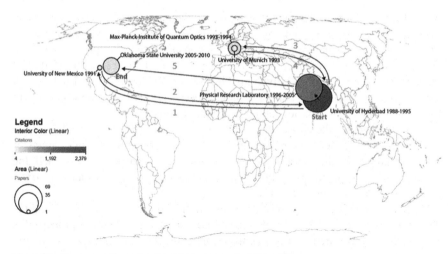

Fig. 14.7 Career trajectory for Dr. Girish Agarwal, 1988–2010

the number of all citations received up to 2009 by the papers published in a specific year (red line). As expected, the bulk of the paper production tends to happen at one "home" institution.

Next, the geocoded files for each author were visualized in Sci2 to show the world-wide career trajectories over geospatial space. Specifically, the "Proportional Symbol Map" algorithm[37] was applied to the physicist with the highest combined number of papers and citation count and the resulting map was saved as PDFs and further edited in Adobe Illustrator to add the directed edges connecting the institution symbols. The resulting map for Dr. Girish Agarwal is shown in Fig. 14.7. Each circle corresponds to an institution and is labeled with the institution's name and the date range in years that the physicist is associated with the institution. The symbols are sized proportional to the number of papers produced at each institution and colored proportional to the number of citations those papers have received. A green "Start" label was added to the first institution associated with the physicist in the dataset, and a red "End" label was added to the last. Green numbers indicate the sequence of transitions between start, intermediary, and end institutions.

As expected, the majority of the paper production occurred at the institutions where Dr. Agarwal spent the most time, University of Hyderabad and Physical Research Laboratory, where he was the director for 10 years. The papers produced at these institutions also have the highest citations counts, as these papers are much

[37] http://wiki.cns.iu.edu/display/CISHELL/Proportional+Symbol+Map

older than the others in the dataset. Visualizing the dataset over a geographic map gives a sense of the geospatial trajectory of Agarwal's career, but the limited scope of the dataset results in a somewhat misleading visualization. Simply by looking at the map, one might assume that Dr. Agarwal started his career at Hyderbad University, but he actually received his Ph.D. from the University of Rochester in 1969. It is highly likely there are other publications at other institutions that, due to the limited date range, are not captured by this visualization.

Next, to visualize the topical distribution Dr. Agarwal's publications, his ISI file was mapped using the "Map of Science via Journals" using the default parameter values. Figure 14.8 shows the topic distribution of Dr. Agarwal. As expected, the majority of his publications occur in the fields of Math and Physics and Electrical Engineering and Computer Science.

Finally, to analyze the connections between the ten different physicists and the 112 journals in which they publish, a bipartite network was created. A property file was used to add the total citation counts to each node. Figure 14.9 shows the graph of all ten physicists, where the circular nodes represent the authors and the square nodes represent the journals. The author nodes are sized according to their out-degree, or the number of journals in which they publish, and the journal nodes are sized by their in-degree, or more popular publication venues within the context of this dataset. The nodes are colored from yellow to red based on citation count, with all author nodes labeled and the journal nodes with the highest citation counts also labeled.

The author with the most diverse publication venues is Vyacheslav Yukalov who published in 38 unique journals. The author with the highest citation count in this dataset is Girish Agarwal, with 3,917 citations to 163 papers. The journal with publications by the greatest number of authors in this dataset is *Physical Review B*, but the journal with papers that have the highest total citation count is *Physics Letters B*, with 4,744 citations.

14.4 Discussion and Outlook

The Sci2 Tool is one of several tools that use the OSGi/CIShell framework (Börner, 2011, 2014). Other tools comprise the Network Workbench (NWB) designed for advanced network analysis and visualization (http://nwb.cns.iu.edu); the Epidemiology tool (EpiC) that supports model building and real time analysis of data and adds a bridge to the R statistical package (http://epic.cns.iu.edu); and TexTrend for textual analysis (http://textrend.org) (Kampis et al., 2009). Thanks to the unique plug-and-play macroscope approach, plugins can be shared between the different tools, allowing individuals to customize the tool to suit their specific research needs.

Current work focuses on the integration of the smart local moving (SLM) algorithm (Waltman & van Eck, 2013) into Sci2 making it possible to run clustering on any network file (Fig. 14.10). This algorithm detects communities based on the

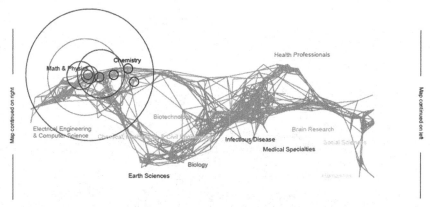

Legend

Circle area: Fractional record count
Unclassified = 1
Minimum = 1
Maximum = 277
Color: Discipline
See end of PDF for color legend.

Area

29.09
16.19

2.8

How To Read This Map

The *UCSD map of science* depicts a network of 554 subdiscipline nodes that
are aggregated to 13 main disciplines of science. Each discipline has a distinct
color and is labeled. Overlaid are circles, each representing all records per
unique subdiscipline. Circle area is proportional to the number of fractionally
assigned records. Minimum and maximum data values are given in the legend.

▨ Chemical, Mechanical, & Civil Engineering

1 sensor letters

■ Chemistry

1 advanced functional materials
1 journal of nanoparticle research
1 journal of physical chemistry a

▨ Electrical Engineering & Computer Science

1 applied physics b–lasers and optics
2 ieee journal of quantum electronics
4 journal of the optical society of america a–optics image science and vision
14 journal of the optical society of america b–optical physics
38 optics communications
14 optics express
22 optics letters

■ Math & Physics

1 acta physica polonica a
1 advances in atomic molecular and optical physics
1 applied physics letters
1 europhysics letters
1 fortschritte der physik–progress of physics
1 foundations of physics
2 international journal of modern physics b
1 international journal of quantum information
17 journal of modern optics
3 journal of optics b–quantum and semiclassical optics
4 journal of physics a–mathematical and general
8 journal of physics b–atomic molecular and optical physics
1 laser & photonics reviews
1 laser physics
2 modern physics letters a

▨ Math & Physics

9 new journal of physics
1 optics and spectroscopy
2 physica scripta
234 physical review a
5 physical review b
28 physical review letters
7 physics letters a

Multiple Categories

2 physical review e

Unclassified

1 pure and applied optics

Fig. 14.8 Topic distribution of Dr. Agarwal's ISI publications on the Map of Science and
discipline specific listing of journals in which he published. Note the one journals in "Unclassified"
that could not be mapped

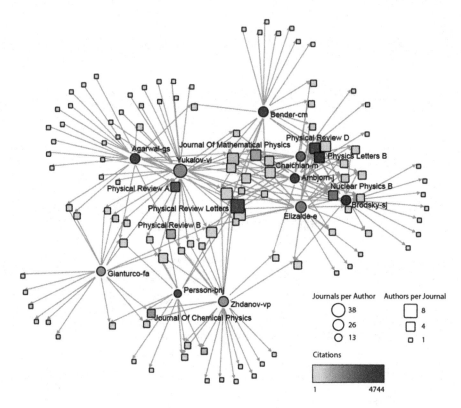

Fig. 14.9 Bi-modal network of all ten authors and the 112 journals in which they publish

relative number of links to nodes in a network. Future development will incorporate the full community detection capabilities of VOSviewer, which allows users to specify the level of granularity for clustering, resulting in fewer larger communities or more smaller communities (Van Eck & Waltman, 2010).

In addition, the Sci2 Tool can be run as a web service making it possible to request analyses and visualizations online and to execute more computing intensive jobs in the cloud. A first interface will soon be deployed at the National Institutes of Health RePORTER site. Eventually, Sci2 as a web service will be released publicly with full build support so that users can build and deploy Sci2 to the Web themselves. Furthermore, algorithms developed for the desktop version will be compatible with the online version. Ideally, users will be able to use the workflow tracker combined with the Sci2 web service functionality to create web applications that read in data and output visualizations.

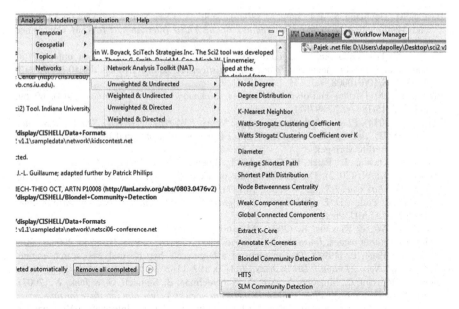

Fig. 14.10 Sci2 menu showing the SLM Community Detection algorithm

Acknowledgments We would like to thank Vincent Larivière for compiling the dataset of the top-10,000 physicists with the most affiliations and Robert P. Light and Michael P. Ginda for comments on an earlier draft of the paper. The Sci2 Tool and web services were/are developed by Chin Hua Kong, Adam Simpson, Steven Corenflos, Joseph Biberstine, Thomas G. Smith, David M. Coe, Micah W. Linnemeier, Patrick A. Phillips, Chintan Tank, Daniel AKM Halsey, and Russell J. Duhon; the primary investigators were Katy Börner, Indiana University and Kevin W. Boyack, SciTech Strategies Inc. Sci2 uses the Cyberinfrastructure Shell (http://cishell.org) developed at the Cyberinfrastructure for Network Science Center (http://cns.iu.edu) at Indiana University. Many algorithm plugins were derived from the Network Workbench Tool (http://nwb. cns.iu.edu). This work was partially funded by the National Science Foundation under awards P01 AG039347 and SBE-0738111 and National Institutes of Health under awards P01 AG039347 and U01 GM098959.

References

Adar, E., & Kim, M. (2007). *SoftGUESS: Visualization and exploration of code clones in context.* Paper presented at the 29th International Conference on Software Engineering, Minneapolis, MN.

Andrienko, N., Andrienko, G., Voss, H., Bernardo, F., Hipolito, J., & Kretchmer, U. (2002). Testing the usability of interactive maps in commonGIS. *Cartography and Geographic Information Science, 29*(4), 325–342. doi:10.1559/152304002782008369.

Anselin, L., Syabri, I., & Kho, Y. (2006). GeoDa: An introduction to spatial data analysis. *Geographical Analysis, 38*(1), 5–22.

Bastian, M., Heymann, S., & Jacomy, M. (2009). Gephi: An Open Source Software for Exploring and Manipulating Networks. *Proceedings of the Third International ICWSM Conference* (pp. 361–362).

Blondel, V. D., Guillaume, J.-L., Lambiotte, R., & Lefebvre, E. (2008). Fast unfolding of communities in large networks. *Journal of Statistical Mechanics*, P10008. doi:10.1088/1742-5468/2008/10/P10008

Börner, K. (2011). Plug-and-play macroscopes. *Communications of the ACM, 54*(3), 60–69. doi:10.1145/1897852.1897871.

Börner, K. (2014). *Plug and play macroscopes: Network Workbench (NWB), Science of Science Tool, (Sci2), and Epidemiology Tool (Epic)* (Encyclopedia of social network analysis and mining). Berlin: Springer.

Börner, K., Klavans, R., Patek, M., Zoss, A. M., Biberstine, J. R., Light, R. P., Larivière, V. & Boyack, K. W. (2012). Design and update of a classification system: The UCSD map of science. *PLoS One, 7*(7), e39464. doi: 10.1371/journal.pone.0039464

Börner, K., & Polley, D. E. (2014). *Visual insights: A practical guide to making sense of data.* Boston, MA: MIT Press.

Boyack, K. W., & Klavans, R. (2008). Measuring science-technology interaction using rare inventor-author names. *Journal of Informetrics, 2*(3), 173–182. doi:10.1016/j.joi.2008.03.001.

Brin, S., & Page, L. (1998). The anatomy of a large-scale hypertextual web search engine. *Computer Networks and ISDN Systems, 30*(1–7), 107–118.

Buono, P., Plaisant, C., Simeone, A., Aris, A., Shneiderman, B., Shmueli, G., & Jank, W. (2007). *Similarity-based forecasting with simultaneous previews: A river plot interface for time series forecasting.* Paper presented at the 11th International Conference Information Visualization (IV '07), Zurich, Switzerland.

Cronin, B., & Sugimoto, C. R. (Eds.). (2014). *Beyond bibliometrics: Harnessing multidimensional indicators of scholarly impact.* Cambridge, MA: MIT Press.

Hochheiser, H., & Shneiderman, B. (2004). Dynamic query tools for time series data sets, timebox widgets for interactive exploration. *Information Visualization, 3*(1), 1–18.

Kampis, G., Gulyas, L., Szaszi, Z., Szakolczi, Z., & Soos, S. (2009). *Dynamic social networks and the TexTrend/CIShell framework.* Paper presented at the Conference on Applied Social Network Analysis (ASNA): 21. http://pb.dynanets.org/publications/DynaSocNet_TexTrend_v2.0.pdf

Kleinberg, J. M. (2002). Bursty and hierarchical structure in streams. *Proceedings of the Eighth ACM SIGKDD International Conference on Knowledge Discovery and Data Mining* (pp. 91–101). doi:10.1145/775047.775062

Light, R., Polley, D., & Börner, K. (2014). Open data and open code for big science of science studies. *Scientometrics*, 1–17. doi:10.1007/s11192-014-1238-2

Milojevic, S. (2014). Network property and dynamics. In Y. Ding, R. Rousseau, & D. Wolfram (Eds.), *Measuring scholarly impact: Methods and practice.* Berlin: Springer.

Nooy, W. D., Mrvar, A., & Batageli, V. (2011). *Exploratory social network analysis with Pajek* (2nd ed.). Cambridge: Cambridge University Press.

Saito, R., Smoot, M. E., Ono, K., Ruscheinski, J., Wang, P.-L., Lotia, S., . . . & Ideker, T. (2012). A travel guide to cytoscape plugins. *Nature Methods, 9*(11), 1069–1076.

Scharnhorst, A., Börner, K., & van den Besselaar, P. (Eds.). (2012). *Models of science dynamics: Encounters between complexity theory and information science.* New York, NY: Springer.

Sci2 Team. (2009). Science of Science (Sci2) Tool: Indiana University and SciTech Strategies. Retrieved from http://sci2.cns.iu.edu

Song, M., & Chambers, T. (2014). Text mining. In Y. Ding, R. Rousseau, & D. Wolfram (Eds.), *Measuring scholarly impact: Methods and practice.* Berlin: Springer.

Takatsuka, M., & Gahegan, M. (2002). GeoVISTA *studio*: A codeless visual programming environment for geoscientific data analysis and visualization. *Computers & Geosciences, 28* (10), 1131–1144. doi:10.1016/S 0098-3004(02)00031-6.

Van Eck, N., & Waltman, L. (2010). Software survey: VOSviewer, a computer program for bibliometric mapping. *Scientometrics, 84*(2), 523–538.

Van Eck, N., & Waltman, L. (2014). Visualizing bibliometric data. In Y. Ding, R. Rousseau, & D. Wolfram (Eds.), *Measuring scholarly impact: Methods and practice*. Berlin: Springer.

Waltman, L., & van Eck, N. J. (2013). A smart local moving algorithm for large-scale modularity-based community detection. *European Physical Journal B, 86*(11), 471–485.

Waltman, L., & Yan, E. (2014). PageRank-inspired bibliometric indices. In Y. Ding, R. Rousseau, & D. Wolfram (Eds.), *Measuring scholarly impact: Methods and practice*. Berlin: Springer.

Index

© Springer International Publishing Switzerland 2014
Y. Ding et al. (eds.), *Measuring Scholarly Impact*,
DOI 10.1007/978-3-319-10377-8

Printed in the United States
By Bookmasters